POISONED PROFITS

*The Toxic Assault
on Our Children*

PHILIP SHABECOFF and
ALICE SHABECOFF

 RANDOM HOUSE / NEW YORK

Published in the United States by Random House,
an imprint of The Random House Publishing Group,
a division of Random House, Inc., New York.

RANDOM HOUSE and colophon are registered trademarks of Random House, Inc.

Library of Congress Cataloging-in-Publication Data
Shabecoff, Philip
Poisoned profits: the toxic assault on our children /
Philip Shabecoff and Alice Shabecoff.
p. cm.
Includes bibliographical references and index.
ISBN: 978-1-4000-6430-4
1. Environmentally induced diseases in children—United States.
2. Pollution—Health aspects—United States.   3. Industrial toxicology—
United States.   4. Children and the environment—
United States.   I. Shabecoff, Philip, 1934–   II. Title.
RJ383.S53 2008   618.92'98—dc22   2007032250

Printed in the United States of America on acid-free paper

www.atrandom.com

1 2 3 4 5 6 7 8 9

First Edition

Book design by Susan Turner

For William
And for all the children

The test of the morality of a society
is what it does for its children.
—DIETRICH BONHOEFFER

Contents

Introduction

THE TOXIFICATION OF THE ENVIRONMENT BY INDUSTRIAL AND COM-
mercial activity has been a fact of modern life for decades. But this
plague of pollution is so insidious, like the slowly heating pot of water
that boils the unsuspecting frog, that its true dimensions have crept up
on us largely unheeded. So has its impact on the health of our children.

There have been warnings, of course. Rachel Carson's *Silent Spring*
sounded what should have been an arousing alarm nearly half a century ago. Scientists, physicians, and environmentalists have told us of
the danger. Some initial but ultimately ineffectual steps were taken by
government to slow the tide of poisons into the environment.

For most of us, however, the threat has seemed abstract, a problem
for other places, other families, other children. Preoccupied with what
we regard as more immediate concerns, we tend to ignore the degradation of our habitat and its toll on our children or assume that someone
else—the government, the medical community, industry—is correct-
ing the problem. It is a false assumption.

As we looked around, we found that a surprisingly large number
of children were suffering from chronic illnesses. In one of our grand-
sons' neighborhoods alone, a quarter of all the young boys, by our
count, were afflicted with some sort of cognitive or behavioral problems of varying degrees of severity. And as we began to probe more

deeply, to study the data, we found what we consider to be clear, alarming evidence that there has been a steep increase in the incidence of a variety of serious chronic childhood illnesses over the past half century. These include childhood cancer, asthma, birth defects, and a range of neurological problems. The data also underscored that Americans were experiencing growing difficulty conceiving children.

This sharp rise in chronic childhood diseases has been paralleled by an increase in the volume and range of toxic substances into the environment that we perceive as astonishing in magnitude. These substances pervade our habitat—our air, our water, our soil, our homes, our schools, and our places of work. They not only come from toxic waste sites, industrial sites, power plant smokestacks, automotive tailpipes, and pesticide-sprayed fields, but can also lurk in our food and many (if not most) of our commonplace consumer products such as cleaning products, cosmetics, plastic bottles, and clothing. As far as we are concerned, the link between these substances and chronic childhood illness is inescapable. There is not a human on earth who is not exposed to toxic pollution. But it is the children who are most vulnerable.

We undertook this book because we felt it our duty to do what little we could to end this toxic assault on children. While we may try to rationalize and ignore the impact of toxic contamination of the environment, we cannot ignore the health and welfare of our children. Perhaps the information we present here will persuade some Americans of the seriousness of the problem and the need to act.

Over the long run, the toxification of the environment will probably be understood as as serious a threat to human welfare and the future of life on earth as most of us now understand global warming to be. Warnings about human-induced climate change and its consequences have been issued periodically by the scientific community for several decades now. Only recently, however, have a majority of Americans been persuaded of the reality of the threat and the urgent need to address it. We hope and believe there also will be an awakening to the dangers of an increasingly poisoned environment. It cannot happen too soon.

As we gathered material for this book, it was quickly apparent that any solution to the problem must involve not only industrial and commercial activity and a scientific and medical response, but also adjustments to our economic system and a change in the direction of current politics. All of society will have to be involved in protecting the children. As we try to demonstrate, the roots of the problem permeate the institutions and systems of American life.

We have cast our book as a crime story and adopted the voice of a prosecutor presenting a case to a jury and pointing at those responsible for the toxic assault on our kids and those who abet them. We do not mean that these crimes are the kind punishable under our current criminal justice system, although in some cases they may be. Our terms are a metaphor, not a factual allegation. What we are saying is that what is happening to our children as a result of toxic substances in the environment is criminal—in the sense that it is reprehensible and should not happen. We are saying that those who assault the health of children with chemicals or heavy metals or nuclear wastes should be held accountable for their actions and made to stop. But it is not just those who make and sell these poisons and products that contain them who are on trial. It is also the government officials, from the highest level on down, who do little to stop or even cooperate in the crime against our children. It is members of the scientific community who see no evil, either because they are in the pay of the polluters or convince themselves, contrary to evidence, that there is no harm. It is all of us who remain passive while our children grow ill who are also in the dock.

Brookline, Massachusetts
May 2007

POISONED PROFITS

One

INQUEST

THE SMALL CITY OF DICKSON IS IN THE MIDDLE OF TENNESSEE, BUT IT could be anywhere in America. It is reached by taking Interstate 40 some thirty-five miles west from Nashville and then exiting onto a busy multilane state highway that bisects the city. The highway itself is drearily familiar, lined with McDonald's, Burger King, Wendy's, Taco Bell, and every other fast-food joint imaginable and by new- and used-car dealerships, body shops, modular-home sales lots, tiny churches with prefabricated steeples, army and navy recruiting stations, convenience stores, gas stations, motels, and a "Save-a-Lot" supermarket. The only out-of-the-very-ordinary feature on the road is a large structure with a bulbous dome, called the Renaissance Center, a multipurpose gathering place built by one of Dickson County's wealthiest and most influential families. The road is heavily trafficked by speeding cars and trucks, and a pedestrian trying to get from one side to the other is living dangerously.

Once off the highway, however, the city appears to be a pleasant place to live. And, in fact, it was attractive enough to increase its population by nearly 40 percent in the 1990s, and it now stands at nearly thirteen thousand. Founded by General Ulysses S. Grant during the Civil War as a railway depot to bring munitions and supplies to the Union Army of the Tennessee, it is an old-fashioned-looking town of small but well-kept frame houses, neat lawns, old trees arching over the streets, pretty little parks, and clean parallel roads. The old downtown, basically a single street three blocks long, is now a bit decrepit, an economic casualty of the standardized commercial activity on the big highway. Some of the shops are boarded up, and what had been a movie house now is a store selling general merchandise. A quarter of a mile or so outside of the downtown area, the landscape turns rural: undulating fields with rich green grass, wooded hills, pastures where horses and cows graze indolently, and large, newer houses with carefully tended gardens.

It is also a well-watered countryside, with springs, creeks, rivers, and lakes throughout Dickson County, sweet-tasting waters that prompted some of the current residents to buy land and build homes here.

NOT GOD'S FAULT

On a Friday early in May 1998, Judith Cude and her husband, Gary, accompanied their daughter Jenny to the hospital in Nashville where Jenny was to give birth to a son. Mrs. Cude, an attractive middle-aged woman and "Miss Judy" to all, runs the Honey Tree Christian Day Care Center in Dickson. She was looking forward to the birth of her third grandchild. She and the baby's father and the father's mother were in the delivery room, and all seemed to be going well. Gary Cude waited in the hall. The attending physician kept up a cheerful, reassuring stream of talk as he assisted Jenny with her labor. "Peyton came out facedown. When Dr. Booker turned him over, he stopped talking," Judy recalled. "He had his back to us, but when the nurse gave him a shove, he turned around and he had tears running down his cheeks."

The baby's face was badly disfigured with a cleft lip and a bilateral cleft palate. And though they did not know it immediately, Peyton also had a damaged heart, a valve that failed to close properly. "I had never seen this defect except after it had been fixed," Judy said of Peyton's cleft palate. "My heart was in my throat. My husband walked into the room and put his arm around me, and we went into the hall. The first thing I said was, 'Why would God do this to me, as much as I love children and worked with them all my life? Why would He do this?' My husband said, 'Honey, this is not God's fault.' "

ELEMENTARY EPIDEMIOLOGY

Two weeks after Peyton was born, Jenny was given the name of another mother in Dickson whose child, born a couple of months earlier, also had a cleft lip and palate. Then a woman called Judy at her day care center and asked if she could accommodate children with special needs because ultrasound tests found that her child was about to be born with a cleft palate. "That made three," Judy said. She and the other mothers kept a tally. Soon they had counted six. Judy placed a newspaper advertisement asking families with similar defects to contact her. And, as it turned out, nineteen children had been born in Dickson with a cleft lip and palate in a little over two years. The odds against such a series of identical birth defects were almost certainly too high to be coincidental. In a city the size of Dickson, perhaps as many as two cases of bilateral cleft lip and palate might be expected in the same period. Clearly, something was happening.

Cleft palates were not the only ills afflicting Dickson's children. Within a brief period, four babies were born with a rare brain malformation, where the two hemispheres of the brain are not connected. There have been a large number of cases of hypospadias, a condition in male children where the urethra is inverted. There also has been a high incidence of heart problems in Dickson babies, as well as childhood leukemia, the families reported.

"When I realized how many children had defects, I called the Dickson County Health Department," Judy said. "The registered nurse there said, 'We'll bring this up and call you back.' They never did. I'm still waiting six years later. Then I called the Centers for Disease Control [and Prevention (CDC)] in Atlanta, but they said, 'This isn't what we do.'" Finally, on the advice of a public health official in Nashville, she contacted Betty Mekdeci, who runs a nonprofit organization in Orlando, Florida, that monitors birth defects around the nation. Mekdeci told her that there were as many babies with cleft palates born in Dickson during that period as in the entire state of Wisconsin and nine times the national average. Following Mekdeci's advice, Judy got a map of Dickson County and put an "X" down to mark the location of each family with a cleft palate baby. What she found was surprising. Most of the families lived in the southwest quadrant of the county, where the Dickson County landfill is located.

THE SOURCE?

The landfill had opened in 1968 as the Dickson City dump. A decade later, the county bought the property and expanded it for use as a sanitary landfill; though the Tennessee Department of Public Health found the area suitable for use as a sanitary landfill, it recommended that no liquid wastes be disposed of there. Nevertheless, the landfill soon began accepting industrial liquid wastes from manufacturing facilities in the area, including Scovill-Schrader, Inc., which made automotive parts. It would be another ten years, however, before tests were conducted to determine if water beneath and around the fill was contaminated, and then only after a nearby resident contacted the county to voice a suspicion that a spring on her property might be contaminated.

It was. Tests conducted by private contractors working for the county and state, and later by the federal Environmental Protection Agency (EPA), found that a brew of chemicals from the landfill had made its way to the groundwater under the dump and was spreading out through the karst rock, a geologic foundation riddled with countless cracks, that underlies much of Dickson County. The pollutants included toxic chemicals such as benzene, toluene, xylene, and, most ubiquitously, trichloroethylene (TCE), an industrial solvent widely used for degreasing machine parts and in the production of other chemicals. TCE had been heavily used and then dumped in the landfill and elsewhere by the Scovill-Schrader plant and other manufacturers in the area, Judy Cude and other residents said. TCE is known or suspected of causing a number of chronic illnesses, including several forms of cancer and birth defects. There is evidence that it can be a specific cause of cleft palates, although the available data is limited.

DUMPING AND DENIAL

By 1975, the Tennessee Department of Public Health said that no more liquid wastes should be disposed of in the landfill, but Scovill continued to dump "trailer loads" of liquids into the facility, according to local residents. Residents told Judy Cude about the barrels carted away by private contractors and buried on farms in the area. One worker confessed to her, "I buried this shit all over the county." Lynn Agee, a lawyer representing several of the families in lawsuits against Scovill, said that discovery had produced substantial documentation of dumping.

The state ordered the landfill closed in 1988 because it had reached its capacity, not because it was contaminating Dickson's groundwater.

At the end of the 1980s, the county, state, and federal EPA, the U.S. Geological Survey (USGS), and the CDC finally began to monitor the groundwater, wells, and streams in the area. While there were fluctuations, test after test revealed high levels of TCE and other toxic contaminants in the water, including in wells used by families for drinking and bathing and in at least one well that fed Dickson's public water supply. In 1994, the Tennessee Department of Solid Waste Management sent a letter to the manager of the landfill saying that it was in violation of the state's groundwater protection standards. There were no repercussions, however. The department merely recommended steps be taken to correct the violations and even left the landfill wiggle room, inviting its management to prove the contamination came from other sources. The EPA did not recommend the landfill be classified as a Superfund site, which would have required a thorough cleanup of the area. It did classify the Scovill plant as a Superfund site, but the designation was changed after Scovill undertook its own cleanup effort.

Scovill-Schrader pulled up stakes and moved to North Carolina in 1984, resuming a company history of leaving a community and its workers when local conditions proved unpleasant. Ostensibly, the move was because of poor labor relations with the United Auto Workers (UAW), which represented employees at the plant. But Don Corn, a UAW official in Tennessee, contends that "Schrader left the Dickson plant because the heat was on. The state environmental agencies were starting to look into their habits, and they were running out of space to put their industrial waste." Confirming what Judith Cude had heard, Corn adds, "They had filled up the area behind their plant and the county landfill and even contracted out to private firms to bury the TCE. Once the cat got out of the bag, many barrels were dug up and hauled to Emelle, Alabama."

Dickson families who are trying to sue the company for the injury inflicted on their children are finding that they are chasing a will-o'-the-wisp. The company claimed bankruptcy and then was sold and sold again. Dickson County and the city of Dickson finally did sue a successor company to Scovill for damages and settled in late 2006. Under the settlement, the families of the kids will get 75 percent of the award. But since the successor company, Saltire Industrial, Inc., filed for bankruptcy, the families could end up with zero money, one county attorney said.

William Andrews, former CEO of Scovill-Schrader, now chairman of the Corrections Corporation of America, a private company that builds and manages prisons and other detention facilities for federal, state, and local governments, claims the lawsuits are "ridiculous." In fact, he noted, much of the dumping occurred before there were federal rules and guidelines governing such activities. The problem, according to Mr. Andrews, is that "lawyers are stirring things up just to make money for themselves. Everybody in Dickson and the surrounding area knew there was a problem with the landfill. Now they are coming back twenty or thirty years later and saying there are grounds for a suit.

"The point is that I can say that Scovill, whether I was there or anyone else, would not have done anything wrong if they knew it was wrong," Andrews asserted.

Nor are the families getting much support from federal, state, and local governments. While EPA and CDC officials acknowledged that the many children with cleft palates constituted a cluster of illnesses, they could not verify that the TCE or any other contaminant was the cause—there was just no firm scientific evidence to do so, they told the families. Christie Piland, whose son Luke was one of the kids born with the cleft palate, said that when she called the health department, the CDC and "everybody" told her there was no problem in the county. "When we finally did get a visit from the CDC, they made me feel it must have been something I did wrong. And that was the hardest thing." Christie's voice grew loud and harsh as she imitated the CDC investigators asking her questions: " 'Did you smoke? Did you use drugs? What medicines did you take while you were pregnant?' It wasn't, 'Oh, there might be a problem in this county.' "

Early in 2003, Jimmy Cagle, Christie's father, wrote to former senator Bill Frist of Tennessee, then Senate majority leader, to ask for help in getting federal agencies to act on the birth defects in Dickson and to urge the senator to meet with the families affected by the birth defects. Senator Frist, a medical doctor who had expressed outrage on moral grounds for removing Terri Schiavo from life support, never replied, Cagle said.

The city has shifted some families from well water to the municipal water supply. But the families of the afflicted children think that state and local officials have not responded with any sense of urgency to what has happened to their children and what might happen to other kids yet to be born. Lynn Agee, one of the lawyers whom several of the families have retained to seek damages from the city and the companies who dumped the chemicals, said that "their posture is, 'Maybe it will just go away.' " A letter sent by an environmental specialist in the Tennessee Department of Public Health to Mike Apple, head of the state's Department of Environment and Conservation, seems to sum up the prevailing attitude: "Talking with my Director," the specialist wrote, "he has discouraged any public presentation of cleft issues at this time. He is worried that the cleft lip and palate information is a 'can of worms' that could reopen emotional issues distracting from the environmental concerns."

Doug Jackson, a tall, athletic-looking lawyer with a self-assured manner, represents Dickson County in the Tennessee State Senate. His family, which numbers lawyers, doctors, judges, and businessmen in its ranks, is one of the county's oldest and wealthiest. It owns a lot of land in Dickson and surrounding counties and built the Renaissance Center. Senator Jackson is a Democrat with a reputation as one of the few members of the legislature who cares about the environment. He authored a Tennessee law that prohibits out-of-state operators from building landfills without local approval. But some of the families are furious with him, convinced he has failed to come to their assistance or to deal with the pollution of the county.

Senator Jackson, however, said there is nothing he could have done. "It's very hard when, as a legislator, you get a call and people are concerned because their child has cancer or something. And they'll say, 'We live not too far from the landfill.' But trying to prove causation from environmental contamination is very, very difficult and legally very challenging. We know there's been leakage of TCE in Dickson and that there are people who have relied on groundwater as their water source, who live below the landfill along the plume of pollution, and who have consumed TCE. Did that cause someone in the family to become ill? I don't know. I've had a couple of conversations with a doctor from the EPA. He indicated that there is a cluster of health problems in Dickson, but he lectured about the difficulty of trying to connect a cluster of cleft palates to a particular source of contamination.

Those families have every right to be upset. No one can deny there is a cluster, so I have to assume there's a cause. But trying to connect that cause is difficult."

He also suggested that trying to address the problem by state or local government would be tilting at windmills. "We have to understand," he said, "that we're a capitalistic society. To try to implement something that might be good for the public but that could create a burden, embarrassment, or hardship for the business community creates a real challenge for you if you are trying to pass legislation. The general population is not well represented as opposed to special interests and big business. Come up here on the days when committees are working hard. The hallways become jammed with lobbyists. One company may have multiple lobbyists."

Senator Jackson, who grew up on a farm near the landfill, has himself had a bout with cancer. But he insists that chemical contamination was not the cause. His cancer, he said, was caused by his work cutting pine trees as a youth.

The families' neighbors, employers, and community institutions of Dickson have not given much support, either. Many express disbelief that chemicals in the water are a health threat. Others worry that stories about birth defects and other illnesses will decrease the value of their homes.

Jenny's niece came home from school one day and reported that her math teacher had said there was no problem in Dickson County but that "the problem was that all the mothers were on drugs during pregnancy." A local pediatrician who is a member of the Dickson County Health Board said that most of the cleft palates could be attributed to genetics, and others have said the problem was that the families were "inbred."

Hundreds of families who draw their water from contaminated sources still cannot get municipal water.

THE CHILDREN'S STORIES

Meanwhile, the children and their families continue to suffer. On a hot May morning, several of the families gathered in a windowless, stale-smelling meeting room of the town's Holiday Inn to tell us their stories.

Jenny Casteel, Judy Cude's daughter, who is a real estate agent and an attractive, seemingly self-possessed young woman, said that she was so distraught the night her son, Peyton, was born that the hospital put her on Valium. The doctors would not even let her hold her new baby. "But the next day I started to pull myself together and they let me have him, and after that we went home. You learn to live with it."

So far, Peyton has had five surgeries and five different doctors, including a plastic surgeon and a cardiologist because of his heart defect. "He'll have to go this summer and have a retainer brace put in to spread his teeth apart on the top; then in two years he'll need a bone graft to fuse the jaw and his top lip together. Then at fourteen he'll need a complete nasal reconstruction. Altogether he will need five to seven more surgeries. Peyton has got a great personality. I couldn't ask for a better child. He goes along with the plan. He knows he is different, but he seems to be okay with it. He has a severe speech impediment and didn't begin talking until he was three years old. He's been in intensive speech therapy and will have to continue for all his life." (Peyton and a number of the other cleft palate children have been taught sign language to help them communicate with others.)

Amy Woods, a small, pretty, dark-haired, sweet-tempered young woman, had been warned by an ultrasound test while pregnant that her baby daughter would have a cleft lip, although she did not know about the cleft palate until after delivery. Lauren, her daughter, has had five surgeries so far, including an operation to remove part of the middle of her palate, which was growing up into her sinuses. Her daughter learned sign language and, Amy said, has to have speech therapy. Lauren's surgeries were conducted without charge at the Shriners' Hospital in Chicago, as were the surgeries of other children with birth defects.

At that point in her story, Amy started to cry, the tears rolling down her cheeks as she tried to gain control of her emotions. Almost immediately, several of the other women started to shed tears as well.

"The thing that really upset me," Amy said, "was that because of the insurance I was not able to get the best doctors for my baby. It's not right not to be able to give your child the best care. It's not fair to have to sit in a waiting room with people who are there just so they can look better for society while my child is trying to look like a normal child."

At the time of our meeting, Amy was again pregnant—and apprehensive. She had waited five years before deciding to have another child. The pregnancy had been difficult. "It's hard," she said. "I'm still not confident."

When Kim DeLoach's son Paxton was born, he had only a hairline scar on his lip, and the hospital made no diagnosis of a cleft palate. They sent Kim and her baby home without any instructions for caring for the child. "For the next three months, he gained no weight," Kim related. "I kept carrying him back to the pediatrician, and then we switched pediatricians, but that didn't help. At the six-month mark, when he couldn't sit up, couldn't roll over, couldn't eat, and was losing weight, they decided we were mistreating our child and they were going to call the state social agency. So we switched to a third pediatrician, who, thankfully, realized there might be something wrong here and after a little bit of further investigation realized the palate was malformed. The palate is not fully open. It is jammed all the way up Paxton's nasal cavity, so he couldn't suck his bottle. That's why he couldn't eat, and when he did eat it would come out his nose or he wasn't able to ingest enough. He was in pain. That's why my child screamed for the first six months. We took shifts at night and we'd get, on average, about two hours of sleep. He would only sleep on our chests because he had stomach pains from hunger. He was nine months old before he doubled his birth weight, eighteen months before he crawled."

Cathy Flake moved to Dickson in 1999 and became pregnant a month later. One of the reasons the Flake family chose their property was that it included Bruce Spring, a pretty pool of clear water bubbling out of the ground at the bottom of her hill. The water tasted so sweet that Cathy was thinking of starting a business bottling and selling the water. When her son, Spencer, was born, he seemed completely normal. But soon after birth something seemed to be happening to one of Spencer's eyes. It began to swell more and more and to turn black and blue and, Cathy said, "veins started to show up like a spiderweb all the way down the side of his face. As a little boy, he was always sick. His eyelid was shut and hanging halfway down his face. He lived on antibiotics. Never one time did a doctor say, 'There may be something in the environment that affects your child.' " At one point, Cathy related, a neighbor called the state's Department of Children's Services to come and investigate her, on suspicion of hitting her son.

Spencer was finally diagnosed, Cathy said, as having a condition where pockets around the orbit of his eye would fill with lymphatic fluid until, Cathy said, "they would swell up like a water balloon." The odds against a baby having that disease, she found out, are 1 in 50,000. But soon thereafter, the anchor and investigative reporter for a Nashville television station, Demetria Kalodimos, who had done a series of reports on the pollution in Dickson, located another child, the same age as Spencer, living in the same area, with the same affliction.

It was not until four years after their child was born that the Flakes learned that Bruce Spring was contaminated with TCE far above levels recommended as safe by the EPA. It was the TCE that gave the water its sweet taste. Although the spring water had been tested and found contaminated before they purchased the land, the Flakes had not been told about it, Cathy said.

One mother who did not come to the motel meeting was Sheila Holt-Orsted, an African American woman whose family has farmed 150 acres close to the landfill for many years. We spoke to her by telephone, and she said that contamination from the landfill had affected three generations of her family. "I've got breast cancer, my dad has prostate cancer, my sister has bleeding polyps in her colon, and my daughter has a speech impediment—TCE causes that. It's been thirty-eight years that this chemical has been dumped into the ground, so I've been drinking this water all my life." Her family is also suing the company and the county, and their suit includes allegations of racial discrimination, because, she asserted, they told white families in the area that their wells were contaminated but they did not tell hers. "We were trusting them. They said, 'Oh, there's nothing bad in the water.' "

The Flake family is leaving Dickson. In a letter to us several weeks after our meeting, Cathy wrote: "I just can't do it anymore. I have been so worried day and night about my kids. I hope and pray they do not get cancer or other problems. One day we were having a good day and putting some of this out of our minds when I found my daughter sitting at the spring with her feet in the water. I instantly yelled, 'Get away from the spring.' Then I thought, What in the world am I still doing here?" But the Flakes will not be able to leave all their troubles behind. Spencer continues to have health problems. And then there are economic issues. "While the big corporations get out of their debts and restructure, we will be making two house payments, one on a home we can never sell or regain our investment on. We will be spending our children's college money to rent a home because we can't buy a new one now. We saved for seventeen years to buy this property."

Amy's new baby, Hailey Brooke, was born in July 2005. She is perfect. Neither Amy's family nor any of the other families with damaged kids are still using contaminated groundwater. Amy Wood and her family are also moving from Dickson. But she has not given up fighting. She lifted a binder bulging with papers and, tears flowing again, said, "I want to be able to document for Lauren what happened to her and why. I want her to know we fought for her, that we didn't take the easy way, that we stood up for her."

Two

INDICTMENT

DICKSON'S CHILDREN ARE VICTIMS OF A CRIME.

It was, no doubt, a crime without criminal intent. Nobody deliberately set out to deform these children. No one plotted to cause their families grief and economic hardship. To some degree, the crime may have been committed within the margins of the law. Much of the dumping of TCE and other toxic substances into the ground and then into the water was done before it was formally illegal to do so—although at some point the dumpers must have known what they were doing was wrong. Government agencies and medical institutions can say with accuracy that there is no absolute proof that the poisons in the water and air were responsible for what happened to the children. It is rarely possible to determine with the certainty required by the standards of science that a specific pollutant caused a specific illness in a specific person or group of people. The factors surrounding the illness—the chemistry, genetics, multiplicity of exposures to toxic substances, lifestyles, social contexts, and other factors—are so complex as to defy definitive proof, although that may change with advanced diagnostic techniques in the coming years.

Often, as in the case of Dickson, the scientific evidence is cloudy. Perhaps, despite the clear association of TCE with the cluster of birth defects, it was not that chemical but some other that actually *caused* the problem. Maybe a combination of chemicals damaged the babies in their mothers' wombs—emerging evidence suggests that some illnesses are caused by multiple hits of different chemicals. It is possible that the many chemicals in the environment sensitized the fetuses to render them susceptible to the agents that could cause birth defects. Despite the long odds, the cluster of cleft palates could have been a statistical fluke unrelated to the toxic environment in Dickson.

But to us, as well as to the mothers and families of the babies with cleft palates, and to many in the scientific, medical, public health, and environmental communities, the kind of tragedy that occurred in Dickson bears the hallmarks of a crime. There are the victims: the deformed children who bear the pain and stress of their affliction and the families who suffer severe emotional distress and economic hardship. There is evidence of an assault on the children from chemicals known to cause birth defects and other illness. These assaults occurred when the children were in their mother's wombs, the most vulnerable period of their development. There are perpetrators, the industrial facilities that used and disposed of these chemicals in ways that exposed residents of the area to their dangers. There is a motive: the desire of corporations to operate at lowest cost, to turn the highest possible profit. There are co-conspirators: legislators and government officials and institutions that impose lax laws and rules to regulate pollution and then laxly enforce those rules and seldom even bother to keep track of illness patterns among children.

What is missing is justice for the children.

THE SEA OF CHEMICALS

Dickson could be anywhere in America. It looks like so many other places. The health and well-being of children anyplace and everyplace in this country are under assault from a vast and largely uncontrolled array of hazardous, human-created substances in their environment. The impact of these chemicals on the bodies and minds of millions of children is, in cumulative effect, an act of violence against the most innocent and defenseless of our citizens.

Our children are conceived, born, develop, live, and sometimes die in a sea of pollutants. TCE is one of the more ubiquitous of these poisons. It was found in Woburn, Massachusetts, of *A Civil Action* notoriety, where a disproportionately high number of children sickened and sometimes died from diseases such as leukemia. It is found in Toms

River, New Jersey, where there is an even larger incidence of childhood cancer. It is found in Brick Township, New Jersey, a community with an unexplained cluster of children with autism. It is found in Camp Lejeune, a North Carolina military base with elevated childhood cancer rates. TCE is the most prevalent contaminant in Superfund waste sites and one of the most common toxics in the water supply across the nation. But though we know that the chemical is toxic to humans, that it can cause cancer, low birth weight, fetal death, and birth defects, among other ills, our nation has yet to come up with a program for protecting citizens from TCE's dangers.

We and our children are exposed to toxins invisibly, stealthily, without our knowledge. Trillions of pounds of tens of thousands of toxic chemicals pour into the environment and into the products in our homes, workplaces, and schools, and their number is growing daily. There is no place to hide from them. Our children are not safe in their homes, in their yards or playgrounds, or in their schools. They are not safe in their mother's womb: A study of blood from the umbilical cords of newborns found their blood tainted with traces of nearly three hundred synthetic chemicals. They are not fully safe at their mother's breast: The milk of virtually every mother on earth contains high levels of dozens of chemical pollutants. They are not safe even at the moment of their conception, because toxic substances within their parents' bodies may damage the seed or egg. Recently emerging data suggests that genetic change or damage caused by these poisons can even be carried through succeeding generations. Poor children whose families live near contaminated areas or breathe polluted air or drink tainted water, most often African American, Latino, or new immigrant families, are most at risk. Children of the poor are doubly damaged because they often lack access to decent medical care and adequate nutrition and are subject to violence within their communities. But no child can escape exposure to these poisons, no matter how wealthy their parents, how large their homes, how exclusive their schools, how clean and well-ordered their communities.

It is a bleak prospect for the children, but it can be changed. The first step in stopping the toxic assault on our kids is to understand what is happening—acquiring knowledge about what substances are threatening them, who is putting them into the environment and why, and what the practical steps we can take are.

DIMENSIONS OF THE PROBLEM

How do pollutants that we are totally unaware of, that we are never directly exposed to, find their way into our lives and bodies? Through complex pathways, unseen, unexpected, undetected. Toxicants are part of the fabric of our daily lives, released from industrial plants, traffic, natural events, power generation and heating, industrial farming, wastewater discharge, and our chemical-laden household products. The molecules dissipate, travel the globe on airstreams, rain down onto the soil, seep into surface water and groundwater, and evaporate back to the air we breathe, into dust, food, and drinking water.

There is literally no place on earth that is not contaminated, from the farthest reaches of the Arctic to the depths of the ocean. Synthetic chemicals and pesticides are found in salmon, birds, whales, bears, in all wildlife. The Inuit people, who live near the Arctic Circle far from the sites of manufacture and sources of release, register some of the highest burdens of persistent toxic chemicals.

Many of these pollutants have been found to be toxic, to cause or contribute to what is increasingly understood as an epidemic of serious, chronic childhood illnesses. Among environmentally linked health problems are birth defects, several forms of cancer, asthma, and a range of neurological illnesses that include lower IQs, learning disabilities, attention deficit/hyperactivity disorder (ADHD), and autism. The rising incidence of reproductive problems, including infertility and miscarriages, are also understood by many scientists and physicians to have environmental roots.

Today's synthetic products are created from chemical combinations that do not exist in nature, so their effects cannot be anticipated in advance. Yet most of the chemicals that are poured into the environment every year have never even been tested for the full range of toxicity to humans. Chemicals, unlike medications, are considered safe until proven otherwise. As Dr. Bruce P. Lanphear, director of the Environmental Health Center at Cincinnati Children's Hospital Medical Center, stated, "We have erred on the side of protecting industry rather than protecting children for the past one hundred years or so. The regulatory system to protect children from environmental toxins is precarious. Under current regulations, manufacturers of commercial chemicals (excluding pesticides) are not required to supply (the full range of) toxicity data before selling their products. Nor are pesticide manufacturers obligated to supply basic premarket toxicity and exposure data necessary to ensure that children will be protected from exposure and potential harm," especially neurotoxic harm to their developing minds.

Those chemicals that have been tested have been examined chiefly for the impact on adults, not children. Much of whatever testing has been carried out has been done by the manufacturers themselves or by scientists in their hire. Overworked and overwhelmed government agencies, handicapped by flawed legislation, rely heavily on whatever data industry chooses to give them. These agencies, moreover, have been increasingly subject to political and ideological pressures in recent years, with many of the appointees who run them coming straight out of the industries that are supposed to be regulated and as likely as not returning to those industries. Finally, there has hardly been any attempt at all, up until now, to assess the effect on children of the multiple exposures of the hundreds of chemicals in their daily environment.

The chemical-manufacturing industry is both the single largest manufacturing sector and the single largest industrial source of chemicals that can harm our children's bodies and brains.

Even when regulatory agencies do try to act to reduce or eliminate dangerous substances from the environment, they are frequently thwarted or, at least, delayed by powerful, well-funded challenges by industry and its satraps in the offices of lawyers, lobbyists, and public relations firms. A whole new industry has been created to "manufacture doubt" about the dangers of industrial products. The chemical, energy, automotive, and other industries pay for scientific studies to contradict evidence that shows a chemical causes harm. The companies hire public relations firms, lawyers, and product defense specialists to sway public opinion, harass critics, win lawsuits, and ward off regulation. They give large sums of money to fund the campaigns of politicians who support them in legislatures and state houses, Congress, and the White House. Many of the tactics currently employed originated with the tobacco industry, first to hide and then to deny the dangers of smoking. With sales of more than $600 billion a year, the chemical industry has enormous resources with which to overwhelm its critics.

The burden of proof that environmental pollution causes sickness and sometimes death rests with the victims of pollution, not with the economic interests that produce and sell them. Those who continue to assault our kids with chemicals go largely unpunished. Eric Schaeffer, former head of enforcement for the Environmental Protection Agency, lamented, "You look at a case in California where a guy convicted of stealing a pizza under the state's 'three strikes and out' law gets thirty years, and then you look at plants that are killing children and nothing is done—well, it does seem wrong."

Much of the nation's scientific community remains reluctant to demand action to stem the flow of these hazards in the absence of definitive proof of harm. Medical schools devote little time to training doctors on the relationship between environmental hazards and illness and even less on how those hazards can affect children. Research and medicine in this country continue to devote most of their energy and expertise to treating chronic illnesses rather than preventing them. The commercialization of the nation's health care system encourages drugs and heroic interventions rather than prevention, which is less profitable. "Insurance companies often refuse to pay $150 for a diabetic to see a podiatrist; but nearly all of them cover amputations, which typically cost more than $30,000," Paul Krugman commented in a *New York Times* column.

Caveat emptor! Except in this case, it is our children who are at high risk. Yet we nevertheless continue to entrust their health, their safety, and their lives to rules of the marketplace.

Dr. Philip Landrigan and Dr. Herbert Needleman, physicians and research scientists who have done pioneering work in environmental pediatrics, have this to say about the impact of hazardous chemicals in the environment: *"We are by default conducting a massive clinical toxicological trial. And our children and their children are the experimental animals."*

THE LEARNING CURVE

In the second half of the twentieth century, the United States did make a serious attempt to come to grips with its polluted, degraded environment. In the 1970s, Congress passed in rapid succession a series of landmark laws to lessen the impact of air and water pollution, industrial chemicals, pesticides, and other threats to our habitat and to human health. President Nixon created the Environmental Protection Agency by executive order, and many other environmental institutions at all levels of government sprang into existence. For a period, some substantial progress was made in cleaning our habitat and safeguarding health. Unfortunately, however, in recent years administrations and Congresses have rolled back many of the gains in protecting the environment achieved in the last century.

Children must run a gauntlet of threats to their health and welfare other than toxic substances in their environment. Violence in their homes and neighborhoods, traffic on their streets and highways, drugs, tobacco, and alcohol in their schools and playgrounds, poor diets, and sedentary lifestyles lurk ominously in the shadows of American childhood. But that these threats are real and omnipresent should not lead us to ignore or minimize the equally real dangers of a degraded environment, as some ideologues and corporate apologists would have us do. Protecting our children should not be a zero-sum game. It cannot be any game at all.

Our high-tech, consumption-driven civilization, floating in a sea of chemicals, powered by fossil fuels and nuclear energy, has brought innumerable, often precious benefits to us all, including our children. The scourge of infectious disease that claimed so many children in past centuries has been dramatically reduced by improved nutrition and hygiene, clean water supplies, antibiotics, vaccines, and other medical advances. When we were growing up in the 1930s and 1940s, our parents lived in dread of polio, which paralyzed and killed many children. That disease and the fear of it have been eradicated in this country. Other childhood diseases, such as measles, scarlet fever, tetanus, smallpox, and dysentery, have been eliminated or substantially erased—although we must beware of a new generation of infections, including those carried by disease vectors that are expanding their range as the globe warms. In most parts of the country, children receive medical care that ranges from adequate to superb, provided their parents can pay for it. Recent generations of children can look forward to a longer life span and enjoy far more material benefits than past generations. If their families are well clear of the poverty line, they will have access to ample (if sometimes unsafe) food throughout their childhood and will be adequately sheltered, warmed, and clothed.

As Dr. Landrigan points out, "Patterns of illness have changed dramatically among children in the United States. Today the most serious diseases confronting children in the United States are a group of chronic, disabling, and sometimes life-threatening conditions termed 'the new pediatric morbidity.' The protection of children against toxic chemicals poses a major challenge to modern society. The problem is not going to go away."

Not all chronic childhood illness can be wholly or directly attributed to poisons in the environment. Family lifestyles and habits and stress can lead to sick children. Pregnant mothers who smoke, drink alcoholic beverages, or take recreational drugs place their babies in grave danger. Secondhand smoke in the home can create long-term health problems for children who live there. Tobacco, in fact, has been found to be one of the more serious threats to children. We do not devote much attention to the dangers of smoking, alcohol, and so-called recreational drugs in this book because these hazards are now common knowledge. An inadequate or improper diet is a factor in many childhood illnesses.

Nonetheless, there is abundant evidence that the trillions of pounds of hazardous pollutants that have been poured into the environment are, in all likelihood, responsible for much of the sickness, suffering, and, too often, death of America's children.

An early reference to an environmental assault on children's health came out of Australia a century ago when physicians found an epidemic of lead poisoning among children who had ingested lead paint. The vulnerability of the young to environmental hazards did not rise in public consciousness again until the 1950s, when pregnant women who had eaten mercury-laden fish from Japan's Minamata Bay gave birth to babies with distorted limbs, mental retardation, cerebral palsy, and other horrible disorders. In the same period, concern over the impact of ionizing radiation on children, spurred by the fallout from nuclear weapons testing, led some in the medical community to focus on that danger. Herbert Needleman's research, beginning in the 1970s, on the heavy toll lead was taking on children's neurological health was a turning point.

It was not until the 1990s, however, that a major breakthrough was made in creating a special layer of protection for children who might be threatened by environmental hazards. A 1993 report by the committee of the National Academy of Sciences, headed by Philip Landrigan, entitled *Pesticides in the Diets of Infants and Children,* was, according to the American Academy of Pediatrics, "instrumental in highlighting environmental hazards unique to children and the relative paucity of information connecting environmental exposures and child health." The landmark project marked the first time a major scientific study looked at the special vulnerabilities of children and the way one aspect of their lives, their diet, differed from that of adults. Scientists began to realize that children's immature bodies with their unformed protective systems combined with their childish behavior make them other than just little adults.

The study led to congressional passage of the Food Quality Protection Act (FQPA), which for the first time required that the regulation of pesticide residues in foods had to reflect the special vulnerability of children. In 1995, EPA administrator Carol Browner directed the agency to create a new policy requiring that the vulnerability of children be considered when assessing environmental risks. The federal government also launched an initiative that funded thirteen new centers across the country to research children's environmental health. The first was opened at New York City's Mt. Sinai Hospital in 1993, under Dr. Landrigan. So the field of environmental pediatrics was born.

Remarkable advances in science and research technologies in recent decades have created powerful new tools for understanding human physiology and how it is affected by chemical pollutants and other environmental contaminants. The decoding of the genome, new molecular biology, advanced toxicology, supercomputers, and many other feats of wizardry are being pulled out of hats in laboratories and medical facilities across the country, enabling scientists and physicians to better understand what is happening to the children and why.

The new science has revealed that the effects of chemicals on fetuses, newborns, and very young children may be completely different from, and often far more powerful than, effects on adults. Small amounts during windows of vulnerability can change lives permanently. The equivalent of a single drop in 118 bathtubs of a mixture of chemicals called polychlorinated biphenyls (PCBs) can cause a lowering of IQ and a rise in attention disorder in a child exposed in the womb. Chemical mixtures can exert harmful effects while one chemical at a time may not. A toxic substance can cause multiple illnesses in one victim, and it can cause different illnesses in different children.

Science's new understanding of the genome has proven that few illnesses stem from genes alone. What is inherited is the manner of re-action of our genes to a given environment. Because of genetic varia-tions in susceptibility, some mothers in Dickson may have showered in and drunk chemically tainted water while pregnant yet produced a normal baby. They were recipients of a lucky roll of the genetic dice. A saying we heard repeatedly from physicians and scientists was "Genet-ics loads the gun, but the environment pulls the trigger."

The epidemic of chronic illness among our children need not be. There are or can be more benign alternatives to many, if not all, of the likely unsafe chemicals now in commerce, and a motivated industry could come up with many more. Indeed, the world managed to get along quite well without these dangerous substances, most of which came on the scene after World War II. Since the 1930s, the number of synthetic chemicals put into commercial use has doubled every seven to eight years.

As Dr. Martha Herbert, a physician and research scientist at Har-vard Medical School and Massachusetts General Hospital who works with neurologically damaged children, asserted, "A whole generation of kids has been trashed by causes that are preventable."

Change is possible. Indeed, it is a central premise of this book that this wave of environmentally induced illness among our children is preventable. More and more parents are demanding relief from a de-graded, dangerous environment. Unless the rest of us join them in their anger and activism, however, decisive change will not happen. Unless the rest of us think about the ways we consume and the ways we live, unless we abandon our lethal passivity in the face of the toxic jug-gernaught, unless we become intelligent voters who elect governments that represent true, vital interests of our families, not their own and their moneyed patrons, then we are all accomplices in this crime against our children.

As a nation, we can make different and better collective choices.

Three

VICTIMS

PUBLIC HOUSING PROJECTS ARE ALMOST ALWAYS DUMPED INTO THE worst part of town, and in Port Arthur, Texas, a scruffy sacrifice zone of oil refineries and petrochemical plants, silent streets, empty houses, and shuttered stores, the worst is a vision of industrial purgatory. The project here, rows of two-story concrete buildings containing tiny apartments, is separated by a small playground from a massive refinery complex. Beyond a chain-link fence are acres of squat, round storage tanks, cracking towers and flare ducts, and seemingly endless miles of pipes and wires. Around that complex are more refineries and chemical plants.

"So many of the children here have health problems, you can find a sick kid in almost any one of these houses," said Hilton Kelley, who organized and runs a citizens' group seeking to reduce and obtain compensation for the pollution that plagues so many of Port Arthur's residents. "Sulfur dioxide, benzene, butadiene, volatile organic chemicals, particulate matter—we have it all."

In fact, the first apartment we approached was occupied by a young mother whose small son was afflicted by asthma.

Latisha Montgomery, a pretty, well-spoken twenty-four-year-old woman, shares her two rooms with her son, Jobori, a chubby toddler with soft curls. When we visited, he was curled sleepily in the lap of a friend who had dropped by.

"He has breathing problems, especially when that funky smell is in the air," Ms. Montgomery said. "He also has allergies. He has to use a nebulizer a lot." A nebulizer is a device that delivers misted medication to the airways of people who have trouble breathing. She went into the back room to bring out the apparatus to show to us. As soon as she came back and Jobori saw what was in her hands, he climbed down from the friend's lap and walked unsteadily toward his mother, his face lifted up to the device.

Sometimes, as in Port Neches, a town not far from Port Arthur, the effect of a childhood exposed to industrial pollution may not show up for years.

In Port Neches, as in much of small-town America, communities dote on their high school football teams. The Friday night games of the Port Neches–Groves High School Indians are high points of the town's weekly social calendar, and the stands at the sidelines are always packed. The Indianettes, the drill team that performs at halftime, is an extra attraction at the games.

A teenager named Jane was a member of the Indianettes in the 1970s and a conscientious one. She would often spend hours by herself on the football field, practicing baton twirling and other routines of the drill team. As it happened, the field is immediately adjacent to a large facility that manufactures synthetic rubber using a compound called butadiene. Federal agencies have found butadiene to be a potent human carcinogen.

The field was also near a plant that had been the biggest producer of butadiene in the country. The plants had been there since World War II, when the U.S. government set them up to produce synthetic rubber to replace the natural rubber that was increasingly in short supply. After the war, the plants were turned over to private corporations, and operated by the B. F. Goodrich and Texaco-Uniroyal companies. Gulf Oil owned a rubber plant in Port Neches that provided the butadiene feedstock. Air monitoring outside the plants found levels far above safety standards. There are also oil refineries and chemical plants in the area.

Jane graduated from high school, married, and had children. At age twenty-nine, she was diagnosed with acute lymphocytic leukemia. She died at age forty-one.

Jane was not the only student at Port Neches–Groves High School to contract leukemia, lymphoma, and related cancers. In the years 1963 to 1993, twenty-six cases were verified by medical records or death certificates, according to Dale Hanks, a lawyer in Beaumont, Texas, who represents families of kids sickened by pollution. Of those young people, fourteen died. So notorious was the cancer rate among graduates of the school that some local residents refer to it as "Leukemia High."

The Texas cancer registry is inadequate, so other cases of cancer in the area may well have gone unrecorded. But even the incomplete Texas registry showed that cancer rates in Port Neches, Port Arthur, and Beaumont were significantly higher than the state average, particularly among children. Much of the Gulf coasts of Texas and Louisiana, which are carpeted with polluting refineries and chemical plants, experience high rates of cancer, respiratory disease, and other illnesses. It is not the best environment for the developing bodies of children.

Children are vulnerable to cancer-causing agents not only because of their developing bodies, but because of the long latency period of many forms of cancer. Dr. Richard Jackson, former director of environmental health programs for the Centers for Disease Control and Prevention, calls this phenomenon "the long shelf life of children." The growth of cancerous cells can be a slow process that takes place over years or even decades. Children and adolescents, with many years of life ahead of them, have ample time to let this insidious process proceed.

A study commissioned by some of the polluting companies found that there was only a small excess number of cancers among male students at the high school and a less than average number among female students. The study was conducted by a firm headed by Kenneth Rothman, a respected epidemiologist who is frequently hired by corporations as an expert to defend the safety of their products and operations. But as Dale Hanks pointed out, the firm's study included only those students who had died from cancer, not all those who had contracted the disease yet were still alive, and only students who had attended the high school during the years Rothman chose for the study. And as the study itself noted, there was also a substantial number of unexplained nonmalignant tumors among the young people who had attended the high school.

In response to an e-mail query, Jim McGraw, a spokesman for the International Institute of Synthetic Rubber Producers, the industry trade group, at first replied that he knew of no research into the health effects of butadiene. But later, in a follow-up contact, he said that the institute had conducted an occupational epidemiological study in 1975, which it continues to update, and found elevated levels of leukemia associated with high exposures to butadiene. In actuality, butadiene is one of the chemicals whose track record of carcinogenicity is the longest and most public. The University of Texas School of Public Health's 2007 study concluded that children living in areas with increased emissions of butadiene from petrochemical industries have an increased risk of leukemia.

There is no definitive proof that butadiene caused the illnesses at the high school, but the evidence suggests that had these children not gone to school in an environment laden with toxic chemicals, they would not have been at such high risk of falling victim to leukemia. Dr. Jonathan Ward of the University of Texas Medical School in Galveston noted that the highest recorded levels of butadiene had been recorded in the Port Neches area. There is not enough data, he said, to prove that the pollution actually caused the leukemia of the students there. "But any sense of caution should tell you that you certainly shouldn't emit high levels of a carcinogen."

The high school students were not the only children in the Port Neches and Port Arthur area assailed by cancer. Ann Tillery, an angry grandmother whose flame red hair looks like a walking danger flag, has been keeping track, often by obituaries in local newspapers, of children in her community who have fallen to cancer. One of those children was her own grandson.

"Justin got sick when he was twelve years old," she told us. "He was diagnosed with glioma, a very rare brain cancer. They said it was inoperable; he would be dead in six months. I took him straight to Ronald McDonald House, and he had six weeks of radiation. Then I took him to his home, gave him herbal remedies, music, and touch therapy. I learned how to play pool with him. I stayed at his house every day; I slept on the floor. I asked God, 'Where is my miracle?' He lasted forty-two months. I suppose that was my miracle."

She showed us a photograph album, on the cover of which was embossed "In Celebration of the Life of Justin Lee O'Neill." The pictures at the beginning of the album show a smiling, fat baby. In the middle was a husky young boy. At the end was a thin, bald, sad-eyed teenager.

GROWING A CHILD

For much of history, children were treated simply as smaller people. Physical and mental defects were attributed to inheritance alone or infectious disease and sometimes to improper care by parents. Until very recently, environmental regulation in this country failed to account for the vast differences in the physiology and behavior of children. It is largely still failing to do so. But one mantra we heard again and again as we gathered material for this book was, "Children are not little adults."

"Children are not just small adults" is the fundamental principle of the first international report, by the World Health Organization, on children's special susceptibilities to chemical exposures at different periods of their growth. The WHO scientists make it clear that methods to evaluate risks of chemicals do not yet adequately deal with these differences between adults and the young.

The conception, development, and growth of the human mind and body is an enormously complex and delicate process that involves the creation and evolution of trillions of cells at precisely the right time, at precisely the right rate, and in precisely the right sequence. Some have compared this process to an intricate dance. If so, the dance is an elaborate ballet where the dancers on a crowded stage must time their steps with exquisite accuracy to the music and to the movements of all the other dancers and where all the instruments of the orchestra in the pit must be together, in pitch and in tempo. If any of the dancers or musicians misses a step or a cue, it can confuse and disrupt the other performers, and the ballet can go irreparably awry.

The embryonic and fetal periods of new life contain numerous critical windows of vulnerability during which they are especially susceptible to the toxic action of many chemicals. The nature of that vulnerability changes with each stage of development, almost on a day-to-day basis. Interference at different stages and from different substances will lead to different outcomes. Even temporary disruption at the wrong moment can cause lifelong disabilities.

Fetuses are not necessarily more vulnerable to toxic exposures at every point. There are some periods where they are less susceptible and more able to resist and recover than adults. For the most part, however, new life is far more defenseless in the face of toxic assaults.

In the first weeks after conception, before major organs begin developing, the new life can be killed by a single "hit" of a toxic substance, causing a spontaneous abortion. As pediatrician and children's health advocate Dr. Philip Landrigan noted, "Most chemicals do their damage and then disappear. The mom might not even be aware that she was exposed to the chemical." In the later stages of embryonic development, cells begin to specialize under hormonal messages and migrate to different zones that will in turn originate different tissues of the body. The mother's production of thyroid hormones, along with many other factors, triggers fetal brain formation in the first few weeks of gestation and continues to control many of the stages of brain formation.

Birth defects such as a cleft palate, or other attacks on organs or major biological functions, are generated later in the development of the fetus while the organs are being differentiated, although some occur sooner in the development process.

Toxic exposures can cause "irreversible structural and/or functional abnormalities." In the fetal stage, new life takes on human shape. All of the major organs and vital function, including the nervous, immune, reproductive, respiratory, cardiovascular, gastrointestinal, excretory, and endocrine systems, emerge during this period. Skeleton, skin, and limbs become fully formed. During early fetal life, all children are alike, basically female in organization. As the testes develop, they release the hormone testosterone, which launches the reorganization of the fetus leading to maleness. Hormonal drugs and synthetic chemicals can interfere with this process.

As this process goes forward, there is constant messaging among the various components of the developing organism that instructs the DNA (deoxyribonucleic acid) when and how to act. DNA is the fundamental human genetic material; it contains the genes, handed down from generation to generation, that direct the construction and functioning of every cell and every organ in the body. Humans have at least thirty thousand genes in each human cell—although not all are functional.

It is hard to exaggerate the complexity of this process. The nervous system alone, for example, contains over one hundred billion cells for transmitting information and over a trillion connecting cells. They regenerate less readily than other cells, such as liver cells, after a toxic assault. Many of these cells have to move from one place to another to form synapses, or connections, differentiate, and get covered with a protective myelin sheath. Interestingly, the fetus contains more brain cells than the adult human, and normal development requires many of these cells to die off as a way of fine-tuning neural circuits. This delicate cascade of elements places the brain at special risk of a "whoops" factor that can forever diminish the child. Pollutants that can cause permanent damage to the developing brain—"neurotoxicants"—include, among others, mercury, lead, pesticides, and PCBs. Chemicals that have scant or no effect on the adult brain can produce permanent changes in a child's.

Information about the connection of environmental exposures to parents and the impact on fetal systems is emerging at a rapid clip. For example, Dr. Paul Ashwood at the M.I.N.D. Institute in Sacramento, California, a research center and clinic for children with neurological illnesses, reported that a new study suggested that asthma among expecting mothers is associated with an increase in autism.

The nervous system is especially vulnerable to toxic harm. "If something happens to the brain at development, you don't get a second chance," said Dr. Philippe Grandjean of the Harvard School of Public Health, who has extensively studied the effect of toxic metals on the brain. Heavy metals disrupt the normal workings of neurotransmitters, the molecule messengers of the brain and nervous system. Neurotransmitters leap by electrical discharge from the end of one neuron to the next, bearing signals. Among the fifty or so neurotransmitters are dopamine and glutamate. After their signals have been transmitted, enzymes normally rush in to wipe away any excess neurotransmitter molecules. This step ensures that the neurons do not continue to send excessive signals, which would cause excitability.

In their book, *Generations at Risk*, Dr. Ted Schettler and his colleagues emphasize that "any of the organs or processes whose coordinated function is essential for normal reproduction and subsequent development of the fertilized egg is a potential target of toxic exposures." They go on to note, "Normal reproductive function requires timing, balance, properly set feedback loops and communication among cells and organs from the time of conception through the reproductive years. Normal fetal, infant and child development depends on genetic makeup, a healthy environment and interactions between the two."

Environmental hazards have many means to disrupt the development of an embryo or fetus. Poisons in the mother's system can damage or destroy the new and rapidly proliferating cells. They can cause mutations in the genetic structure. Toxic substances can also cause serious harm to the developing child without changing the basic structure of the genes, but by chemical modifications that change the way those genes express themselves—the way they turn on or off. Gene expression is the process by which a gene's information is converted into the structures or functions of a cell.

In normal development, messenger RNA (ribonucleic acid) reads the genetic code in the DNA for making proteins, the basic building blocks of life; the RNA copies and carries the message to the cytoplasm, which then produces the proteins that the gene had coded for. Some of these proteins are enzymes that cause specific chemical reactions, including detoxification and gene repair. The unique pattern of proteins in each and every one makes us who and what we are.

John Peterson Meyers, coauthor of *Our Stolen Future,* notes that the pattern of gene expression "has a huge impact on a person's life. When working properly, those changes give you male or female sex organs, a functional brain, and an immune system that defends against disease. They regulate weight and protect you from cancers. In short, the right changes in gene expression as the fetus is growing are essential to a person's health and quality of life." But, as Peterson notes, genes can be "hijacked" by exposure to even minuscule amounts of a toxic substance, changing their normal expression and leaving the body vulnerable to disruption of health and quality of life.

Some toxic hits as small as a molecule can cause the RNA to deliver the wrong message or deliver it at the wrong time, causing too little or too much protein or another malfunction. The wrong message could, for example, interfere with the process of programmed cell death, which in turn could lead to physical deformities of the infant or to childhood cancer. Michael Lerner, who directs the Collaborative on Health and the Environment, compared gene expression to the music roll on a player piano. The piano still has exactly the same notes, but a new roll determines how the notes are played, in what order, and at what tempo.

Gene expression is "emerging as the main and so far missing link between genetics, disease and the environment that is widely thought to play a decisive role in the etiology of virtually all human pathologies," states the Sanger Institute, a renowned genome research institute.

Many environmental toxicants can affect gene expression by disrupting the hormone system of the developing baby. The system consists of a number of glands that secrete hormones. These glands include the thyroid, the adrenals, the pituitary, the female ovaries, and the male testes, among others. The system performs many essential functions in the development of the fetus, including controlling metabolism, governing behavior, determining sexual development, and a broad spectrum of other biological services. In recent years, scientists have found close interconnections among the nervous, immune, and endocrine systems.

Any disruption of the hormone system of a child in the womb can lead to devastating consequences. From the 1950s through 1970s, a large number of pregnant women in the United States took DES (diethylstilbestrol), a drug that mimics estrogen, to prevent spontaneous abortions. A significant percentage of their daughters were afflicted with cancer of the vagina and cervix and also had trouble conceiving their own babies. Some studies suggest that even their granddaughters may be at greater risk of developing cancer. Male babies exposed to DES in utero developed malformed testicles, and many had greatly reduced sperm counts. There is now increasing evidence that these endocrine disruptors can masculinize female brains, feminize male brains, and cause other disorders.

FRAGILE DEFENSES

What had been thought of as built-in defenses for the developing embryo and fetus have been found in recent years to be fragile and permeable. It was formerly believed, for example, that the placenta filtered out harmful substances before they could reach the developing child. That belief was dispelled tragically when children born to mothers who had taken the drug thalidomide in order to prevent nausea during pregnancy emerged with shrunken, deformed, or missing limbs. Further, the "blood-brain barrier," which shields the brain from harmful substances in the blood, is undeveloped for babies in the critical embryonic and fetal stages and is not, in fact, fully formed until the child is six months old.

ENTERING A HAZARDOUS WORLD

The newborn child is by no means safe after it emerges from its mother. While the windows of vulnerability are narrower for a baby and child than for an embryo or fetus, the new life remains extremely vulnerable.

Although children live in the same world as their parents, because of their size, metabolism, eating and playing habits, continuing physical and neurological development, and inability to defend themselves from whatever is in their surroundings, they inhabit a very different environment.

Children spend most of the day at a different altitude from that of their parents and other adults. Whether crawling or standing, they live closer to the floor, where chemical residues can linger for weeks or longer in dust and on carpets. Concentrations of some toxic substances such as pesticides are four to six times higher near the floor than at adults' breathing level. Children are also out of doors more than adults and are exposed to outdoor pollution: They play on the grass, which may be contaminated with insecticides and herbicides, or on the playground or dirt, which may be tainted with arsenic or other poisons. Infants, who are inside most of the day, are exposed to indoor air pollution, now recognized as a potentially serious health problem. Small children also notoriously engage in constant hand-to-mouth behavior. Anything they can grab—and today much of what they can grab contains toxic ingredients—can go into their mouths and the contamination into their systems. In their book, *Toxic Nation*, Fred Setterberg and Lonny Shavelson told of an unguarded chemical dump site in Columbia, Mississippi, that local children used as a playground. They noted that the ground there is coated with a viscous mud infused with benzene, which can cause leukemia, and quoted a six-year-old boy as reporting enthusiastically, "There's gooey stuff and you can squish and play with it."

Kids eat more food, drink more water, and breathe more air pound for pound than do adults. Their diets, which include a much higher percentage of fruits and vegetables, juices and milk, expose them to a substantially higher intake of pesticides than the diets of most adults.

Dr. Richard Jackson recalled that a study by the American Academy of Pediatrics in 1991 found that children were getting forty times more pesticides from foods per pound than adults. Moreover, children's brains and nervous systems continue to develop through adolescence and remain vulnerable to assaults of toxic substances.

Immature bodies have lower levels of defenses against assaults on their bodies. Because their lungs are still developing, their airways are narrower, and because they have a more rapid rate of metabolism, kids need more oxygen and breathe more rapidly and are far more sensitive to air pollution. Immature bodies and organs, moreover, have less of an ability to get rid of many invasive toxics. They absorb a higher percentage of the pollutants to which they are exposed. They retain, for example, about half of the lead they ingest, while adults absorb only one-tenth of that lead. The liver, kidneys, and other cleansing agents in the bodies of the young are much slower to eliminate foreign substances than those in adult bodies. The developing immune system can be significantly more sensitive to toxic assaults.

Children are further lacking in the capacity to repair oxidative stress. Oxidative stress is an outcome of the body's perpetual struggle to rid itself of environmental contaminants and the by-products of normal cellular metabolism; in its struggles, it generates oxygen compounds in the cells known as "free radicals." As free radicals build up, they create oxidative stress, which interferes with the enzymes that should clear away excess neurotransmitters. To prevent or detoxify these injurious free radicals, cells of mammals have developed an army of antioxidants in their immune system. Even plants under environmental assault can suffer oxidative stress. However, from conception through infancy, all children have low levels of the most powerful antioxidant, glutathione, a small molecule found inside every cell.

The fetus and young child are further handicapped in the capacity to detoxify substances because the levels of protective enzymes that would enable them to break down the toxics do not fully develop until adulthood. In addition, their immature immune and nervous systems can be damaged more easily and have a lower capacity to make repairs.

Because of these vulnerabilities, children are, on average, ten times more vulnerable to cancer-causing chemicals than adults and accumulate half their lifetime risk of cancer by age two, according to EPA estimates. The EPA further estimates that chemicals that cause cancer by interfering directly with DNA are up to sixty-five times more potent for infants and toddlers than for adults.

Children's health is also sensitive to their social environment. Such factors as exposure to mental illness or high anxiety in the family, a missing father, large family size, poverty, inadequate nutrition, or minority status can have an effect on the mental and physical well-being of the child. The harm to a child's intellect from exposure in utero to a pesticide is more pronounced if the mother has experienced deprivation of shelter, food, or clothing at some point during her pregnancy, it's been discovered.

Beyond the issue of toxic substances, children lack the ability to protect themselves from the countless other dangers that lurk in their physical environment. They cannot recognize and avoid poisoned foods, household products, and toys; they cannot tell when water is tainted or air polluted. Their parents and teachers can shield them from many of these assaults up to a point—but only up to a point. The stresses of their combat with ever-present toxic chemicals will diminish the ability of children (as well as adults, animals, and plants) to deal with the effects of global warming, and global warming will bring new diseases that will put further stresses on their bodies.

But why do some children get sick from toxic exposures and others do not? The decoding of the human genome has only recently begun to unravel this mystery. It has revealed that in addition to inherited variations that are genetic mutations, there are inherited variations that are slight differences in the sequence of DNA in the genes. These variations, called "polymorphisms," are the reason a child is born with green rather than brown eyes or blood type A rather than O. They predispose a person, from the womb throughout life, to be more or less vulnerable to environmental insults such as bacteria, viruses—and toxic chemicals. Yet a child might have a predisposition to asthma or autism that might never surface without the toxic assault. One of these variations, for example, can affect the ability of a fetus or child or adult to metabolize and excrete heavy metals such as lead. As Dr. Jill James, a professor of pediatrics at the University of Arkansas for Medical Sciences, discovered, the autistic child's immune system is lower in the stress-fighting antioxidant glutathione, as well as lower in the amino acid cysteine, which synthesizes glutathione, so that his or her system is inept at detoxifying pollutants such as heavy metals and pesticides. A recent study has confirmed that children with autism suffer from oxidative stress, as shown by the reduced amount of blood reaching their brain.

Another minor gene variation can affect the level of enzymes available to protect against carcinogenic damage from pesticides or air pollutants.

These variations mean that a fetus or child might well be doubly vulnerable: first, by the defenselessness of his or her immature body; and second, if the fetus or child carries a variation that predisposes him or her to a lowered level of defense to specific toxic substances.

DIFFERENCES OF OPINION

Not all physicians and scientists subscribe to the view that children are particularly vulnerable to toxicants in the environment. Writing in the journal *Pediatrics*, Dr. Robert Brent of the Alfred I. duPont Hospital for Children and Dr. Michael Weitzman of the University of Rochester School of Medicine and Dentistry asserted that there is a "polarization" over the issue in the scientific community and that some "scientists and lay individuals believe that environmental risks have been grossly exaggerated."

Still, it is difficult to explain away the growing mass of evidence linking the rising incidence of chronic childhood illness with the flow of hazardous pollutants into the environment.

Four

EVIDENCE

DISABILITY, DISEASE, AND DYSFUNCTION AMONG OUR NATION'S CHIL-
dren have reached epidemic proportions.

Of America's 73 million children, almost 21 million, nearly 1 out of
3, suffer from one chronic disease or another.

Cancer threatens the lives of 58,000 children. Almost 2.5 million
live with disfiguring, debilitating birth defects. Those whose bodies
and minds are poisoned with lead number 310,000. About 6 million
children suffer and some of them die from asthma. Twelve million
have some form of developmental disorder, from autism to ADHD
and serious learning disabilities that cloud their minds and torment
their behavior.

These numbers mark today's reality. It takes these stark, aggre-
gated statistics to grasp the extent of the epidemic of childhood chronic
illnesses.

A SAD TRAJECTORY

Even more telling are the rates of increase in illnesses from the past to
today. These rates show a steep upward trend, a trajectory that
presages a distressing future if we do not reverse it. These rates parallel the swelling volume of toxic chemicals invading our children's lives
from one generation to the next.

This trajectory began with the baby boomer generation, those children born after World War II to parents whose bodies were fairly free of man-made toxins. The boomers' birth and childhood coincided with the onset of our nation's unthinking, revolutionary experiment with "Better Living Through Chemistry." The boomers had lindane scrubbed into their hair to get rid of lice and cavorted in the mists of DDT sprayed over their neighborhoods. They were the first generation exposed from infancy to a massive array of synthetic toxic products; as they matured, new, more powerful toxins were created, seeping ever more deeply into their bodies.

Twenty-five years later, the baby boomers began to give birth to the next generation. These children, exposed to pollutants in the boomers' wombs, began to exhibit unprecedented rates of chronic illnesses and disabilities. Now, as these boomers' kids start to have children, or struggle to do so, the numbers and rates of increase in illnesses drive further upward. (As for the boomers, as they age, they are experiencing their own epidemic of the kinds of environmentally triggered illnesses that take years to show up—breast cancer, prostate cancer, Parkinson's disease, Alzheimer's disease—in rates considerably higher than those experienced by any generation before.)

Childhood cancer was once a medical rarity. In two generations, from 1950 to 2001—from the boomers' childhood to their children's—the rate of childhood cancers of all types leapt 67.1 percent, exceeding rates among adults. The United States has the fourth-highest incidence of childhood cancer in the world. Since the 1970s, when tracking such data became reliable, brain cancer among children has increased about 35 percent and acute lymphocytic leukemia over 47 percent. The worst is not behind us. On the contrary, the highest rates for all childhood cancers are for the years 2000, 2001, and 2002, and it's thought the trend will continue. Though treatment has improved almost miraculously, one-sixth of children who have had cancer will eventually face a second bout.

The trends for birth defects are uncertain because registries to track them are inadequate. It is known, however, that of the 4 million babies born in the United States each year, about 150,000 (or 3.5 percent) are born with immediately apparent defects, such as cleft lip and palate, or a hole in the wall between the two upper chambers of the heart, or with partial blockages of the urinary tract. Because defects of even a major nature are not always detected at birth, 8 percent is probably the true rate. Major structural birth defects cause about 70 percent of all deaths before one month of age and about 22 percent of the six thousand deaths each year of infants less than twelve months of age.

Asthma among children, prevalent in rural Iowa as well as in the tenements of Harlem, increased from 3.6 percent in 1980 to 8.7 percent in 2001. A 2006 study from the Centers for Disease Control and Prevention declared that the prevalence of childhood asthma reached a historically high rate. It is the foremost cause of school absenteeism in America, outranking colds and flu; it is the most frequent cause of hospital admissions for children beyond the newborn period. Among children under four, the disease has exploded by 160 percent. The death rate, however, is beginning to decline, except among black children.

Learning and behavioral problems, from autism, dyslexia, ADHD, diminished intelligence, and mental retardation to a propensity to violence, are all on an upward trend. About 16 percent of boys and 8 percent of girls between the ages of five and seventeen have now or at some recent earlier point been diagnosed with ADHD. A national household survey conducted by the CDC found that 6.5 percent of children under the age of eighteen have one or more learning disabilities, that 6.5 percent have emotional or behavioral problems, and that 4 percent suffer from delays in growth and development. In addition, 1.6 percent, meaning approximately 1.17 million children under the age of eighteen, are mentally retarded. About 300,000 American children, or 1 out of 166, have been diagnosed as having autism. The trends show no sign of abating. Other developed nations report similar rising levels of the same illnesses.

There is also strong evidence that children are increasingly disabled before they are born. Such disability shows up in preterm birth and low birth size; the United States has the highest rates of both. Preterm births have increased 23 percent over the past two decades and continue to rise, even discounting prematurity from multiple births. Every year, more than 440,000 babies are born before the end of thirty-seven weeks of pregnancy and are also born dangerously small, weighing less than five pounds. Those rates too continue to creep upward, by 8.2 percent in 2005. Preterm and low-weight babies are more likely to suffer from lower IQ and learning problems, cerebral palsy, mental retardation, autism, asthma, and diabetes. We also have one of the highest infant mortality rates in the industrialized world, ranking forty-third in the survival of newborns among all developed nations.

About half of all pregnancies in the United States result in the baby's death either in the womb or shortly after birth or in an otherwise less than healthy child. Spontaneous abortions, including those that occur before the woman knows she is pregnant, end 30 to 34 percent of all pregnancies. Among fetuses who die, an increasing number are male.

Over the course of modern history, there had been a constant ratio of 100 female births to 106 male—that is, 51.5 percent of newborns in modern history had been male. No longer. That proportion of male to female births has declined since 1950 worldwide. The United States has seen a decrease since 1970 of 1.7 males per 1,000 live births; in Japan, the decrease is 3.7 per 1,000. In the United States and Japan combined, more than a quarter of a million boys are missing since 1970 compared with the number that would have been expected had the sex ratio remained unchanged. Researchers warn that this decline may be a "sentinel health indicator"—a serious shot across the bow of human health, a wake-up call for action.

Further, boys are facing an increase in demasculinization and a complex of reproductive disorders, including a deformity of the penis in which the urinary tract opening is in the wrong place, called "hypospadias," undersized and undescended testicles, called "cryptorchidism," and testicular cancer. These trends appear in other developed countries, at higher incidence in some than in others. Sperm quantity in industrialized countries appears to be declining at an average rate of about 1 percent every year, a remarkable 50 percent in fifty years. The decline, not confined to the United States, is sharper in some countries than in others.

Girls face their own set of reproductive problems. They are beginning to develop secondary sexual characteristics at a younger age, which signals a future problem since earlier bursts of estrogen are associated with increased breast cancer risks later in life. A deformity of the uterus that can cause infertility, known as "endometriosis," now strikes twice as many teenagers as a generation ago. At least 1 in 12 American couples reports problems conceiving, turning fertility treatment into a big business, with more than one million customers and revenues of at least $3 billion a year. As with low-birth-weight babies, women under age twenty-five, the group that should have the easiest time of it, report the most problems in conceiving or carrying a baby to term.

PATHOLOGICAL DENIAL

These reports of illnesses are not without controversy. Some critics argue that the epidemic is more apparent than real, the product of better detection and record keeping, of increased reporting or changes in reporting, or of expansions in diagnostic criteria or such. While these explanations may account for some of the numbers, they fail to explain most of them.

The rising cases of malformation of the penis (hypospadias), for instance, cannot be chalked up to improved reporting and diagnosis because increased rates have occurred for severe cases, which are unmistakable, as well as for mild cases. The enormous rise in asthma cannot be explained away by an increase in bad genes, as more and more people with no family history of asthma are developing the disease.

Some researchers argue that the increase in cancer represents improved technology and detection, such as the use of magnetic resonance imaging (MRI). Other scientists reply that the increase in cases is real, so large that better diagnosis and reporting are unlikely to be the principal explanation, that it is impossible to miss brain cancer because the symptoms are so painfully obvious or to miss leukemia because the tests are so accurate, that the observed increases of testicular and other childhood cancers are too consistent and too large to reflect anything but actuality. Rather, they say, the incidence may be underreported, as a result of reporting delays and errors such as those that have caused upward revisions in trends for adult cancers. Dr. Martyn Smith, a cancer researcher at the University of California at Berkeley who attributes a good portion of the rising statistics to improved diagnosis, adds, "Something is seriously going wrong if children are getting cancer at all."

Learning and behavioral disabilities may be overdiagnosed because medication is so easy (and profitable) a solution, as skeptics claim. Maybe these are just the "bad boys" who were kept in line through remedial beatings in the old days. "Baloney," responds Jo Behm, a registered nurse who has been a longtime leader of the advocacy organization Learning Disabilities Association of California and the mother of a son with dyslexia. "Veteran teachers are definitely reporting a sharp rise in the number of students, especially boys, in their classrooms with learning, attention, memory, and organization problems—on top of other mental health issues. It's tragic when those who should know better casually write this off as better diagnostics."

Debates about autism—what it is and how much of it there is—raise the most dust. Skeptics claim that the change in diagnosis criteria for autism and special ed classification means that children are now diagnosed as autistic who might have been otherwise categorized. Dr. Martha Herbert, a pediatric neurologist at Harvard Medical School, calls this viewpoint "a near pathological denial." If diagnosis and labeling are at issue, where are the studies that have found the previously misdiagnosed autistic people among older Americans? Further, the claim for diagnostic substitution does not hold up in light of California's situation. There, the number of children entered into the autism registry increased by 210 percent between 1987 and 1998. Yet during that increase in autism, reporting for cerebral palsy, epilepsy, and mental retardation remained stable or increased slightly. Nor can the increases in autism reporting be attributed solely or mainly to families moving to the state in order to avail themselves of special services.

Whatever the debate, Dr. Herbert says, "even though we may have neither consensus nor certainty about an autism epidemic, there are enough studies coming in with higher numbers that we should take it seriously. Environmental hypotheses ought to be central to research now."

THE ASSAULT

A heavy and constantly growing weight of evidence supports the thesis that this rising tide of childhood illness is linked to the rapid toxification of the world in which children are now conceived and born.

We live in a society drenched in pollutants. The United States now produces or imports at least fifteen trillion pounds of chemicals a year, which works out to forty-two billion pounds of chemicals per day.

The increase since 1980, when our nation produced two hundred billion pounds of chemicals annually, works out to a 750 percent increase in twenty-five years.

The chemical industry is now the largest manufacturing sector in our nation, a behemoth that grew to revenues of $635 billion in 2006 from $484 billion a year in 2004 and from a relatively modest $2 billion a year in 1962, when Rachel Carson wrote *Silent Spring*. It might more accurately be called the petrochemical industry, since oil is the base for 90 percent of the chemicals. Shell Oil, for example, which transmuted petroleum into the formerly favorite and now banned cancer-causing pesticides aldrin and dieldrin, counted its revenues at $180 million a day in 2007. These products are created from chemical combinations that do not exist in nature.

The diversity of synthetic chemicals has exploded in parallel with its volume. Carson wrote about the two hundred pesticide products in use in the 1960s. Today, there are nine hundred active pesticide ingredients, formulated into about eighteen thousand different pesticide products. Roughly 4.5 billion pounds of pesticides are applied annually by households and agriculture, compared with the 400 million pounds a year used in the 1960s, which translates into an increase of 1,125 percent in pesticide use in twenty-five years.

In addition, there are now eighty-two thousand industrial chemicals, formulated into more than ten million products. Industrial chemicals are those used in the manufacture of other products.

So in sheer volume and variety, the growth is astounding. The assault is also much broader than the world of pesticides that engaged Rachel Carson's attention. The danger spans not only pesticides, but food additives and growth hormones, schools built on toxic waste dumps, and malignant manufacturing processes, along with the environmental hazards of everyday life, from infants' plastic teething rings to fake wood.

Most of these industrial chemicals are captured inside the products they were used to manufacture, then slowly release as the products degrade, transforming into household dust, leaching into food and water, piling up in waste dumps. Still, a substantial portion is left to escape from their manufacturing sites into the air, soil, and water, year after year. Industry must report to the Environmental Protection Agency's Toxics Release Inventory the releases of some 650 of these pollutants, under legislation narrowly enacted by Congress in 1987 to advance citizens' right to know, after the deadly release of a pesticide gas in Bhopal, India, raised citizen anxiety about chemicals here. The most recent report puts the release at at least 4.4 *billion* pounds—1,500 pounds for each man, woman, and child in the United States each year. And that's probably a vast undercount.

FINGERPRINTS

Synthetic chemicals and heavy metals often leave their marks behind. Even the baby in the womb carries these imprints, passed into his or her body from both mother and father, the traces of their lifetime exposures to chemicals. New technology now allows researchers to measure the body burden of those chemicals that enter and linger inside babies in the womb; they find a surprising variety in every fetus.

Not one of these body burden studies, not by the government or by the private or nonprofit sector, has yet to find an unpolluted child.

Ten newborns from different regions of the country, tested in late summer 2004, all carried a body burden of between 154 and 231 chemicals in their umbilical cord blood, meaning that these chemicals had entered their fetal bodies. Among the group as a whole, each of the 287 chemicals tested for was detected. The chemicals included mercury, stain repellents such as Teflon, flame retardants, by-products of vinyl incineration, and molecules from air pollution. DDT and PCBs, once used to insulate electric equipment, were also found in the cord blood, though these two chemicals had been banned before the babies' parents were born.

When the Columbia Center for Children's Environmental Health ran tests on a group of 230 mother-and-newborn pairs, the infants' first stool universally carried fingerprints of the then most popular type of pesticide, chlorpyrifos (CPF, whose brand name is Dursban), while their blood samples revealed that twenty-nine other pesticides had also invaded their unborn bodies.

Tests of the chemical body burden among newborns are still rare and very costly, but testing among children six years and older has become more common. By now, the CDC has run three biannual nationwide studies of children six to eleven years old as well as of adults. In the most recent and largest CDC study, one hundred different substances were found in the bodies of the children.

Other body burden studies come up with parallel findings. When the 3M Company tested 598 children from twenty-three states for traces of Scotchgard, the stain-resisting chemical that was one of the company's best sellers, they found traces of it in every child. The same kind of pesticide, chlorpyrifos, that the Columbia Center researchers found in newborn stool was found in the body of every child the CDC tested and found at levels four times higher than the level the EPA considers "acceptable" for a long-term exposure.

Not only are unborn and newborn babies and infants more saturated with chemicals than one could have imagined, but the concentrations within their bodies are higher than in adults, a surprise even to scientists. Mercury concentrates in umbilical cord blood at levels 1.7 times higher than the level in the mother's blood, a senior EPA scientist, Dr. Kathryn Mahaffey, recently discovered. This higher burden, she explains, probably reflects the fact that the developing fetus does not excrete mercury as efficiently as an adult. Dr. Mahaffey's discovery means that twice as many babies are born each year with unsafe mercury levels than the government had earlier estimated. The number works out to 630,000 newborns, 1 in 6.

School-age children carry a burden of about one and one-half times more than adults of a family of chemicals, called "phthalates," used in plastics and cosmetics, while DEET, the active ingredient in many common mosquito repellents and various other pesticides, showed up in children's bodies in higher concentrations than in adults', sometimes twice as high.

Rowan, a toddler whose body burden was tested in 2005 for an *Oakland Tribune* (Oakland, CA) investigative news series, carried a higher level of flame retardants in his twenty-two-pound body than his parents, and his level was higher than ever found in someone not handling these chemicals for a living.

Women of childbearing age also carry a high burden of synthetic chemicals, sometimes higher than other adults. These toxins can affect conception or a child in the womb. Women have on average a body burden of methylmercury at or above the "acceptable" standard. While every single adult the CDC tested had one particular type of phthalates in their bodies, women of childbearing age carried a higher level of the chemical, suspected of disrupting the fetal hormone system.

Science has not yet perfected a method to measure chemicals that vanish without leaving a trace. Such a test will result in body burden statistics far higher than today's.

PROTECTING CHEMICALS OR CHILDREN?

Science still can rarely link a specific chemical burden to a specific harm in a specific child. As CropLife America, the trade association for the pesticide industry, took pains to point out twice in its brief press release in response to the CDC study findings, "The public should take assurance from CDC's position that just because people have an environmental chemical in their blood or urine does not mean that the chemical causes disease or adverse health effects." This may be a valid point, explains Dr. Brenda Eskenazi, a neuropsychologist who heads a children's environmental health research center at UC Berkeley's School of Public Health and has struggled with the complexity of environmental health for years. "The chemistry is getting so good that we can detect anything at the level parts per billion. The average person thinks that because we can pick it up in the body, it's got a biological effect. It may be that we're just really good at chemistry at this point."

Yet, as Dr. Bruce Lanphear of the Cincinnati Children's Hospital Medical Center cautions, "Saying 'just because you can measure them doesn't mean they are causing harm' assumes that when there is any uncertainty about the toxicity of a chemical, we should protect the chemical rather than human health. Given the numerous examples of other chemicals shown to be toxic, any rational policy would work in the reverse."

Science can in fact connect exposures to specific chemicals to specific illnesses, if not yet in a specific child. The stain repellents in Teflon and Scotchgard, whose innocuous nature the manufacturers DuPont and 3M upheld for decades, turned out in lab tests to be a likely cause of cancer, birth defects, and liver damage.

COSTS

Pollution's assault on children carries economic ramifications that are seldom considered and even more seldom quantified.

The drop among a generation of children of a "mere" five points in average IQ shifts the national IQ from 100 down to 95, almost doubling the number of retarded adults and halving the number of gifted. Instead of 6 million gifted adults, the nation will have only 2.4 million to draw on for leadership. Instead of 6 million mentally retarded adults, the nation will have 9.4 million to support.

Some scientists seriously fear "a generation of intellectual cripples," a downward national slide in intelligence. "Maybe we're all getting exposed to chemicals that will make us too stupid to figure out that we're being exposed to something," says Duke University's Theodore Slotkin, professor of neurobiology (and of psychiatry and behavioral sciences and of pharmacology and cancer biology).

This is not a hypothetical concept; it is current reality. Over half of the most heavily used industrial chemicals are known to be toxic to the brain and nervous system. They diminish intelligence and disturb behavior. Further, the intellectual development of children with asthma, cancer, and birth defects may be compromised by the school days they miss, the pain that diverts their powers of concentration, and the treatments and medication that dull their awareness.

Another way to look at costs is to consider how a nationally diminished IQ will affect productivity. The reduced productivity from just one pollutant, mercury, found in hundreds of thousands of children's bodies at levels that cause loss of intelligence, costs our nation at least $2.2 billion and as much as $43.8 billion annually. If the cumulative effects of environmental toxicants reduce the average American's IQ by just one IQ point, the annual cost to society would come to $50 billion and the lifetime societal costs to trillions.

There is also the cost of caring for sick children. As calculated by a research team led by Mt. Sinai School of Medicine's Dr. Philip Landrigan, that cost is at minimum a staggering $54.9 billion a year in current dollars. This figure covers only four illnesses—lead poisoning, cancer, asthma, and learning-behavioral disorders—and includes only the portion of the illnesses that can be confidently attributed to environmental factors. The $54.9 billion includes emergency rooms, prescription drugs and therapies, special education services and home care, the parents' forgone earnings from days of lost work, and the costs of the children's lost lifetime productivity from death or disability, but it does not include the pain and suffering of the children and their families. The costs of caring for preemies amounts to another $26.2 billion a year.

The costs of lost leadership, reduced productivity, and year-in-year-out care are borne by the children, their families, and American society, not by those who generate environmental toxicants. The manufacturers are relentlessly effective at "externalizing" these costs—having someone else pay for them. "A corporation tends to be more profitable to the extent it can make other people pay the bills for its impact on society," notes economist Robert Monks.

CONCEALED EVIDENCE

How is it possible that our children are in jeopardy from commonplace products, that a toxic chemical in plastic baby bottles and another chemical used in lipstick have become part of the warp and woof of daily life? It is because the evidence of their possible harm has been concealed.

We assume there are systems in place to make sure any product we buy or use has been tested and approved by the government or some surrogate empowered to watch over our lives. After all, we are required to wear seat belts, our water has been fluoridated, and our drugs have been approved by the government.

In the United States, however, industrial chemicals (those used to manufacture other chemicals) are rarely tested for health and safety before they are sold and used. Our country's system for allowing chemicals into production inspires all too many metaphors. It's the fox guarding the henhouse. It's leap before you look. It's Alice in Wonderland: Words have elusive meanings, the rules keep changing, children lose the game.

Two facts explain why. First, industries have gamed the system since the first law addressing chemicals' impact on health and safety was passed. The second fact is that scientific advances have completely changed the picture; previous decades used the inadequate science of the day to examine chemicals, resulting in false assurances of safety.

The most basic information—whether an industrial chemical is toxic or not—is not available for most chemicals. (An analysis made ten years ago found that basic information was missing for 92 percent of the chemicals produced in greatest volume.) That is, it's missing from the public record. The manufacturers withhold a lot of information as "trade secrets." Some companies have even asserted that the location of their plant site was a secret. States are denied access to information they need about chemical plants for antiterrorist planning.

Chemicals that do get tested are examined not to prevent, but to set "acceptable" levels of potentially harmful use.

Industrial chemicals are considered innocent until proven guilty. Contrast this situation with the requirements for a chemical marketed as a drug, which before it can be sold must first pass a series of safety tests on animals and humans, mandated by the Food and Drug Administration (FDA). If a company wants to sell an industrial chemical that acts identically or very similarly, there is no regulatory agency or governmental oversight to ensure the product is safe. "Medicines are the only chemicals that have to be proven safe," explains the Mt. Sinai School of Medicine's Center for Children's Health and the Environment.

There have been some efforts to change this situation, but with scant success. One was the Toxic Substances Control Act (TSCA), passed in 1976 in hopes of heading off harmful chemicals before "the bodies pile up," recalls Jacqueline Warren, an attorney and environmental consultant. But the TSCA was and is weak, and furthermore, an exception immediately rent a huge hole in the law. The sixty-two thousand industrial chemicals in use when the law passed were "grandfathered"—that is, allowed to stay in commerce, exempt en masse from any testing. The EPA has used its authority to require testing of fewer than two hundred of these chemicals. Yet it is these exempt chemicals that constitute the vast bulk of the chemicals we use.

The situation is scarcely more reassuring for new industrial chemicals. The manufacturer must send an application to the EPA to register it for production, but the EPA's assessment is limited to the studies the manufacturer supplies. The EPA does not—indeed, is not allowed to—conduct safety tests. Dr. Michael Firestone of the EPA's Office of Children's Health Protection said, "The first day I came to the EPA, I was astounded to find that the people who are asking to register these products are the ones giving us the data."

The manufacturer is under no obligation to study a chemical's effects on people or the environment. The company is required only to identify the chemical and its structure and to offer an estimate of its production volume, a number they're allowed to change later. Some companies do include information about rudimentary screening studies they may have conducted.

If they have done tests that surfaced any indication of possible harm, companies are supposed to advise the EPA. That requirement is a clear disincentive, since such a heads-up allows the EPA to ask for more information. Indeed, numerous companies that have come across adverse information have hidden the facts. "Companies don't set out to harm their customers," concludes Eric Schaeffer, who as former head of the EPA office in charge of enforcing the laws witnessed the dynamics of corporate behavior. "The problem is that industry has dueling instincts; there's a human tendency to bury the bad news when it conflicts with their self-interest."

The Government Accountability Office (GAO), the independent federal watchdog agency, reviewed the situation and found that companies had provided health data to the EPA for about 15 percent of the industrial chemicals that had been introduced over the past thirty years, since the 1976 passage of the act. Without data from the company, the EPA resorts to computer models of molecularly similar chemicals (often not available) for which toxicity data are available (also often lacking), and these models are not always accurate in predicting if a new chemical will be harmful.

A chemical, once approved, may be put to any other use by anyone else in any quantity without notifying the EPA, with but a handful of exceptions.

Since 1976, the EPA has banned or restricted only five groups of grandfathered chemical substances—including PCBs and dioxin—and the last time the agency tried to ban a chemical was in 1990. The substance was asbestos. The EPA began gathering evidence about asbestos in 1979; over the next ten years, it reviewed more than one hundred studies of the health risks, amassed comments from outside parties, and finally concluded that asbestos at any level brought along the risk of injury or death. In 1989, the agency issued a rule banning asbestos. The asbestos manufacturers fought back. They filed a lawsuit, which they arranged to bring to the Fifth Circuit Court of Appeals, a court long known as strongly conservative. In 1991, that court ruled with the manufacturers, finding that under the TSCA's standards the agency could not demonstrate substantial enough evidence of unreasonable risk and that the EPA's decision was too burdensome for the manufacturers.

Since then, the agency has tried to regulate the chemical industry only by asking companies to pass information on to consumers or to remove a hazard voluntarily. Arsenic, for example, was considered "regulated" when industry agreed to write up consumer information sheets. "Every one of those agreements puts children at risk because of the nature of the negotiated settlement," asserts Jay Feldman, who for decades has headed Beyond Pesticides, a national nonprofit group trying to rid our country of the most dangerous pesticides and find better alternatives.

The asbestos ruling, capitalizing on the TSCA's weaknesses, made it virtually impossible for the EPA to prove that an industrial chemical already in the market is harmful. The agency may request data only if it can first substantiate that a chemical is causing an "unreasonable risk of substantial injury" to human health and, in addition, that the risks outweigh the costs to industry and the lost benefits of the chemical's use. But the EPA cannot gather such substantiation because the law prohibits it from doing its own studies and the data from the companies are mostly out of bounds as "trade secrets." Basically, the agency has to wait for decades until independent scientists, alerted by multiple real-life cases of harm, raise the funds and then take up the study of a toxic product and eventually produce a body of scientific evidence demonstrating toxicity. Even then, however, proving unreasonable risk of substantial injury that outweighs economic profit is very difficult, given the opposing legal, political, and science-for-hire resources available to billion-dollar manufacturing companies.

The real-world result of the decision has been pernicious, motivating industry to protect existing products rather than innovate to find safer ones.

Pesticides used on foods must now hew to a standard stricter than that for industrial chemicals. The Food Quality Protection Act, passed in 1996, gave the EPA the muscle to focus on health, especially children's health, rather than pitting health risks against profits, at least in theory. For both new pesticides used on food crops and the six hundred ones in use before the FQPA was enacted, pesticide manufacturers must turn in results of a set of six EPA-mandated animal tests, using EPA-mandated protocols, to disclose a minimal level of information about basic toxicity. The aim, however, is not to prove safety, but to indicate the levels of exposure at which different toxic reactions take place.

The EPA has the power to decide how much pesticide can be applied and how much residue left from the pesticide on food crops will be tolerated. Notice, however, that the manufacturers perform the tests and that residues are allowed in and on our food at "acceptable" levels. A bit of pesticide comes home on your apples, some blended into the canned creamed corn, some baked into the bread; even if each was safe, the cumulative effect has been ignored. Further, the permitted residue levels may be too high; four states petitioned the EPA in 2005 to reduce them but were rejected.

One major regulatory gap exempts pesticide manufacturers from disclosing what inert ingredients they use to make the active chemicals more effective, though these inerts can constitute as much as 90 percent of a pesticide product and can be active and dangerous. Fourteen states have petitioned the EPA at least to require disclosure on labels.

If the EPA has incomplete information about a pesticide's safety, the agency may apply a tenfold safety factor limiting the chemical's use (to achieve this objective, the agency adds user instructions to the label), intended to protect children and women of childbearing years. But the tenfold factor has seldom been imposed. Some of the weaknesses in the FQPA may be chalked up to industry's effective lobbying to mold the law to their liking. While the law was under debate, 145 pesticide, food industry, and agribusiness groups formed the Food Chain Coalition to lobby Congress on issues including the FQPA, contributing $84.7 million outright to lawmakers from 1987 to 1996, while also treating them and their aides to nearly one hundred junkets. The coalition groups finally bartered their support in exchange for the elimination of a 1950s law they had found particularly anathema, the "Delaney clause," which had prohibited any residues at all in processed foods of pesticides that caused cancer in lab animals. Now small amounts of carcinogens are permitted in our food supply.

Indeed, the EPA has been severely hampered by inefficient rule-making tools. Jay Feldman views its weaknesses this way: "It's 75 percent political that there's inadequate data. There's a culture at the EPA that's predisposed to meeting industry needs. The rest is that there's not enough staff at the EPA because the sheer volume of data is unmanageable." About 1,500 new applications for industrial chemicals and 7,000 applications for pesticides land on EPA desks a year.

MISSING INFORMATION

Whatever the reason, the absence of data is not proof of safety. The absence of data proves only ignorance.

Industry representatives flatly deny that manufacturers do not test their chemicals for the EPA. Chris VandenHeuvel, a spokesman for the American Chemistry Council, the industry's trade association, argued, "Any new chemical that comes on the market today is going through a battery of screens and tests for health and development effects." Dow points to the $1 million it spent on 3,600 studies of Dursban (though it failed to tell the EPA about dangers it had in fact found yet buried and about the dangers it failed to understand).

Industry omits to say that their tests and the tests the EPA requires for new chemicals and for pesticides are flawed and inadequate and still cannot detect most of the hazards facing children. Testing is usually confined to single active ingredients; sometimes that misses the danger. The pesticide Roundup, for instance, is more dangerous in its finished formulation than its active ingredient.

The tests assume we are exposed to one chemical at a time, though in real life we encounter them in multiples. Children can also encounter the same chemical cumulatively, from food, air, and water. Ten years after Congress required the EPA to test certain pesticides for their cumulative dangers, the agency did so, in 2006, but minimized the risks that are unique to developing fetuses, infants, and children. In a déjà vu parallel, more than ten years after Congress directed the EPA to do so in 1998, the agency had not yet begun by the time this book went to press to test seventy-three pesticides for their potential to damage the hormone system; and again the tests it plans are deemed inadequate for understanding risks to the fetus.

The tests for industrial chemicals, when they are done at all, assume that we are all healthy adult males who weigh 150 pounds and live seventy years, though children are so differently vulnerable. The tests fail to measure in utero exposures, the very exposures that matter the most, and fail to capture the late effects of early exposures (such as heart disease, Alzheimer's, or Parkinson's). They fail to assess possible harm to the immune system.

Most tests for industrial chemicals involve giving high doses to a few hundred lab animals for a year or two, while we now know that low doses are harmful in ways that high levels are not. High-dose tests reveal only crude damage, such as death or organ dysfunction, and miss subtle dangers to the fetal or infant developing brain and nervous system, which are particularly sensitive to harmful agents. "If we know there are four or five or six chemicals out there capable of causing brain injury, what about other chemicals that have not been adequately tested?" asks pediatrician Philip J. Landrigan, who recently coauthored an important study of brain damage.

Since EPA hands are tied in getting data about industrial chemicals, the agency has agreed to a project whereby corporations will voluntarily offer up a specific set of basic health data for the most widely produced chemicals (one million pounds or more). Skeptics say that "this is an attempt to preempt effective government" and refer to a 1992 memo written by the chemical industry trade association that came to light, urging that "voluntary development of health, safety and environmental information will . . . potentially avert restrictive regulatory actions and legislative initiatives." Environmental Defense, a long-established national nonprofit group, agreed to work on this project because it seemed to offer some screening information that might be used to prioritize chemicals for a deeper analysis in the future but gives industry a "D" for its foot dragging.

The safety of cosmetics is even more deeply hidden in shadows. The Food and Drug Administration has legal authority over cosmetics, but it is a hollow power. The FDA website explains, "Cosmetic products and ingredients are not subject to FDA premarket approval authority, with the exception of color additives. . . . Recalls are voluntary. . . . FDA is not authorized to require recalls of cosmetics. Manufacturers are not required to register their cosmetic establishments, file data on ingredients, or report cosmetic-related injuries to FDA."

The FDA is also responsible for overseeing the safety of food additives. Its protocols for testing are strict, but its oversight is blinded by exceptions: As is the case for other industrial chemicals, the manufacturer is the one doing the safety testing; and the roughly two thousand additives in use before 1958 and those otherwise "generally recognized as safe" are exempt from testing. Food colors are supposed to be tested, but again, the testing is left to their producers. The FDA receives a "user fee" from the manufacturer for each pound of food dye it certifies, which certainly has the appearance, if not the reality, of a conflict of interest.

PRICING CHILDREN'S HEALTH AND LIFE

Beyond the constraints in collecting evidence, the EPA and FDA and all federal agencies are further hamstrung because all proposed new rules must be based on "risk analysis," a process only economists could love. In the first step, the agency involved must perform a risk assessment, which combines all the evidence about the hazard of a substance, then from that evidence assesses its potential for harm to people, animals, or the environment. Next follows a cost benefit analysis, in which health is allotted equal weight with "economic costs," which are the assumed lost profits to corporations and the economy if the manufacturers are required to change or withdraw the chemical.

The essential flaw, critics point out, is that risk and cost benefit analysis rest on the premise that a society can and must attach a dollar value to health in order to generate economic efficiency in spending private or public funds. Yet "the process of reducing life, health, and the natural world to monetary values is inherently flawed," declares Tufts University professor Frank Ackerman, coauthor of *Priceless,* a critique of the concept and practice. Risk and cost benefit analysis assume that America is "a world of scarce health resources," as the dean of the Harvard School of Public Health once declared. Who assumes that, and on the basis of what information? What is worth spending money on, and who is at the table when that decision is made?

There are other significant problems. The data used in determining risks and costs are generated and controlled by the manufacturers, who may have the incentive and opportunity to maximize the economic value of using the chemical while minimizing the health risks. Further, those who produce the risk (the manufacturers) are not the ones subjected to its risk (American families and workers). Yet another problem is that the analysis considers one chemical at a time, with no consideration of the cumulative or synergistic effects of multiple toxics in real life. Nor are alternative solutions, such as the substitution of a less pesticide-intensive method of raising a crop, ever included in the analysis.

On top of these problems, the dollar values attached to health are questionable, especially if it's your child. Would you agree that one lost IQ point is worth about $9,000 over a lifetime, which was the dollar estimate the government generated in debating how strictly to control childhood lead poisoning?

In monetizing the value of health and life, children are at a disadvantage because the formula discounts health benefits occurring in the future. The Office of Management and Budget (OMB), the White House agency that watchdogs other agencies' cost benefit analyses, requires a formula that decreases the value of a life by 7 percent per year. So the younger the disabled child, the less his or her future life is worth.

Cost benefit analysis assumes that a certain amount of illness is bound to occur—that, for example, there will be one extra death in a million from exposure to one chemical from one source—and assumes that's acceptable. This kind of approach creates an anonymity that makes judgment by numbers palatable. Lois Gibbs, the mother who organized the families of Love Canal to fight for the cleanup of the more than twenty thousand tons of toxic chemicals dumped next to their backyards for over ten years, then founded and now runs a center to teach other communities how to fight as effectively, has a dramatic way to demolish the acceptability of the unnamed one-in-a-million premise. In a room full of people, she will point to someone, usually a little child or fragile elder, and say, "She is your one in a million, are you okay with that?"

Cost benefit analysis is something we all do every day (buy a new TV versus save for college). But as a nation, we have failed to frame these analyses to ensure that our children's and our nation's well-being takes precedence. Butadiene, for instance, a known carcinogen, continues as a key ingredient in making synthetic rubber. "Because it is such a high production-volume chemical, its removal would probably have a tremendous economic impact," explained Dr. Ronald Melnick, the National Institute of Environmental Health Sciences (NIEHS) expert on this chemical. "It's not based simply on science. Decisions on permissible levels of carcinogens in our environment are based on politically acceptable levels of risk." It's clear he would change this situation if he could to one with greater emphasis on disease prevention. A panel making recommendations to the EPA on regulating butadiene was stacked with so many members with corporate connections that the situation triggered a congressional investigation. Jane became one of the victims, her story told in chapter 3.

THE EVIDENCE WE DO HAVE

Despite the barriers that keep the nation from investigating and reining in the toxics among the fifteen trillion pounds of chemicals in use here every year, science moves forward, researchers pursue clues, parents press for answers, and evidence (though far from as much as it should) finds its way to light.

In 2002, three senior scientists associated with world-class medical centers and nonprofit environmental research organizations across the country sat down to create, for the first time, a summary of the known harmful effects of almost the entire universe of chemical contaminants. As Dr. Gina Solomon, one of the researchers, describes it, they listed all the major human diseases as identified in the leading occupational and environmental medicine textbooks and through additional literature searches, for a total of 198 disorders; then for each disease they cross-referenced the environmental contaminants either known to play or suspected of playing a role in disease. The project took two years. They found that more than 120 diseases have been linked definitively to pollution, while for another 33 illnesses, evidence of a link is judged to be "good." For the rest, there is evidence, though it is "limited." This database is continually updated, and there have been other such "meta-analyses," enabling scientists to create similar databases, by now covering virtually all the illnesses with an environmental element that beset children.

It takes the concept of a database, or grid, to grasp the body of knowledge connecting illnesses and toxicants. Across the top are a series of columns in which you note the illnesses; running down the side, you list the substances.

Start with a column for cancer, a childhood illness for which a substantial amount of evidence has accumulated. Much of this evidence finds its way into the hands of Dr. William Jameson. He directs the center within the National Toxicology Program that evaluates the potential carcinogenicity of chemicals, one by one. Sitting in a conference room in the handsome brick building on the Raleigh, North Carolina, campus of the NIEHS (one of the program's sponsors), he described to us the careful, apolitical, open procedures he and his staff follow to reach judgments. This task culminates in a congressionally mandated biennial *Report on Carcinogens*. The latest report lists 228 chemicals as either "known human carcinogens" or "reasonably anticipated to cause cancer in humans." The state of California, with stricter standards, considers 475 chemicals as carcinogenic.

After Dr. Jameson completed his description of how chemicals are evaluated, he said quietly, as if thinking aloud, "We are paying for the sins of our fathers, who introduced these chemicals, not knowing or caring how bad or hazardous they were."

The first inkling that external agents could produce malignant change dawned in the mind of a London physician in 1775. He was Sir

Percivall Pott, who deduced that the scrotal cancer so common among chimney sweeps must be caused by the soot accumulated in their bodies. The science of his time could not proceed further. Fast-forward to the present. The Columbia Center for Children's Environmental Health has now discovered that particles of hydrocarbons inside soot initiate the cancerous DNA and chromosomal changes that likely felled the chimney sweeps . . . the very changes found among babies in Harlem exposed to these pollutants in the womb.

Scientists began to look at the causes of cancer only in the 1950s, with the publication of a handful of studies of smoking and lung disease. The "war on cancer" has trudged along since its launch by then president Richard Nixon in 1971, but the research has focused largely on cures, not evidence of causes, not prevention. From the start, this national campaign was blocked from dealing with causes of the disease by folks who had major economic interests in seeing this happen. Only in the past ten years have studies begun to look at the possible connections between cancer and about twenty synthetic chemicals, notably pesticides, PCBs, dioxin, the plastics softener called bisphenol-A, and solvents. Now, in this relatively short time and despite lack of adequate research support, the evidence mounts. "The vast majority of those studies show direct links to cancer, or increased susceptibility across almost every body system," according to a comprehensive analysis of current scientific studies by the University of Massachusetts's Lowell Center for Sustainable Production.

"The growing incidence of cancer among children offers some of the most convincing evidence of the role of toxic pollutants in causing this dreaded disease," concludes Boston University epidemiologist Richard Clapp, who directed the analysis. He points out that though tobacco remains the single most significant preventable cause of cancer, it has been linked neither to the majority of cancers nor to the many cancers that have increased rapidly in recent decades, including children's cancers.

Now, back to your grid. Switch the focus down the list of chemicals, to identify those associated with cancer. Among those toxics with the highest number of entries in this hypothetical grid will be "pesticides," a term that encompasses insecticides, herbicides, fungicides, miticides, defoliants, rodenticides, and fumigants. Here are some of the entries:

Plotted geographically on a map of the United States, cases of non-

Hodgkin's lymphoma are clearly clustered in agricultural areas likely subjected to pesticides; plotted temporally, the cases creep upward in frequency over the years as the use of pesticides increases. Of all Americans, farmworkers and their children are at greatest risk, with the least protection. Among the numerous pesticide studies, one looked at both household and occupational hazards and found that the mother's and father's prenatal exposures to pesticides, through farm or forest work or through using insecticides to battle bugs in the household and fleas on the family pet, carry greater risks of cancer among their children, especially for brain and central nervous system cancers, than a child's direct exposure.

Using pesticides during childhood creates substantial risk as well. How many parents would use weed-and-seed garden herbicides if they knew that the five most popular varieties are associated with non-Hodgkin's lymphoma? While rates of this illness are on the rise here, they are on the decline in Sweden, where certain pesticides have been outlawed. Pesticides also find their way into the drinking water supply, in air and household dust, as residues on fruits and vegetables, and into the fatty tissues of animals.

Then cross-reference "pesticides" with the other illnesses that exposure can trigger. Asthma is one. For instance, a study of middle-income families in California found that children who had been exposed to pesticides before the end of their first year were four to ten times more likely to suffer from early persistent asthma than nonexposed children. Weed-killing herbicides had an even more severe effect, putting them ten times more at risk of developing asthma. Pesticides are also blamed for the sharp rise in premature births in the United States and other nations; they end up in water during the spring–summer spraying season, with parallel seasonal patterns in prematurity.

Birth defects are another casualty of pesticides. Dr. John Harris, director of the California Birth Defects Monitoring Program, the largest in the nation, concludes that pesticides used at home are associated with birth defects, including limb abnormalities and cleft palate. The program's tracking finds that 20 to 30 percent of California households do buy pest products, such as Monsanto's Roundup, the herbicide used on home lawns and on genetically modified crops. "Names like Roundup make them sound cute and harmless," Dr. Harris adds with more than a touch of anger.

As is true for cancer, maternal or paternal exposure (before or during conception) to pesticides at work can raise the risk of birth defects more than home use. Farming families stand a greater risk than average of giving birth to a child with anencephaly, missing a major portion of the brain, skull, and scalp. The babies born to Gulf War soldiers who fought amid a miasma of pesticides are two to three times more likely than other soldiers' children to suffer from birth defects, from heart valve defects to hypospadias, a penal urinary tract deformity.

Miscarriages are yet another casualty. Miscarriages, as well as birth defects, are higher than average among American farming families who used conventional pesticides from three months before conception to three months into a pregnancy, an analysis by the Ontario (Canada) College of Family Physicians finds.

If you look at the entries for birth defects, you will see they are linked not only to pesticides, but also to solvents. Mothers exposed to solvents, according to research such as the multiyear Baltimore-Washington Infant Study, and the wives of men exposed to solvents bore a higher than average number of children with birth defects. Solvents also increase the risk of spontaneous abortion, as do hormone-disrupting chemicals, from phthalates to bisphenol-A, from dioxin to PCBs.

Among the body of evidence are conclusions of a competing nature. Some studies will claim they have uncovered harm, countered by other evidence claiming benignity. Such is the case surrounding phthalates, accused as an agent of reproductive deformities among male babies—an accusation disputed by industry. Though the family of phthalates was invented in the 1930s, the federal government began to support research into their effects on human reproduction only in 1999.

In the eye of the storm stands Professor Shanna Swan, a professor of obstetrics and gynecology at the University of Rochester Medical Center. Trained also in epidemiology and statistics, she recently launched the Study for Future Families to track reproductive health over the years to come. She is one of the scientists who definitively documented declines of 1 percent a year in sperm count since World War II in the United States and Europe. It was her research that discovered that lower sperm quality in U.S. men can vary from one area

and environment to another and that lower sperm quality in the agricultural Midwest may be due to exposure to pesticides.

In 2005, Dr. Swan published the results of a landmark study of
phthalates funded by the EPA and the National Institutes of Health.
She and her colleagues found that among the eighty-five mother-son
pairs they studied, women whose urine during pregnancy carried
higher levels of traces of phthalates were more likely than the others to
have baby boys with genital conditions, including smaller penises and
a shorter distance between the genitals and the anus. In studies of male
rodents, the distance is usually about twice as long as in females, so a
shorter distance is thought to indicate the "undervirilization" of males.
Boys with this shorter distance were also more likely to have incompletely descended testicles.

The American Chemistry Council's Phthalate Information Center, a group representing the manufacturers, counters that phthalates
"make everyday life more convenient, colorful and fun" and that
human exposure is well below the EPA safety margin. The head of
the council's phthalates panel argues that differences in genitalia have
no known significance in humans and could be caused by natural variability.

The National Toxicology Program's Center for the Evaluation of
Risks to Human Reproduction, refereeing the debate, concluded that
evidence of harm to adults was still insufficient, then expressed "concern" for infants and "some concern" for toddlers, strong language for
this research body. It stated that if infant and toddler exposure were
severalfold higher than that in adults, male reproductive tract development would be adversely affected. Then the center added information showing that's exactly the situation: "Exposure of children one to
six years old has been established to be several-fold higher" because of
their smaller bodies and immature metabolic systems. The most recent
CDC survey of Americans' body burdens in fact found that eleven of
twelve phthalates tested were higher in children than in adults.

Phthalates are not alone as suspects of demasculinization. A Dutch
study, started in 1990 to track a group of children exposed shortly before birth to hormone-disruptive PCBs and dioxin, found the children's play to be less sex-appropriate. By age seven or so, the PCB levels

in boys were associated with less masculine play, while PCBs made girls more masculine than normal. Dioxin turned the play behavior of both boys and girls more feminine. Clear evidence from creatures in the wild shows that exposure to hormone disruptors has demasculinized them.

Chemicals that disrupt the fetal hormone system also factor in the disruption of the male-female birth ratio. The most vivid real-life story begins back in midsummer 1976 in the town of Seveso, Italy, when an explosion in an herbicide factory showered an immense cloud of dioxin over nearby towns. Researchers were wise enough to take thirty thousand blood samples from the townspeople shortly after the explosion and have since followed them, human guinea pigs, to track the outcomes of this chance experiment. During the first seven years after the accident, forty-six females compared with only twenty-eight males were born to exposed parents, even though the amount of dioxin that found its way into their parents' bodies was the equivalent of 1 drop of dioxin in a string of 7,400 bathtubs. The fewest male children were born to the most highly exposed fathers. Overall fertility was also markedly reduced. By 1985, the male proportion of births and the townspeople's overall fertility returned to normal.

In Canada, living immediately adjacent to several large petrochemical and chemical plants in Ontario, across the river from Detroit, the Aamjiwnaang First Nation has experienced a plunge in the proportion of male live births since 1990, statistically verified by scientists with whom the community is collaborating. Even in the Arctic, seemingly far from industry, twice as many girls as boys have been born, and high levels of hormone-disrupting chemicals have been found in women's blood in villages from Russia to Greenland.

Many of the toxic substances that cause cancer, asthma, birth defects, and reproductive problems also trigger mental and behavioral damage, independent researchers find time after time.

Lead continues to carry the curse as the single most significant mental health threat facing American children today. Even after decades of struggle to clear it from their lives, it lingers on. Though successes in lead cleanup, including its removal from gasoline and paints, have lowered lead levels found in children's bodies over the past three decades from an average of 15 to 2.2 micrograms per deciliter today, that level is still too high. A ten-year study of children in affluent Cambridge, Massachusetts, concluded, "You might not notice lead effects in this group because the kids look okay. No matter where they start out, lead exposure pushes children toward lower performance," and, as just discovered, impaired language ability.

Mercury, another heavy metal, stays in the mother's bloodstream for months and is only slowly excreted. It can be passed by the same routes as lead, across the placenta or through nursing. Once the fetus or child has absorbed it, mercury accumulates in and irreversibly damages the developing brain, though the effects of exposure to mercury can take months, even years, to appear.

When asked about the high levels of mercury that body burden studies find in hundreds of thousands of woman of childbearing age, Dr. David Carpenter, of the School of Public Health at the State University of New York at Albany and a national authority on mercury contamination, states unequivocally that "the science is really clear now that at those levels the child almost certainly would have a permanent IQ loss," typically a five-to-seven-IQ-point loss. Prenatal exposure to mercury is estimated to cause 1,566 more children to be born mentally retarded each year. The mercury emitted from American coal-fired power plants is said to be responsible for reduced productivity because of lowered intellect at a national cost of $1.3 billion each year. The researchers authoring this analysis then add, "In contrast to the costs of controlling pollution, which are onetime expenditures, these costs last a lifetime and will recur in each year's birth cohort until emissions are reduced." This well-established connection is disputed, however, by a scientist working for the corporate-supported, conservative American Enterprise Institute; he flatly declares, "Mercury does not reduce children's IQ."

To test the thesis that heavy metals diminish intelligence, a researcher from Harvard University traveled to the small town of Miami, Oklahoma, located near an abandoned site where mining had taken place nearly one hundred years ago. There he snipped off samples of hair from thirty-two of the town's fifth- and sixth-grade schoolchildren. He put the children through a battery of problem-solving, visual, spatial, verbal, and memory tests and, back at Harvard, ran the hair strands through a mass spectrometer to look for heavy metals. The correlation was unmistakable. The higher the combined concentration of the metals manganese and arsenic in the child's hair, the lower his or her verbal intelligence test scores.

The air around us is one route of exposure. A Texas study found that for every one thousand pounds of mercury released in the environment, autism rates rose by 61 percent and the demand for special education grew by 43 percent. Texas with its nineteen coal-fired power plants—five of them the nation's most polluting—releases more mercury into the air than any other state. A similar correlation was found in a 2006 California Department of Health Services study of the San Francisco Bay Area, a typical urban setting. Children with autism disorders were 50 percent more likely to be born in neighborhoods with high amounts of several air contaminants—mercury, cadmium, nickel, vinyl chloride, and the solvent TCE.

Mercury may enter a child's body through vaccinations. A contentious debate rages about whether there is a link between autism and the mercury used as the base of the antimicrobial preservative known as thimerosal, developed by Eli Lilly and used by Merck and other manufacturers in federally mandated childhood vaccines. The unprecedented increase in autism coincides with the 1990s' sharp escalation of the number of vaccines, from eight to twenty, mandated for children under age two. The FDA revealed in 1999 that some infants who had received multiple vaccines containing thimerosal could be exposed to a cumulative dose that far exceeds federal health guidelines. Some parents of children with autism hold thimerosal responsible for their children's illnesses and severely criticize the Institute of Medicine's report that rejected any connection between thimerosal and autism in 2004.

Animal studies seem to give credence to such a connection. One study comes from Dr. Mady Hornig, a Columbia University researcher. She and her colleagues injected thimerosal at levels similar to that in typical vaccinations into a strain of mice bred to have a compromised immune system, like the compromised immune systems found in children with autism and other brain-body disorders. The mice reacted with autismlike symptoms: They behaved oddly and had brain abnormalities and delayed growth. Two other strains of mice, without problem immune systems, showed no such effects.

Manufacturers removed the preservative from most vaccines in 2001, though without admitting harm. It still remains in flu shots. Even if found not to be a determining cause of autism, where is the

wisdom of vaccinating a baby a few days or months old with a substance known to be neurotoxic?

Pesticides, like mercury, can also bring about oxidative stress, especially to a child carrying a genetic variation that leaves him more vulnerable. A 2007 study found that 28 percent of children born to women
living while pregnant near agricultural fields sprayed with organochlorine pesticides were autistic; a 2006 study had identified the genetic variation that predisposed a person to these very pesticides.

The damage that PCBs inflict upon children's brains has also been
tracked over the years. The most infamous examples of harm are
found in communities around the Great Lakes. For decades, manufacturers such as Shell Chemicals and Occidental Petroleum as well as
shipping companies dumped millions of tons of persistently hazardous
wastes, featuring PCBs, into the Great Lakes, while families and public health agencies warned of the coming ill effects. By the mid-1980s,
a study by a husband-and-wife research team showed that children
born to mothers who ate the most fish from Lake Michigan in the years
before becoming pregnant (though they ate only three fish meals a
month) weighed less, had smaller head circumferences, and exhibited
more startle responses than average. Tested at age eleven, these children had IQ scores more than six points below those of their peers, as
well as memory and attention disorders, and were twice as likely to be
at least two years behind in reading comprehension.

Cousins to PCBs in their chemical makeup and action are polybrominated flame retardants. Though there are no human data on
their health effects yet, animal studies show that, like PCBs, they can
do damage by acting on the thyroid. The common antibacterial triclosan, packed into so many household products, is another chemical
cousin. Research keeps piling up indicating that these substances could
be contributing to autism, ADHD, and other mental and behavioral
disorders. It is called "a silent pandemic" by two leading scientists, Dr.
Philippe Grandjean of Harvard's School of Public Health and Dr.
Philip Landrigan of Mt. Sinai Hospital.

ENOUGH EVIDENCE FOR ACTION

The accumulated evidence points to one conclusion: Chemicals have
been tested enough just to know that they are toxic, but not tested
enough to plumb in full the casualties they wreak.

Linda Gillick, the mother of Michael, one of the dozens of young cancer victims in Toms River, New Jersey, put it bluntly: "These are our children, they are our future, and if we can't protect them, we're in big trouble. The atrocities we've done to the environment for years and years, to make our lives easier and better, in the long run they've cut the life expectancy of our children big-time. We're just seeing the tip of the iceberg of what's coming. This is just the beginning."

This tragedy is of our own making. But in that very fact, then, we all hold the power to end the tragedy. Whatever part of the illnesses and deaths are triggered by man-made poisons can be prevented. We now have the science to understand enough of the causes and so too the science to eliminate them.

Five

SCENE OF THE CRIME

It's said that crime occurs most often in the home, and that can be true of environmental assaults on children. The home, which should be a haven of comfort and safety for our kids, is often instead a minefield of toxic hazards. Here is what you may find in a walk through your home and a day spent with your children.

THE GAUNTLET

The baby awakens in her room, arguably the most toxic room in the home. If you had decided to spruce it up just before her arrival, you probably painted it with petroleum-based paints, which continue to release their long-lasting, polluting volatile organic compounds (in this use, "organic" means a carbon-based product, and "volatile" means they dissipate into the air). Her crib mattress contains flame retardants if it's an Air Flow Sleep Positioner, though you may have chosen a Sealy Baby Soft Classic Mattress, which does not. If she's chewing on a Little Teethers teething ring, she's ingesting some phthalate molecules, a chemical used to soften plastics, though if it's a Soft Freezer Teether, it's free of phthalates. If the baby's furnishings or floors are not made of solid wood, then you can bet they are made of pressed wood, glued together with formaldehyde. If baby takes along her rubber ducky or the Splish-Splash Jesus bath book to her bath, both expose her to another dose of phthalates. (Phthalates, like flame retardants, disrupt the body's hormones.)

In the bathroom, the distinctive, unpleasant smell of the new vinyl shower curtain means it's offgassing more phthalates and other unwanted chemicals. Air fresheners may be emitting a lemony and pine smell; the scent derives not from lemons or pine needles, but from a manufactured chemical conglomeration of multiple volatile organic compounds, benzene, formaldehyde, and more phthalates. (Air fresheners are a $1.72 billion-a-year industry.)

Getting ready for the day, as you apply makeup and perfume or shave, you encounter more phthalates, there to disperse the scent evenly so that it lasts longer. After you rub a product with phthalates into your skin, the chemical will pass straight into your body and, just two hours later, show up in your urine.

If you are like the average American, you use ten personal care products a day, made of 126 different ingredients, some toxic, others benign. If you are pregnant or intending to conceive, many of these products could put your unborn baby at risk. Nail products contain solvents and formaldehyde. Revlon Moondrops lipsticks contain phthalates, as do two-thirds of all personal hygiene products; hairsprays, deodorants, nail products, and hair mousse usually contain two. The nonprofit research organization Environmental Working Group shopped at a local Rite-Aid store, cross-referenced the products against patent office records, and found one or another phthalate in thirty-seven popular nail polishes, top coats, and hardeners, including products by L'Oréal, Maybelline, Oil of Olay, and CoverGirl. The Avalon line of personal care products is formulated without phthalates. Phthalates are recognized as toxic substances under the major U.S. law controlling industrial chemicals, but "companies are free to use unlimited amounts in cosmetics," the group points out.

If your baby was born prematurely, the hospital's intensive care unit may well have used a medical device for her care, such as intravenous drip tubes and catheters, made with one of the chemicals from the phthalate family. If so, she could have been exposed to twenty times more of that chemical than the FDA standard. Dr. Ted Schettler, science director of the nonprofit research group the Science & Environmental Health Network, notes that cost-competitive medical products without phthalates are readily available and already in use by a number

of prominent medical supply manufacturers and hospitals, including
Catholic Healthcare West and Kaiser Permanente (about which more
later).

As you dress for the coming day, you slip on some freshly dry-
cleaned clothes that promptly release vapors of perc (perchloroethylene, or PCE), a chlorinated solvent related to TCE and a probable
human carcinogen. Underfoot, the synthetic rug was probably manufactured with a bonding agent concocted of (carcinogenic) styrene-
butadiene, applied to the underside to hold together the carpet yarn
and backing, and most likely impregnated with Stainmaster, another
stain resister like Teflon. Your children breathe these vapors.

In the kitchen, bacon is browning on a pan coated with Teflon, a per-
fluorochemical that DuPont brought to market in 1945. If there's a pet
bird down the hall, it might keel over dead in its cage from fumes that
a human can't see or smell. (These chemicals are thought to cause cancer and delay or harm development.) Perfluorochemicals are used as
water, stain, and grease repellents for food wrap, carpet, and furniture.
The Stainmaster and Scotchgard formulations stainproof fabrics, from
sofas to ski clothes, and coat fast-food packaging.

In the refrigerator, food is probably stored in plastic containers, yet
another everyday household item often made of polyvinyl chloride
(PVC) with phthalates. After your ten-year-old daughter microwaves
leftover mac-and-cheese in a plastic bowl covered with a plastic wrap,
she may ingest a double portion of phthalates in her quickie meal. Her
Barbie doll is made of PVC plus phthalates. If you are expressing
breast milk to give later on to your baby, the pump probably contains
one of the phthalate chemicals.

The problem is that phthalates can leach out because they are not
chemically bonded to the PVC plastic polymer. They have even been
found in deep-sea jellyfish more than three thousand feet below the
surface of the Atlantic. If a container is microwaved or refrigerated,
the molecules may migrate into any high-fat food, such as milk, meats,
and cheeses, stored there. Dr. Donald T. Wigle, epidemiologist and author of the textbook *Child Health and the Environment,* says he even
slices off a thin layer of whatever food came into contact with plastic
wrap. Some plastic wraps are made with phthalates or another type of

softener that leaches, some are not. Glad Cling Wrap contains no softener, and S. C. Johnson reformulated Saran Wrap without any phthalates.

Getting ready for school, your daughter puts on a T-shirt with a shiny printed design and fills her bright blue Nalgene water bottle, which she tucks into her backpack. The snazzy-colored Nalgene bottle and other hard (#7) polycarbonate plastic bottles harbor a different plastics ingredient, a molecule called bisphenol-A that was originally invented in 1936 as a synthetic estrogen. (We now know the damage that can be caused by hormone replacement estrogen.) This plastic is used in most polycarbonate baby bottles, in toddler sippy cups, and in some reusable and microwavable food and drink containers, such as Rubbermaid's Dip 'n Snack Tray. The five-gallon jugs that hold "pure spring water" are made of polycarbonate. Bisphenol-A is also usually found in the plastic resins that line food cans, supply pipes, and even dental sealants used on kids to prevent cavities. The chemical, produced in excess of 6.4 billion pounds a year, generates $6 million a day for businesses in the United States, Europe, and Japan.

Like phthalates, the chemical bonds that create the polycarbonate polymer are not stable. They degrade in water and heat, when exposed to alcohol, soaps, or acids, and simply with age. As they break up, bisphenol-A leaches out.

As for your daughter's T-shirt, the design on it, as well as the waterproof coating on her backpack, may have been made out of PVC, as may the pipes in your house, the siding, and the roofing. At a production level of fourteen billion pounds a year, PVC and its basic compound, vinyl chloride, are everywhere, in children's toys, garden hoses, wallpaper, office supplies, and tablecloths. It's in the shrink-wrap of the packages the supermarket sets out in its cases.

The PVC used in all these products is fabricated from petroleum on a base of chlorine, embellished with various toxic additives, including lead. Some vinyl lunch boxes, for instance, are made this way. Just one bottle of some kinds of PVC can contaminate a recycling load of one hundred thousand otherwise recyclable bottles. That means used

PVC items end up in landfills or dumped from barges into the seas. Even at the end of their lives, these chlorine-based products present hazards, because when vinyl chloride breaks down in a landfill, it can transmute into other toxic substances, including TCE, the toxin in Dickson's water supply. The manufacture of PVC products releases thousands of pounds of carcinogenic vinyl chloride gas into the air and into the community water supply, endangering the families in surrounding neighborhoods. The federal Agency for Toxic Substances and Disease Registry has characterized fetuses, infants, and young children as "highly susceptible populations."

Your twelve-year-old son is doing some last-minute homework on the computer in the family room. The computer, the TV housing, the chair, and the sofa, like the other items made of PVC or polyfoam in your home, may have been impregnated with polybrominated flame retardants, another petroleum-based chemical, intended to help the product withstand high heat. One manufacturer might use a brominated flame retardant, while another, such as IKEA, might use something else that works just as well, but since these elements are not labeled, you can't tell.

Like phthalates and bisphenol-A, the flame-retardant molecules, trapped inside but not bound to the plastic, detach themselves from products and disintegrate over time, then enter humans, wildlife, and the natural environment. Levels of these chemicals are higher in U.S. breast milk, blood, and dust by orders of magnitude than in Europe, leaving researchers "flabbergasted" by their quick buildup.

While cleaning up after breakfast, you may stop to wonder why dishwashing liquid contains the warning "Keep Away from Children." Perhaps it's because, as Colgate-Palmolive's consumer affairs staff explained when asked about the ingredients of their Palmolive Oxy Plus Marine Purity dishwashing liquid, "some ingredients combine to release formaldehyde." Perhaps even more problematic are the dishwashing liquids that contain the antimicrobial agent triclosan, which disrupts the body's hormone system. It also reacts with chlorinated water to produce chloroform, a possible carcinogen. Triclosan is also used in toothpastes, deodorants, and hand soaps in the United States, though other nations, including China, have removed it.

Household and bathroom deodorant products (such as toilet bowl deodorizers, air fresheners, and surface disinfectants) are often formulated on a base of chlorine and benzene. Spot removers contain benzene and solvents such as perc, the dry-cleaning solvent. Children whose mothers used the most household cleaners were found to be more than twice as likely to wheeze and eventually develop asthma.

To dramatize the unsuspected toxicity of many household cleaners, Professor William Nazaroff, an environmental engineer at the University of California at Berkeley, demonstrates with a few capfuls of Pine-Sol with water. Pine-Sol consists of 15 to 20 percent terpene, a relatively harmless hydrocarbon. But if he mixes these cleaners in a room where there's smog, even a room ventilated with an open window, the elements interact and turn into formaldehyde and other carcinogenic compounds.

Labeling is not required for household consumer products, which are the jurisdiction of the Consumer Product Safety Commission. The recipes are considered confidential business information.

THE INSIDE STORY

Because of the increasingly complex mix of chemicals in your cupboards, appliances, and furniture, and because outdoor air has improved, the air inside your home may be more polluted than air outdoors. No government agency, however, regulates indoor air. A test of homes on Cape Cod detected sixty-six chemical compounds in the dust and fifty-two in the air, while another study, covering homes in seven western states, found thirty-five toxic industrial chemicals, with Teflon and other perfluorochemicals topping the list.

The air inside your home is also tainted with chemicals you'd never expect because you did not knowingly bring them in. Many of the substances banished decades ago, such as DDT and PCBs, have not quietly faded away. PCBs are still present in virtually every household dust sample ever tested and in your food. From 1929 until banned in 1978, PCBs were thought of as a miracle product, manufactured and widely used in the United States as insulators in electrical transformers and capacitors, as ingredients in plastics and adhesives and incorporated into household products like wood-floor finish. Over those decades,

PCBs pervaded the environment through careless disposal, or as they leaked from industrial facilities and waste disposal sites, or as the products in which they were used disintegrated with age.

Lead is another virtually indestructible toxin. Though dramatically reduced in many settings, it remains a threat in the thirty-eight million homes built before 1950, where paint is deteriorating year by year, warns Dr. Herbert Needleman, the original champion in the decades-long fight against its once pervasive uses. Drinking water continues to carry lead from old plumbing. In urban areas, lead from years of vehicle traffic gets scuffed up with street dust, then ends up in kids, raising their blood lead levels above health standards.

Our children carry one thousand times more lead in their bones than pre–industrial age children.

Like the chemicals of former generations, such as DDT and PCBs, today's phthalates, bisphenol-A, flame retardants, PVCs, and dioxin repeat this cycle, finding their way into your home. Brominated flame retardants have recently been discovered in the North Pole.

Like the robot boy in the movie *AI* (also manufactured for human enjoyment), these chemicals, it seems, will outlast generation after generation of life.

SCHOOL DAYS

With breakfast and cleanup chores behind you, it's time for school and work. One parent drives the baby to the child care center. Your new car has that new-car smell, which is actually the odor of leaching phthalates and other volatile chemicals. The plastic and foam of the car are probably full of flame retardants, unless you're driving a Lexus or Volvo.

Your two older children ride off to school in a cheerful yellow school bus (one of the 505,000 other buses transporting twenty-four million other kids, using up a total of three billion hours of their lives per year!), inhaling polluted air. Diesel school buses, some of the oldest and dirtiest vehicles on the road, spew forth diesel exhaust, made up of forty toxic chemicals, including benzene, butadiene, formaldehyde, and various harmful particles. Inhaling all these fumes, your child faces eight times more pollution than his or her classmate walking to school. When they arrive at the school, buses often idle there, emitting even higher concentrations of particulates than when moving. In New

York City, idling school buses spewed 1.3 million tons of soot into the city's air, not to mention 60 tons of nitrogen oxides, in just one year.

In their eight hours in school, your children may be plagued with the same unhealthy kinds of exposures as at home, but more so. The water from hallway spigots may contain high lead contamination, a nationwide problem, but you'll never know because schools are exempt from federal drinking water regulations and need not test their drinking water. The indoor air may be saturated with residues from industrial-strength cleaning products and from the pesticides routinely sprayed in school kitchens, cafeterias, athletic fields, playgrounds, classrooms, and offices, usually without parents having been notified in advance. A law requiring schools to use safe pest control methods was squashed in a congressional maneuver. Of the forty-eight pesticides most commonly used in schools, twenty-four are carcinogens and twenty-five cause learning disabilities. When the *Journal of the American Medical Association* published a peer-reviewed article about pesticides in schools, the pesticide makers' trade association, CropLife, dismissed their findings as "scaremongering." According to a GAO investigation, half of the nation's schools have unhealthy indoor air. Asbestos may be lingering inside the walls and ceilings of many schools. Our nation's schools are in worse shape than our jails, while the poorest children attend schools in the worst repair. An amendment offered in the U.S. Senate in 2003 for $1 billion to states to help schools fix health and safety hazards failed to pass. If the playing field is synthetic, it is probably formulated of ground-up recycled rubber tires mixed with heavy metals as well as carcinogenic chemicals, all of which transform into hazardous air particles.

Building on the cheapest piece of land is one way to limit the line item for school budgets. That means, more times than you might believe, that schools get built on or next to some of the most poisoned places in America. In Pittsfield, Massachusetts, a grammar school was built abutting land where a mountain of PCB-laden soil had been dumped. In the neighborhood around the Houston Ship Channel, the César Chávez Magnet School was constructed less than a quarter mile from five huge petrochemical plants that spew five million pounds of toxics into the air annually; ironically, that school is supposed to specialize in environmental studies.

In Marion, Ohio, a middle school and a high school were built on

an old military depot; at their ten-year reunion, former students discovered they shared an unusual number of leukemias and rare cancers.
In New Jersey, there's a law that says you can't build a dump within a
half mile of a school, but nothing in the law says you can't build a
school within half a mile of a dump or even on top of one. A number of
states have similar rules. In a sample survey of five states, the nonprofit
Center for Health, Environment & Justice found that 1,195 schools had
been sited within a half-mile radius of a hazardous waste site.

THE NOT-SO-GREAT OUTDOORS

After school is out, your daughter joins her teammates in soccer practice on the schoolyard. Here, as her breathing accelerates (and pound
for pound, children breathe more air than adults), she inhales some of
the witches' brew of air pollutants that plague almost every American
neighborhood.

More than half of the nation's population live in or around areas
that violate clean air standards, though the Clean Air Act of 1970 did
impel some improvements in some parts of our nation. Beverly Hills,
for all its tinseled affluence (median family income is over $155,000 a
year), is among the top 10 percent of the nation's dirtiest counties for all
varieties of air pollutants. Every county in the United States has air so
polluted that it could cause an additional case of cancer among one
hundred thousand people exposed over a lifetime, which is the EPA
benchmark for hazardous pollution. Furthermore, air pollution health
standards are set for adults, not for children.

Children virtually everywhere inhale and ingest two of the worst
pollutants, smog and soot, constant companions of human ingenuity
since the beginnings of the industrial era. Smog, which has doubled
since 1900, occurs when sunlight interacts with the nitrogen dioxide
emissions and volatile organic chemicals from tailpipes, smokestacks,
and everyday activities such as dry-cleaning and housepainting. Because sunlight is the key to smog formation, levels are highest at mid-
day and the early afternoon in the hot summer months, just the time
when children are most likely to be playing outdoors.

If you live in New England, your family is breathing the secondhand smog that forms each summer over cities across the nation, particularly in the Midwest, and then moves across the country to the
Northeast, picking up additional pollutants as it travels.

If you live in a city, your lungs and eventually your children's will be blacker than the lungs of country dwellers. Your family is inhaling soot, the mixture of particles rising from the inefficient combustion from diesel motor vehicles, residential heating, tobacco, power plants, and incinerators. Some particles of soot measure no more than 2.5 microns, less than one-hundredth the width of a human hair, so small they can lodge deep in the lungs if inhaled. These particles also accumulate in the food chain, so a child can ingest as well as inhale them.

The greatest sources of air pollution, of toxins in the air, in sheer weight—some 42 percent of the total—are the nation's 1,100-plus coal-fired power plants. Out of their tall stacks spew some four hundred thousand tons a year of lead, mercury, hydrochloric acid, chromium, and arsenic into the air, as well as about 60 percent of the nation's sulfur dioxide (SO_2) emissions, and they are second only to automobiles as the largest source of nitrogen oxide (NO_x) pollution. About forty-eight tons of mercury rise from these plants into the air each year, more than 40 percent of all airborne mercury from industry. Coal-fired power plants can be built to be high in efficiency as well as low in pollution and on the greenhouse gas emissions linked to global warming—one is up and running successfully in Tampa, Florida—but they cost the plant owners more up front.

Globally, the load of mercury in heavily industrialized countries is increasing over time, in parallel with increases in neurodegenerative diseases such as autism and Alzheimer's disease.

Plants that manufacture chemicals, paper, metal, and plastics or other petrochemical by-products do not make healthy neighbors either, nor do printing plants, which emit the highest levels of toxins that affect developing bodies and brains.

When your baby outgrows his or her rubber ducky, which is not rubber but plastic, it ends up in a landfill in the company of bleached paper towels, outmoded computers, and old tires. Or it will be incinerated and transformed into dioxin.

Dioxins are a family of supertoxic pollutants that can be neither seen nor smelled. (They can cause cancer and a host of reproductive problems, including miscarriage.) They are not made by man but, like an evil genie let loose by a thoughtless act, are created and released any time organic materials containing chlorine, such as chlorine-based pesticides and plastics like polyvinyl chloride, are burned, and then find

their way into the food chain. The smokestacks that once were part of
many medical centers sent plumes of dioxin into the air as they incinerated common hospital medical supplies made from PVC, such as IV
bags and tubing. "When I ran an HIV clinic for women, we thought
the more we incinerated, the safer we were making everyone. Eventually we found out that the very bags we burned came back as dioxin in
breast milk," Charlotte Brody recalls. She went on to help found
Health Care Without Harm, a campaign to replace PVC medical devices with safer alternatives. "Today there are fewer than one hundred
medical waste incinerators in the country, down from four thousand
just ten years ago."

Thirteen southern states, stretching from North Carolina to New
Mexico, receive the major portion of all dioxin releases. Of the twenty-
five counties in the United States that account for the greatest burden
of air-polluting chemicals that can damage children's bodies and
brains, fourteen are home to African American populations.

For a taste of really bad air, visit Houston's East End neighborhood, a place that looks and smells like hell or perhaps Mordor. Juan
Parras, a social worker and teacher at Texas Southern University's
Thurgood Marshall School of Law, has worked for years with his son
Brian to organize this poor Latino community of little run-down
houses on dirt streets, hiding in the interstices of land between one
hulking oil refinery after another, built here because the area abuts the
Houston Ship Channel. Texas Petrochemical, Goodyear, Shell, Monsanto, Exxon, and Mobil are here, fifteen miles of flaring smokestacks,
domes, train yards, and storage sheds. The only rise in the landscape
are the mounds of concrete pebbles churned out from a crushing facility, adding toxic particles into the air to join the acrid, nauseating
fumes of butadiene, styrene, sulfur dioxide, and volatile chemicals.
Children living within two miles of this area have a 56 percent increased risk of contracting acute lymphocytic leukemia.

PERIL IN THE GRASS

While your family was away from home for the day, the lawn company
came by for the scheduled application of Weed B Gon, the best-selling
herbicide. Its main ingredient is a chemical by the name of 2, 4-D, a
member of the same family of phenoxy chemicals as Agent Orange,
both produced by Dow. The herbicide remains active for days, a poten-

tial hazard if your children and pet romp on the grass. If your pet is wearing a flea collar, that's another dose of a different class of pesticides. On the weekend, you and your older children may plan to play golf on greens probably groomed with more pesticides per acre than any other landscape, including croplands.

Kids inhale pesticides in their homes, schools, and day care centers, in parks, playgrounds, and malls. They breathe these chemicals while their homes are being sprayed, then inhale them again after the pesticides seep into their plush toys, carpets, and soft furniture, lingering there for up to two weeks. Food and water are the second source of children's exposure, and third comes absorption through their skin. Children who live in crowded, old, and run-down housing, often children of color, are especially likely to be exposed to heavy doses of persistent pesticides. Farmworkers' children suffer from the pesticide dust their parents bring home on their clothes and from the drift from sprayed fields, but their parents seldom speak up, fearing retaliation.

These exposures add up, day after day. By the age of five, children have accumulated half their entire lifetime burden of pesticides.

If you are like most Americans—in fact, 82 percent of us—you are quick to reach for the pesticide bottle, pleased when bugs and weeds disappear. Use in our homes and gardens has doubled since 1982, and families use more than agriculture does. The number two favorite household pesticide is Roundup, which receives rave reviews on the website e.opinions. Roundup's ads picture it as a cute white bottle with muscled arms, the nozzle a masculine head. Americans spent more than $11 billion on pesticides in 2001, the latest year for which figures are available, and are responsible for 33 percent of total world expenditures on pesticides, for more than 40 percent on herbicides, and for more than 33 percent of world expenditures on insecticides.

The most heavily used pesticides in the United States and the world today are formulated from the organophosphate class of chemicals, based on phosphoric acid. German scientists at IG Farben invented these chemicals during World War II when sea blockades cut off their farmers' traditional nicotine-based pesticide. The nerve gas sarin, which Japanese terrorists piped into Tokyo subways in 1995, is a variation on the German invention, as is the poison gas that Iraq's Saddam Hussein used to murder five thousand Kurds. Derivatives of these chemicals of mass destruction, highly diluted, are now the basic arsenal in our country's agricultural offensive against insects and weeds. Residues show up in the urine of nearly every child. Pesticides were known as "economic poisons" in 1947, when a law governing their use was passed, until PR came along to give pesticides their cozy names and cute bottles.

In the twenty years following World War II, before organophosphates took center stage, besides using arsenic and nicotine, farmers began to apply pesticides formulated out of chlorine-based synthetic chemicals. DDT was one such organochlorine, introduced in the 1940s. These pesticides were championed as saviors, which they may be for countries battling malaria- and yellow fever–carrying mosquitoes. They were also championed as safe, which they were not. They persist for decades in the environment and accumulate in living tissues up the food chain, causing deadly harm to birds, fish, and animals, as Rachel Carson's book *Silent Spring* so elegantly relates, and, as science has since found, harm to humans. So most, though not all, of these pesticides were eventually banned, decades after their harm was first recognized.

It turns out, however, that the organophosphate replacements (for example, malathion and diazinon, once used in Spectracide, and carbamates such as Sevin), while less persistent, are more acutely toxic to both pests and humans. Sevin and malathion products were reapproved for continued use in 2006; while the EPA no longer permits diazinon for home use, it is allowed for some agricultural uses.

Local governments that try to regulate pesticides more strictly than federal agencies, as they did successfully for smoking, often find they cannot. Forty-one states, acceding to pressure from an industry front group called the Coalition for Sensible Pesticide Policy that visited every state legislature, have enacted laws forbidding localities to pass ordinances regulating the use or sale of pesticides. These state "preemption laws" often lifted wording directly from the model supplied by the coalition.

IS THE WATER SAFE?

When everyone is back home for the evening, your soccer-stained daughter decides she needs a shower, your baby is ready for her bottle. That turns your attention to the water flowing from your taps.

Children drink more water than adults. In the first six months of their lives, they consume seven times more water per pound of body weight than the average American adult. In baby formula, liquid foods, or soups made with tap water, your child ingests unwanted chemicals as part of the recipe. Drinking is not the only source of exposure. A ten-minute shower in contaminated water, easily absorbed through the skin and through breathing, is greater than the exposure from drinking two quarts of the same water.

Some of what's in the water supply are your cosmetics, household cleaners, and food preservatives, which live a "toilet to the tap" life cycle. Since most do not biodegrade easily, they persist in wastewater even after conventional water treatment and eventually find their way back into your community water supply, then into your taps. Some of these products cannot be detected even through current technology.

Trace amounts of thousands of these chemicals float around in company with the components of our nation's vast supply of medications. Rivers and streams across the country are full of as many drugs as a medicine cabinet because pharmaceuticals are excreted incompletely metabolized. The antidepressant Prozac frequently turns up in streams draining high-income suburban communities. Water engineers cannot match these chemicals with specific pharmaceutical products since drug ingredients are protected as trade secrets, so the chemicals elude understanding and treatment.

When the U.S. Geological Survey, which monitors water quality across the country, tested a sampling of waterways in thirty states, it found ninety-five components of treated human sewage. These included antiseizure drugs, testosterone, estrogen, naproxen, the mosquito repellent DEET, antibiotics, and the persistent breakdown products of cigarettes. Though each of these drugs and household chemicals is present in tiny amounts, there are 130 of these "emerging contaminants," enough to be labeled "sentinels of potential health effects" by the USGS.

The virtually indestructible chemicals used in Teflon and Scotchgard have migrated into water, soil, and sewage sludge; traces of the compounds were detected in two remote lakes in Minnesota and in Lakes Ontario and Erie and will probably be found wherever researchers look. The solvent TCE (trichloroethylene), the pollutant in Dickson's water supply, is a top contaminant of water in our nation; it poses a serious threat on more than 1,400 military bases and their nearby neighborhoods. It also vaporizes and has become an indoor air pollutant in thousands of homes across the nation.

The Safe Drinking Water Act, passed in 1974, required the EPA to set legal limits for contaminants and allowed the states to enforce these limits if they so chose. Not until 1991 were pesticides and herbicides added to the act and a maximum contaminant level set for them. In a landmark survey covering 1992 to 2001, the USGS discovered that over 80 percent of urban streams and more than 50 percent of agricultural streams contained pesticides, almost always as complex mixtures. However, as the lead researcher Robert Gilliom adds, "the water-quality benchmarks used in this study have been developed for individual chemicals," stressing that regulation fails to consider the mixture of chemicals flowing from our taps.

Furthermore, as biologist and author Sandra Steingraber points out, "these numbers have been arrived at through a compromise between public safety and economics. Maximum contaminant levels are not solely a health-based standard. Instead they also take into consideration cost and the ability of available technology to reduce contaminants."

Some of the standard practices intended to improve our water supply have come into question in the past few years. Fluoridation, first opposed by citizens who alleged it was part of a Communist plot, is now suspected as having adverse impacts on children's intelligence and behavior. (The silicofluoride chemicals used for fluoridation are derived from industrial waste and have never been FDA-approved for human ingestion.) The recent, first national analysis of tap water concludes that 137,000 miscarriages a year may be attributed to the by-products of chlorinated drinking water, called trihalomethanes, which are also found to be responsible for smaller than average size and neural tube defects among babies and bladder cancer in adults.

THEY ARE WHAT THEY EAT

Now you face the question "What substances hide in the food I've chosen for dinner?"

Meals for preschool children present a special challenge because they eat much more of many fewer foods than adults, and their favorite foods often contain the highest residues of pesticides. Young kids are voracious herbivores, favoring fruits and juices first and foremost, followed by vegetables. For instance, children aged one and two drink thirty times more apple juice for their weight than the average American and sixteen times more raisins. Of the nineteen foods children eat the most, one-fourth had detectable levels of cancer-causing pesticides and one-third had detectable levels of pesticides that can harm their brains and behavior, says the EPA, citing numerous studies.

Oversight of the nation's food supply is a crazy quilt, with the emphasis on "crazy." Under the 1996 Food Quality Protection Act, the EPA is in charge of setting the limits, called "tolerances," for pesticides used in conventional farming and left as residues on food. The Food and Drug Administration is supposed to enforce those tolerances for all food except meat and poultry, which are the jurisdiction of the U.S. Department of Agriculture (USDA). The FDA regulates eggs in the shell; the USDA is in charge of dried eggs. The FDA also sets levels for allowable residues on foods of contaminants other than pesticides, including lead, mercury, and PCBs.

To complicate matters further, it is the USDA's job to monitor residues. Every year, the USDA buys samples of food from ordinary supermarket shelves, washes them, then analyzes them for pesticide residues. They vary that list year to year. In recent rounds of testing, the USDA found nearly all the butter contained pesticide residues, mostly from breakdown products of the DDT banned decades ago, and pesticide residues on 33.7 percent of the tested items, mostly on fruits and vegetables, the staples of infant and toddler menus.

The residue levels are not health standards. They are a compromise between health and the power of lobbyists, the same kind of trade-off made in controlling contaminants in water. Each standard is set for a single chemical and not for the total amount of similar chemicals found in the entire diet. The FQPA allows traces of carcinogens in every food item.

Companies that manufacture pesticides outlawed in the United States continue to produce and export these chemicals, which are then applied overseas to crops imported back into the United States, in what's been named "a circle of poison." A bill that would have required country-of-origin labels on food was killed in 2004, opposed by food processors and grocery chains.

The level of pesticides in children who eat predominantly organic diets is six to nine times lower than in kids eating conventional foods.

Whether that means their health will be better over time or not has not yet been studied.

Even if regulators could wave a magic wand to eliminate pesticide residues tomorrow, no regulation can remove the toxic products arriving in your food supply as unanticipated and uninvited by-products of decades of "Better Living Through Chemistry." After pouring forth from smokestacks or evaporating from water, mercury travels thousands upon thousands of miles, then falls to earth borne by raindrops or attached to particles in the air and sinks into the seas, lakes, and streams. There, bacteria in the environment convert it to a more toxic form, methylmercury (also known as "organic mercury"), which is taken up by algae; the algae in turn are eaten by fish, so methylmercury eventually accumulates up the food chain into humans.

As the top rung of the fish chain, tuna (especially white tuna) and farmed salmon carry a particularly heavy burden of methylmercury. Although tuna and salmon are a fine source of protein and omega-3 fatty acids, cautions about eating them mount year by year, especially cautions addressed to pregnant women.

Dioxin repeats the cycle described by mercury. After it falls from the air onto the land and pastures, cows eat the grass and store the dioxin in their body fat and milk. When you drink cow's milk or eat fatty meats, you ingest the dioxin. Ninety percent of human exposure to dioxin occurs through your diet of meat, dairy, eggs, and fish, leaving you with twenty-two times the maximum dioxin exposure suggested by the EPA, declares the nation's preeminent dioxin researcher, Dr. Arnold Schecter from the University of Texas School of Public Health at Houston. Nursing infants imbibe thirty-five to sixty-five times the recommended dosage, as if any dioxin at all could be recommended.

Consumer Reports found that an average jar of meat-based baby food contains more than one hundred times the EPA's so-called daily limit for dioxins. That limit is based on cancer risks and does not take into account brain damage or other dioxin-linked health effects.

Dr. Schecter, a scientist of many interests, also looked into the presence of flame retardants in the food supply. He pulled thirty-two everyday items off the shelves of three Dallas supermarkets. His lab sliced, diced, and mashed the items into a pulp, then washed, vaporized, and shot them into a high-resolution gas chromatograph. Dr. Schecter found flame retardants in every food containing animal fats, with the highest levels in fish, followed by meat, then eggs and milk. Market basket studies in Japan and Spain have also uncovered these chemicals, though at much lower levels.

Pesticides like PCBs, dioxin, and flame retardants accumulate in the fatty tissues of the animals that consume the sprayed plants. In addition, all the organochlorine chemicals—DDT, aldrin, dieldrin, and chlordane—banned decades ago, linger in the food chain. Because steer and cows live relatively long lives in comparison with fish and other mammals, they accumulate a substantial amount of these pollutants in their fat. As do humans.

The compounds most likely to survive the trip up the food chain are those that are lipophilic—that is, they seek, dissolve in, and accumulate in fatty organs faster than they are ridded from the animal's body. Humans are replete with fatty organs. Because humans are at the top of the food chain, we get the full benefit, if one can use that term, of bioaccumulation. The embryo and, later, the suckling infant receive an even larger dose because of the parents' lifetime accretion.

AT THEIR MOTHER'S BREAST

Though our imagination and the world's great art portray the nursing baby and mother as purity incarnate, reality intrudes. Women's bodies contain as much as 10 percent more body fat than men's, and the breast especially attracts and stores fatty substances. Any fat-soluble chemical in the food chain will end up in mother's milk.

Multiple pesticides, today's and those long banned, accumulate in the breast, as do the industrial chemicals that crowd our homes. Of recent vintage is the rocket fuel perchlorate, found in breast milk at levels five times higher than the average in dairy milk, a Texas study discovered. PCBs' concentration in breast milk is higher than the level that would trigger U.S. regulatory action if found in cow's milk. Breast milk would be too contaminated with dioxin to be legally sold as a food in three European countries.

The product DDT turns into as it is metabolized is found in the breast milk of virtually all women in the United States today; in some, the level is six to seven times higher in her breast than in her blood. Mercury and lead are also found in breast milk, as are the fluorinated compounds that coat nonstick cookware and the lining of popcorn bags.

Flame retardant chemicals were found in the breast milk of twenty first-time American mothers at an average of seventy-five times higher than levels in Europe. A recent, first nationwide study found that two of the women had the highest levels ever reported in the United States and among the highest ever detected worldwide—1,078 parts per billion (ppb) for Katelyn of Raytown, Missouri, and 178 ppb for Angela of Gainesville, Florida. Older women tend to have higher levels of contaminants because their bodies have had a longer time to accumulate them.

Soy-based infant formula is questionable as an alternative because soy contains a natural environmental estrogen (called "genistein") not present in human milk. Even small amounts of soy, comparable to those in soy-based formula, when fed to pregnant lab mice during the early days of fetal life resulted in a variety of adverse reproductive effects in female offspring. Furthermore, because of the high levels of the heavy metal manganese in soy, soy formula can contain as much as eighty times the amount of manganese that occurs naturally in breast milk.

To top it off, soy is one of the crops typically grown with the full armament of industrial agribusiness tools.

No matter what substances are found in breast milk, the universal recommendation is to breast-feed because it passes on antibodies that protect the baby against intestinal and respiratory infections, childhood cancers, asthma and eczema, diabetes, and heart disease and increase a baby's IQ. New evidence indicates that breast-feeding can also protect the mother from breast and ovarian cancer and osteoporosis. Breast-feeding itself appears to overcome some of the harmful effects of high fetal exposure to persistent chemicals. How much better it would be if the milk were uncontaminated.

BITTER HARVEST

Industrial agriculture, its advocates say, is the reason Americans spend a smaller percentage of their disposable income for food than anyone else in the developed world. But that's economics viewed through a narrow lens, the supermarket cart. The price skyrockets when the accounting includes the real costs of contamination of the food supply, the pollution of air, land, and water in abutting communities (industrial agriculture is the nation's greatest polluter), ill health among farmworkers, and taxpayer-financed subsidies, which amounted to $143 billion over the last decade.

The average factory farm now envelops 14,000 acres, with 250,000-acre spreads looming in the near future. These agribusinesses specialize in commodity monoculture: one type of animal, vegetable, or fruit as far as the eye can see. The animal operations have a technical name, "concentrated animal feeding operations" (CAFOs), each raising more than 1,000 beef cattle, 2,500 hogs, or 100,000 hens. Industrial growers of vegetables and fruits raise one crop at a time over thousands of acres.

The CAFOs depend on the routine use of antibiotics and growth stimulants. The monoculture of produce requires intensive use of pesticides.

Industrially grown chickens, hogs, and cattle spend their lives in cramped feedlots, often indoors, squeezed together from birth to death in disgusting conditions. These practices generate high levels of stress and disease among the animals, which have to be countered with daily doses of antibiotics in their feed. Some years ago, when farmers coincidentally discovered antibiotics also accelerate animal growth, dosing became part of standard feeding practice. About 70 percent of medically important antibiotics in the United States—twenty-five million pounds a year, roughly eight times the amount given to humans—are fed to farm animals. This unremitting use ends up making bacteria resistant to the drugs that doctors depend on for treatment. "And because bacteria readily swap genes, that resistance can be transferred to a vast array of bacteria," reports the nonprofit organization Environmental Defense. The organization adds that the Centers for Disease Control and Prevention ranks antibiotic resistance "a top concern." And these growth-promoting antibiotics are bad business. An analysis of seven million Perdue antibiotic-fed chickens discovered that, because of the costs of the drugs, producers ended up losing money by using them.

As much as 80 percent of the antibiotics fed to poultry passes unmetabolized straight through their bodies into their litter, which is often applied as fertilizer on nearby fields. Poultry litter is also used as feed for other CAFO animals. Though most food producers wean animals off antibiotics a few weeks before sending them to market, some low levels of the antibiotics remain, along with antibiotic-resistant bacteria, which get passed along to consumers. Since chicken parts are ground up to make chicken nuggets, antibiotics can come along in this favorite food, potentially evolving kids' resistance to these drugs. In factory poultry farms, chicken parts, feathers, bones, and all are ground up as feed for their living brethren. Apparently, mad chicken disease is not considered a possible threat.

Antibiotics are not the only surprise. Arsenic is an approved animal dietary supplement, added since the 1970s to chicken feed to promote growth and to kill the intestinal parasites that plague fowl raised in crowded conditions in huge farms. Arsenic occurs in trace amounts in nature. But the arsenic in chickens is far above trace; further, in the chicken, it converts into its more toxic inorganic form. The chickens excrete the compound into their litter. After the antibiotic-arsenic-laced litter is spread on farmland, some of it runs off into nearby water. In 2001, the National Academy of Sciences told Congress, "The data indicate arsenic causes cancer in humans at doses that are close to the drinking water concentrations that occur in the United States."

"Americans who consume chicken, such as my son, who appeared to exist largely on chicken wings during high school, may be exposed to arsenic far higher than recommended," and "exposure to arsenic via food may well be a significant and preventable portion of overall exposures to this human carcinogen," worries Ellen Silbergeld, a toxicologist at the Johns Hopkins University School of Public Health. Anyone who eats over three-quarters of a pound of arsenic-tainted chicken in a day far exceeds the level of arsenic the government deems "acceptable," an obsolete measure set decades ago before American chicken consumption tripled.

Tyson Foods said it stopped using arsenic in chicken feed in 2004, around the time McDonald's started demanding arsenic-free chickens. But arsenic is still around. A 2006 study by the nonprofit Institute for Agriculture and Trade Policy found arsenic at varying levels in Perdue chickens, as well as in 55 percent of uncooked chicken samples purchased in supermarkets and in all ninety fast-food chicken products.

Steroids—the very kind outlawed for athletes—are commonly injected into cattle, to boost their growth rate and beef up their muscles. The drug DES used to be approved for this purpose, as it was once prescribed for human mothers to prevent miscarriages; it was later found to cause cancer and genital abnormalities. The FDA currently approves of six growth promoters, including testosterones and estrogens.

Industrial livestock production involves a two-step regimen: Cattle are bred and raised for a few months in one location, where they are weaned, often on blood, then fed slaughterhouse waste and genetically engineered grains. Next, they are sent to feedlots for fattening, where 80 percent of the animals are treated with steroid hormones, either in their feed or via a controlled-release implant in their ears. *ScienceNews* investigative reporter Janet Raloff, working with an animal scientist, ran the numbers: After calculating the cost of the hormones versus the pounds they add to the animal, cattle raisers make an extra $40 per head. Steroid hormones in beef may be one of the reasons for declining male fertility.

In today's industrial dairy farms, cows are kept pregnant through artificial insemination, while about 75 percent of the time they continue to be milked, "not a normal mammalian behavior," explains Dr. Walter Willett, chair of the Nutrition Department of the Harvard School of Public Health. During pregnancy, hormones including progesterone, estrogen, and insulin-like growth factor 1 (IGF-1) normally rise to sky-high levels. Some insulin-like growth factor is needed for normal growth, but not the elevated amounts that come from a pregnant yet lactating cow and are found in the milk our children drink. An injection of growth hormones raises the IGF-1 yet higher, and "that ignites the fire that increases the likelihood of cancer," Dr. Willett adds.

Another problem with industrial cattle is that they are fed grain, an artificial diet for cows, which evolved to eat grass. To be certified as organic, cows must spend at least 120 days at pasture, a standard allegedly circumvented by organic milk factory farms. Beef and milk from grain-fed cows are deficient in beta-carotene and omega-3 and other "good" fats known to help boost the body's production of stress-fighting antioxidants. And, in fact, boys with ADHD behavior, autism, and learning problems are often found to be deficient in these omega acids.

MANUFACTURED FOOD

Prepared foods revamp the products of industrial agriculture into handy packages through the innovations of high technology. Food morphs into "restructured" "novelties" with "improved shelf life," as boasts one of the leading high-tech food sorcerers, FMC BioPolymer, an arm of the FMC Corporation (manufacturer of insecticides, lithium, and the latest food and chemical machinery).

FMC BioPolymer's website enumerates its offerings: "a uniquely broad palette of functional ingredients and technologies based on Avicel microcrystalline cellulose/cellulose gel, FMA carrageenan and Protanal alginate . . . [for] restructured foods, ranging from fruit for filling and restructured onion rings, to formed meat chunks for producing novelties such as meat nuggets and luncheon loaves . . ." and "a range of functionalities, including texture modification, development of structures and rheologies, consistent air cell structure, freeze-thaw stability and enhanced product stability for improved shelf-life." This company polymerizes food the way polyvinyl is polymerized.

Modified, texturized, hydrolyzed, hydrogenated, and manufactured in labs, these foodlike products bring a substantially higher profit margin than real foods. "Most of the money to be made is in processing very inexpensive raw materials in fancy and convenient packaging and making it more tasty, which is also usually sweeter, saltier, and fattier," explains Dr. Anthony Robbins, a medical doctor and professor in the Department of Public Health and Family Medicine at Tufts University's School of Medicine.

The traditional ingredients that give processed foods their increased shelf life include trans fats, preservatives, and emulsifiers. Trans fats, made by converting vegetable oil into a more stable solid form by partial hydrogenation, are found in 40 percent of the products on supermarket shelves; they contribute to clogged arteries and the risk of diabetes. Processed convenience items are often also high in fructose corn syrup, a sweetener used because it's cheaper than sugar. Both have no nutritional value, crowd out healthy drinks such as fruit juice and skim milk, and make their weighty contribution to childhood obesity and probably to diabetes.

Additives, artificial colors, and flavors complete the recipe for kids' foods. The Coca-Cola Company, trying to break into the milk markets, formulated Swerve, a milk-based artificially sweetened canned beverage in chocolate, vanilla/banana, and blueberry flavors. An additive used to make popcorn taste more buttery causes a life-threatening lung disease among workers at popcorn plants.

Most certified food colors are synthetics derived from petroleum. A single artificial flavoring can be made up of anywhere from a few to hundreds of separate chemicals, mostly petroleum based. The chemical recipes for artificial flavors are protected as confidential business information. Here, however, as a sample, is the formula for a synthetic raspberry flavoring: vanillin, ethylvanillin, alphaionone, maltol, 1-(p-hydroxyphenyl)-3-butanone, dimethyl sulphide, 2,5-dimethyl-N-(2-pyrazinyl) pyrrole. Note that there are no raspberries in this raspberry flavoring. Artificial colors and flavorings are the giveaway signs that the product probably contains no real food.

"All these additives in our food are going to turn out to be harmful," warns former EPA administrator Carol Browner. "They are already affecting the maturation of our children; young girls' big bosoms are just one symptom." Additives have also been identified as a factor in ADHD and other illnesses. Though the use of one additive at a time may be relatively safe, research has found that the common practice of the use of several at a time—for instance, two food dyes plus monosodium glutamate and aspartame—produces a synergistic interaction damaging to nerve cells. The Feingold Association, which researches the effect of diet on children's disorders such as ADHD and Tourette's syndrome, suggests eliminating additives and food colorings. Parents who follow the association's recommendations to avoid these ingredients often report that their child's symptoms have vanished or diminished. One theory suggests that artificial food additives may act as "counterfeit" neurotransmitters, causing neurons to fire false, excitable signals.

Processed foods often strip out the calcium, iron, and zinc, which in adequate amounts equip the child's body to rid itself of heavy metals. This leaves poor children, with their insufficient consumption of essential vitamins and minerals, at higher risk from the negative effects of heavy metals, which could be a factor in aggressive behavior and even crime.

Junk food follows children into their schools. Nine out of ten American schools offer snack food or soda to students in direct competition with school lunches, according to two surveys by the Government Accountability Office, the independent investigatory arm of the U.S. Congress. The schools, strapped for money by local budgets, generate fees from these sales. One-third of the schools made somewhat more than $125,000; on average, sales amounted to $18 a year per child. A proposal by Senator Tom Harkin (D-IA) simply to let the federal government develop model nutrition guidelines for foods sold in school vending machines went down to bipartisan defeat, though many cities and states are banning sugary soft drinks and some other junk foods.

The food industry, which spends roughly $10 billion a year marketing to children, applies many of the public relations and science-for-hire strategies as the tobacco industry. The Center for Food, Nutrition and Agriculture Policy, currently based at the University of Maryland, was originally founded by the Grocery Manufacturers of America (GMA), and now executives from GMA, DuPont, Monsanto, Kellogg, and Kraft Foods sit on its advisory board.

SpongeBob SquarePants and other lovable characters peddle "fun" foods. Shrek's portrait attracts kids to his namesake cereal, made of sweetened corn puffs, marshmallow pieces, and one-half ounce of sugar for every ounce of serving. The National Academy of Sciences has found this advertising situation so alarming that it has called for restrictions on the ads. "We have created this monster by allowing trash food marketers to prey on our children and by letting our children disappear into video screens," writes *Boston Globe* columnist Derrick Z. Jackson.

The marketing that pervades modern life has made food a unit of entertainment. It has lost its meaning as the foundation of life. Yet a mother's food choices before and during pregnancy make a critical difference to fetal development. For example, the risk of neural tube defects, cardiovascular disease, and cancer rises among the offspring of women whose diet during pregnancy is low in folate, a nutrient that boosts DNA production. The risk of cancer also increases if a mother, while pregnant, fails to consume enough of the foods, such as meats and beans, that are rich sources of glutathione, needed to repair the oxidative DNA damage wrought by toxins. The iodine deficiency that plagues one-third of American women leaves them deprived of thyroid hormones. So, too, the food that infants and children eat, or do not eat, can aid or add to their propensity toward illness. "The dominance of psychoanalytical, behavioral, and pharmacological approaches to illness have abrogated any major attempts to look for food or chemical triggers for their patients' illnesses," conclude Drs. Nicholas Ashford and Claudia Miller, experts in understanding multiple chemical sensitivity.

AN ALTERNATIVE WORLD

In this walk through a day with your child, you may have noticed that alternatives for toxic products are already at hand. Some change has begun. Toshiba now uses an inherently flame-resistant plastic, polyphenylene sulphide, for the casings of its electronics products. Safer alternatives that companies can substitute for six of the most egregious chemicals from lead to formaldehyde—highly toxic and in wide commercial use—have been identified by the Massachusetts state–funded Toxic Use Reduction Institute in a first such search. The leading manufacturer of salon nail products, OPI Products, has agreed (under pressure from the Campaign for Safe Cosmetics, a coalition of public health, women's, and consumer groups) to remove from its polishes phthalates and the solvent toluene, a carcinogen also suspected as a cause of birth defects, though the company has not yet found "an efficacious alternative to formaldehyde as a nail hardener."

The standard hospital fare featuring red-dyed Jell-O and gray canned peas has been replaced by local, sustainably grown produce from farmers markets for patients and staff at thirty-two hospitals operated by the huge HMO Kaiser Permanente in six states. Kaiser Permanente has also gone through all its medical centers and replaced toxic products with safe ones—no more blood pressure cuffs with mercury, no more PVC-filled carpets, nothing that uses one of the phthalates; only nontoxic janitorial products and only green building materials. And this reformation is actually driving the market because Kaiser Permanente is the largest purchaser of health care products in the United States. Almost one hundred other hospitals are following suit by now.

Organic apparel, nontoxic furniture, and green cleaning products are now hip. Organic food is jumping off the shelves, while organic farming claims ever more thousands of acres.

Technology, pushed by market and citizen demand, has made some progress creating less toxic processes and products. While the EPA has dragged its feet in regulating TCE, the Nakanishi Manufacturing Company in Athens, Georgia, pressured by a coalition of local grassroots organizations, installed a new vacuum technology as an alternative method to degrease the products it manufactures. Before this switch, a 2006 study of the air around the plant registered what is thought to be the highest level of TCE found in any U.S. community.

The architect-designer William McDonough, expressing his goal "to crank the wheel of industry in a different direction, a safe world that our children can play in," has, among numerous projects, helped a Swiss-based textile plant design a fabric safe enough to eat, made through a process that produces no toxic by-products, and it even saves money.

These shifts in consumer, commercial, and industrial response are bellwethers. They indicate that the worst scenario is not inevitable and that parents have the power to move these advances further and further.

Six
_____

FORENSICS

THEOPHRAST PHILIPPUS AUREOLUS BOMBASTUS VON HOEHENHEIM, A
sixteenth-century physician, alchemist, astrologer, and occultist known
as Paracelsus, is often called "the father of toxicology." His dictum that
"the dose makes the poison," meaning a substance is poisonous only at
doses that produce an acute physical effect, has long been holy writ in
the profession of chemistry.

His notion that low doses of a toxic substance are harmless still is
regarded by many as conventional wisdom. Only the brute force of
large quantities of poison could overcome the body's defenses, according to this view. Safety standards to protect public health are built on
this foundation. Chemical companies frequently invoke this article of
faith to defend their products, issuing statements about the huge
amounts of a chemical that would have to be ingested to produce an
adverse effect. An industry fact sheet states, for example, that "if the
dose is low enough even a highly toxic substance will cease to cause a
harmful effect." In defending the "high-dose only is harmful" approach, industry and its supporters also argue the converse: Even
nontoxic substances are dangerous at high doses. Alex Avery of the
conservative Hudson Institute likes to say that in sufficient amounts,
"water will kill you."

In recent decades, however, long strides in scientific understanding of how chemicals affect the body and mind, along with remarkable advances in the technology of scientific research, are rendering Paracelsus's dictum obsolete. What, then, does make the poison?

Foremost is the new understanding that while the immature body may successfully combat a limited assault, chronic exposures—repeated low doses over a period of time to a chemical that has no immediate health impact—will add up to devastating consequences.

Second is the recognition that fetuses and children are both particularly and also sometimes differently vulnerable compared to adults, so that chemicals in amounts measured in parts per trillion or even quadrillion can affect their development. This can knock away the former underpinnings of regulation. Dow claimed the pesticide Dursban and its relatives in the organophosphate family of chemicals were safe based in part on the 1990s explanation about how they affect the body and mind of human adults. It was known that this class of pesticides inhibits a necessary enzyme from cleaning out its corresponding neurotransmitter, so that the neurotransmitter continues or increases sending signals excessively. This neurotransmitter is found throughout the animal kingdom, including in insects; insects' nervous systems get so overexcited by the chemical that they die. It hurts adult humans, but usually within tolerable limits. Subsequently, however, new research discovered that while organophosphates cause some harm in this way in the developing prenatal or newborn human brain, it is not the way they cause the worst harm. Instead, this pesticide family actually reduces the number of brain cells, alters the architecture of the brain, and confuses the normal connections among others, causing permanent structural damage to the immature fetus or child.

These injuries can happen from a small single dose on a critical day of development. In the human child, the result could include behavioral and learning disorders such as ADHD, according to Dr. Theodore Slotkin, the scientist who made these breakthrough discoveries. (Even so, when the EPA finished its ten-year review of organophosphates to examine their cumulative effect, the agency looked only at the neurotransmitter inhibition. Since the difference between immature or adult reaction is scant in this test, chlorpyrifos, which had formerly been assigned a tenfold safety restriction under the Food Quality Protection Act, was relieved of that restriction. If the EPA had chosen to examine the more serious way these pesticides harm fetuses and young children, the outcome would have been more protective.)

The nature of the chemical itself obviously makes a difference. Just as it has been recognized for some years that exposure to even tiny amounts of some chemicals can cause cancer, advances in science show that other toxicants can be provocateurs of other illnesses, even at low levels. For lead and mercury, the level of "acceptable" exposures, especially for children, has been dropping precipitously: A two-year-old's exposure to lead in an amount equal to one drop in six bathtubs can cause a significant decrease in his or her IQ later in life.

Chemicals capable of disrupting endocrine hormones are now understood to be a different kind of toxin. None of them follow the "dose makes the poison" dictum. Even at tiny doses, they can alter the way the immune and endocrine systems operate, leaving the body vulnerable to sickness or developmental damage. Phthalates, bisphenol-A, dioxins, flame retardants, and some pesticides, as well as long-banned chemicals persisting in the environment, such as DDT, are major hormone disruptors.

Different chemicals affect different systems, to interfere with different processes. Because of their influence on thyroid hormones and the neurotransmitter dopamine, PCBs—even at just 5 parts per billion, the equivalent of 1 drop in 118 bathtubs of water—as well as other chemically similar chemicals are under suspicion as responsible for lowered IQ, learning disabilities, and ADHD.

The timing of an exposure can determine what kind of damage (if any) a chemical may cause. One experiment on laboratory animals, for example, showed that different birth defects were induced depending on whether the animals were exposed on day ten or day twelve of gestation! Thalidomide, once prescribed for morning sickness, caused limb defects among children only when the exposure occurred between days twenty and twenty-four of fetal life, researchers subsequently realized. Toxicologist Ellen Silbergeld of the Johns Hopkins University School of Public Health concludes that the same toxins responsible for a mother's loss of fertility and spontaneous abortions can instead cause heart and other defects in the fetus; the difference is the timing—"the windows of susceptibility"—of the environmental insult. Similarly, substantial air pollution can cause fetal death during certain weeks of the mother's pregnancy, while during a different set of weeks it can result in cardiac defects or cleft lips.

The pathway of exposure—if it is by ingestion, inhalation, or absorption through the skin—produces different outcomes (because dif-

ferent organs of the body detoxify in different ways). Chemicals can
shape-shift—first appear in water, as TCE does, then vaporize into air,
presenting a new danger.

Even gender makes a difference. ADHD and autism strike more
boys than girls; boys have twice as many cleft lips as girls, while cleft
palate affects more girls than boys. Boys also fall victim more often
than girls to cancer and asthma. One probable reason that boys make
up 70 percent of all autism cases as well as the majority of children suffering from ADHD and other learning disabilities is that girls have
higher levels of the stress-fighting antioxidant glutathione as well as
higher levels of estrogen, which also protects the body against oxidative
stress. PCBs and flame retardants are known to increase oxidative
stress, as do pesticides and air pollution.

Yet another conceptual breakthrough is the recent recognition that
substances in combination can have more serious effects than they do
alone. Some harmless chemicals can prove toxic in combination with
others that are also harmless. Biologist Tyrone Hayes and his colleagues at UC Berkeley have spent several years exposing tadpoles to
nine pesticides commonly used in growing corn, singly and in combination. They used very low concentrations, as low as 0.1 part per billion, in the tadpoles' water. None of the chemicals by themselves had
any perceptible effect. But when a combination of the nine was added
to the water, the tadpoles fell prey to endemic infection, and those that
survived took longer to mature and ended up smaller in size than tadpoles raised in clean water. Chemicals can also be more toxic and cause
more damage together than individually, a synergistic effect. When a
man's body is burdened with phthalates and PCBs at the same time,
the mobility of his sperm decreases more than twice what would be expected from each chemical by itself, it was discovered recently.

Genetic variability is a major element in response to a low-dose exposure—one child may be impervious to a chemical or metal, another
dangerously sensitive. For instance, injury from pesticides is more
likely in children with a genetic variation that leaves them with an un-
usually low level of certain protective enzymes. This showed up in the
University of California at Berkeley Center for Children's Environmental Health studies of the pesticide Dursban and its fellow
organophosphate diazinon. The researchers identified children who
are up to 164 times more vulnerable to harm from these chemicals than

adults. This extreme vulnerability derived from the combination of their immature bodies plus their genetically low levels of helpful enzymes.

Genetic variations reduce the ability of some children to metabolize and excrete heavy metals. In children with autism, higher concentrations of mercury, lead, and zinc have been measured in their baby teeth, meaning they retain more of the metals than normal. Other studies have found that children with ADHD retain manganese in their hair samples at levels significantly higher than the non-ADHD test group.

Some chemicals that produce no or even beneficial effects in high doses can cause problems in low doses. For example, experiments by Dr. Frederick vom Saal of the University of Missouri have shown that very small amounts of the hormone disruptor bisphenol-A produce changes in the reproductive and urinary system of fetuses. These exposures can enlarge the prostate gland, while large doses have the opposite effect. These discoveries do not mean that exposure to large doses of other toxic substances do not produce serious, acute health problems. Of course they can. But low doses can in some cases be just as dangerous in a different way.

In any event, effects at low levels should not be surprising since chemicals used as drugs often work so. As Ken Cook of the Environmental Working Group points out, the popular antidepressant Paxil is effective at 30 parts per billion, and the chemical in birth control pills works at 0.019 ppb.

This emerging evidence about the dangers of low-dose exposures to chemicals is driving a significant advance in protecting children from toxics in the environment. It represents "a paradigm shift in the way toxicology studies are designed and has increased concern about the safety of such chemicals at ambient environmental levels," according to Daniel M. Sheehan of the National Center for Toxicological Research, an arm of the Food and Drug Administration.

GENES *AND* THE ENVIRONMENT

Scientific breakthroughs have also demonstrated that the old argument pitting genes versus the environment as the cause of a particular illness in a particular human is outdated and misleading. It turns out that very few diseases (cystic fibrosis is one) are caused solely by a defective inherited gene. In most cases of chronic illness, such as cancer or autism, it is both—genes plus environment.

"In thinking about genes and the environment," explained Dr. Irva Hertz-Picciotto, an epidemiologist at the University of California at Davis, "you could estimate that 60 percent of autisms have a genetic component. Does that mean the environment causes 40 percent? No, actually it could cause 90 percent, because among those 60 percent of cases there had to be something environmental that disregulated the gene at a particular point in development and particular tissues and unleashed some cascade of events altering processes necessary for human development."

Even in families with one child with a birth defect, indicating a genetic predisposition in the family, the risk of having another child with a birth defect dropped by half if the mother moved to another town, out of the reach of some toxicant in the environment, a study demonstrated. The lab run by Dr. Jeff Murray, a pediatrician at the University of Iowa and an expert in birth defects, is sequencing genes in one hundred cases of birth defects to try to identify mutations. In his educated judgment, 70 percent of the cases of birth defects are likely to be a mixture of genes and the environment. He adds, "It's difficult and expensive to analyze a rare disease like birth defects because we'd need a large population sample, and the political will has not been sufficient for it to happen in a thorough way."

Many polluters and their representatives continue to insist that it is inherited genes that cause most chronic health problems, not their toxic products and wastes. Dr. Kenneth Olden, former director of the National Institute of Environmental Health Sciences, commented, "The powers that be, the money, want you to focus on genetics. We could spend another twenty years on genetics and I'd predict we'd be no closer to preventing diseases or curing them. We have to understand how genes and environment interact."

GENE EXPRESSION

Genetic mutations—direct changes to the DNA—were once thought to be the necessary initiators of cancer until one of the most startling discoveries of recent years overturned that precept. It's now clear that cancer can also be triggered by a disruption of the body's normal gene expression. Because of the disruption, genes do not turn on, or off, to do what they should when they should; this disruption may cause a carcinogenic change or may prevent a repair enzyme from fulfilling its function. Then, when the DNA replicates itself, the disruptive changes are copied.

Hormone-disrupting chemicals have been shown to work their ill effects by changing gene expression. Arsenic, for example, which can poison at high levels of exposure, can precipitate cancer at low levels. Dr. John Peterson Myers, coauthor of *Our Stolen Future,* explains that arsenic in this case isn't causing illness directly, but instead it interferes with a hormone necessary to turn on genes that make the proteins that would otherwise help suppress tumors, according to lab studies. Hormones that disrupt normal gene expression might also alter the way cells communicate, thus inducing uncontrolled cell proliferation, the very definition of cancer. Among the many chemicals that can affect gene expression to trigger cancer are the herbicide atrazine, the plastic ingredient bisphenol-A, and dioxin (a by-product of chlorine), the most potent family of carcinogens the EPA has ever studied. The Vinyl Institute, the manufacturers trade association, asserts that only some dioxins are carcinogenic.

And cancer is not the only illness attributable to disarray in gene expression. Learning and behavioral problems, deformities of the male reproductive tract, and male and female infertility can also be traced to the disruption of hormones and indeed by the same chemicals. Some families of pesticides affect hormones, leading the Mt. Sinai School of Medicine's Center for Children's Health and the Environment to warn, "Pesticides could become the ultimate male contraceptive."

The brain and behavior are probably, of all the child's developing parts, the most susceptible to chemicals that disrupt the hormone system. That vulnerability extends from the womb into infancy and the toddler years. A recent count tallied sixty-three hormone-disrupting pesticides; 60 percent of all herbicides disrupt hormones yet remain widely used in agriculture. Because the thyroid system is not mature until the third trimester, even a slight difference in the mother's level during pregnancy of thyroid hormone can lead to significant changes to the child's brain development in the womb. "To date the EPA has never taken action on a pesticide because of its interference with the thyroid system," declares endocrine researcher Theo Colborn, coauthor of *Our Stolen Future.*

"We and the world have a tremendous problem, releasing chemicals into the environment without any way to test whether they interfere with thyroid action," warns Dr. Thomas Zoeller, a molecular biologist and endocrinologist. Zoeller runs a lab at the University of Massachusetts at Amherst whose goal is to understand the mechanisms by which chemicals such as PCBs disrupt the thyroid.

His work shows that PCBs cause harm in different ways, depending on which of the 209 different members of the PCB family is active. They can lower the levels of thyroid hormones by too quickly activating enzymes that clear the hormones from the blood; or they can attach to the thyroid hormone receptor, either activating or turning off genes at the wrong times. This process is graphically captured on the series of slides crowding some of the lab's drawers. These slides hold cross sections of fetal brain cortex taken from rats: The higher the dose of PCBs, the stronger the effect on gene expression.

PRENATAL PROGRAMMING

Scientists have also made a breakthrough discovery by identifying the fetal origin of disease, a concept known as prenatal (or fetal) programming. Mounting evidence shows that the fetus is at highest risk to even minuscule amounts of exposure to some chemicals at key periods of its development, and in turn, the environment of the womb influences human development so powerfully that it seems to set a person's cellular code for life and cause illness at any time from conception until old age. Of course, birth defects always occur in utero, but other illnesses also have their roots in fetal or prefetal life.

"Almost all childhood leukemias and many other childhood cancers begin before birth, so we have to look at the exposures of the mother and even the father," explained Dr. Martyn Smith, whose laboratory is part of UC Berkeley's Northern California Childhood Leukemia Study program. "That wasn't really a focus before," he added. Analyzing the newborn blood samples he had stockpiled from many collaborating hospitals, his lab found chromosomal damage is uniformly present at birth in children subsequently diagnosed with leukemia.

However, chromosomal damage does not lead inexorably to cancer. "The development of cancer takes more than one exposure; the first event, perhaps to catalyze an alteration, occurs in the womb," he said, "through the mother being exposed to a chemical like benzene, or by the mother having a bad diet so there's not enough protection against normal damage. Then, the second event takes place after birth to keep the chain of events going. Something interferes with the setting of the immune system. Perhaps different stages of the disease are affected in different ways," Dr. Smith explained. "A virus could start the disease, then a chemical in the environment might make the disease worse. Or the other way around." Multiple chemical sensitivity is similarly thought to develop through a multistep process. Some experts think autism does as well.

Mental and behavioral disorders often begin in fetal life. When it comes to exposure to neurotoxins, timing is particularly critical because the brain and the nervous system are so exquisitely sensitive and parts of the brain and nervous system that control different functions—motor control, sensory functions, intelligence, and attention—develop at different times.

The new theory of fetal programming further posits that illnesses seeded in the womb can wait to emerge until later in life. Exposure to lead in the womb may contribute to high blood pressure in adulthood. Other environmental assaults before birth, such as from pesticides and TCE—perhaps because they destroy neurons or make neurons more susceptible to later insults—now are thought to contribute to Parkinson's disease among the elderly. Exposure to pesticides in the organophosphate family will also affect cardiovascular, metabolic, immune, and endocrine function later in life, Dr. Theodore Slotkin predicts. This theory of prenatal programming would also help to explain the steep increase in breast and prostate cancers and Alzheimer's disease among the aging baby boomers.

There is also evidence that disease and development disorders can be programmed even before conception—that cancer and other illnesses can begin with alteration of the germ cells of the mother and father or grandmother and grandfather. Shock waves rippled through the scientific community when a lab study of rodents discovered that infertility may even be inherited down to successive generations of male offspring. In this study, pregnant mice exposed to hormone-disrupting pesticides experienced reproductive problems, as might be expected; what was revolutionary was that their male offspring, never exposed to pesticides, displayed similar abnormalities on down through the four generations of the study. And the way the damage was inherited, they realized, was not through basic mutations in the DNA, but through genes that had been reprogrammed to behave differently. Changes in gene expression are generally erased with each new generation, but sometimes, as in this case, the affected genes maintain the alteration and are inherited. This study suggests it is plausible that an individual could experience disease that was triggered by his grandparent's exposure to a pollutant.

This is information that prospective mothers and fathers need to consider. Many, perhaps most women still wait to protect themselves from the environmental hazards around them until they find they are pregnant, and men usually fail to understand that their own health has any bearing at all on the health of the babies they conceive.

ONE CHEMICAL = MULTIPLE ILLNESSES

The understanding that a single chemical has the potential to cause a number of different illnesses is another recent insight. One of the reasons Tennessee state senator Doug Jackson said he was skeptical about the complaints of families in Dickson that TCE in their water supply was causing cleft palates, heart defects, urinary tract malformations, and brain malformation was his belief that one chemical could not cause those differing illnesses. In fact, laboratory experiments have shown that the chemical can, indeed, be responsible for these various problems, and the National Research Council has recently declared that while TCE's major threat is liver cancer, it can also cause kidney cancer, reproductive and developmental problems, neurological damage (such as Parkinsonism), and immune system disorders at levels as low as 1 ppb. One family, the Holts, who farmed adjacent to the Dickson waste site, suffered from a variety of ailments, including cancer and skin disease. Amy Wood, whose first daughter was born with a cleft palate, fell ill with toxemia, a known outcome of TCE exposure, during her first and second pregnancies. Jenny Casteel's son was born with a cleft palate and a heart defect. While (as we have already noted) there is no definitive proof that TCE caused all these illnesses specifically in Dickson, evidence indicates that it can.

TCE is not an oddity in its multi-illness powers. It's now known that most chemicals affect multiple sites on different types of cells and, in some cases, even at multiple targets within the same cell type, resulting in different actions on different tissues or organs, each causing a different disease.

Bisphenol-A, a champion disruptor of hormones, seems to be able to alter the expression of several hundred genes. Depending on the tissues it affects and on the timing, it can cause birth defects, reprogram prostate cells, reduce sperm count, bring about miscarriages, trigger obesity, change behavior, and cause degenerative brain disorders. An upset in thyroid hormones can result in harm to the fetal brain and, as recently discovered, harm to the pregnancy itself.

By melding these discoveries, science can understand childhood illnesses as never before, casting aside the basic premise of the past, that single agents disrupt individual organs. For example, testicular cancer, poor sperm quality, and male genital tract malformations—where the urethra is in the wrong place in the penis or where testicles fail to descend—are now seen as related. The likely explanation is that fetal exposure to hormone-disrupting pesticides, phthalates, flame retardants, and other chemicals causes permanent abnormal development of cells within the fetal testes, which disrupts the normal, healthy unfolding of development; the more severe cases give rise to multiple symptoms, less severe cases to only one or two. The mother's body burden of these chemicals turns out to be a better predictor of testicular cancer than her son's. The higher the concentration of flame retardants in the mother's body, the higher the likelihood her son would suffer from undescended testicles.

MIND-BODY DISORDERS

Recent discoveries have led to a greatly changed view of mental and behavioral illnesses. It is now understood that neurotoxins can result in differing consequences: hyperactivity, attention deficit, learning disabilities, aggressive behavior, dyslexia, cerebral palsy, mental retardation, Tourette's syndrome, and the spectrum of autism disorders. Depending on which specific region of the brain and nervous system is affected, a toxin can interfere in different ways with intellect and behavior. These are all "a set of dysregulations that take different forms in different kids," explains Dr. Martha Herbert, a pediatric neurologist who both treats children and pursues research at Massachusetts General Hospital in Boston.

As the landscape of autism has shifted, so too have the other mental and behavioral illnesses come to be viewed differently. In the 1950s, autism was laid at the feet (or, more accurately, breast) of "refrigerator mothers," often hardworking professional women accused of inadequate maternal nurturing. Then, as Dr. Herbert and her colleague, sociologist of medicine Chloe Silverman, trace the history, this opinion changed in the 1980s with the discovery of abnormalities in autistic brains. This discovery freed parents of blame and shame but led researchers to conclude that autism was determined by genes, hardwired before birth. Following this theory meant that autism would be treatable only by behavior modification, and the appropriate research agenda would be dominated by genetics, neuroscience, and psychology. This focus on finding the hardwired genes still accounts for the bulk of research money (bolstered perhaps by the fact that it's possible to develop profitable drugs from genetic discoveries).

But the view of autism as a strongly genetic, inherently incurable, static disease is unraveling. A new rethinking qualifies as a "paradigm shift," a truly new way to think about autism. The same shift applies to other mental-behavioral illnesses.

Not only does this rethinking demolish the focus on genes, genes, genes, it also acknowledges that these illnesses couple the mind and the body; they are not "just in your mind." While the mainstream research and clinical community may still consider autism—or Tourette's syndrome or ADHD or such—something that's happening inside the brain, and perceive, for example, the autistic child's bodily symptoms, such as severe gastrointestinal pain, as incidental and worthy of little attention, the new thinking sees these symptoms as clues to understanding the biological mechanisms that cause the illness and as clues to treatment and to prevention. Dr. Herbert puts it this way: "The widespread changes we're seeing in autistic brains may occur in parallel with or even downstream of (after, or as an outcome) the widespread changes in the body, such as in the immune system." Treating the bodily symptoms may improve brain conditions, as it has for Michaela Blaxill, a child we write about later in this book, and as it has in other ways for children with different mind-body illnesses.

Furthermore, these revisions in thinking about mental-behavioral disorders also directly acknowledge the impact of environmental chemicals. "New evidence is emerging," Herbert and Silverman explain, "that makes autism look more like an environmentally mediated illness."

A NEW SCIENTIFIC REVOLUTION

These revolutionary shifts in knowledge arise out of the scientific advances of recent decades, from analytical chemistry to epidemiology, computer technology, and genetic analysis, among others. Propelled by these advances, we stand now at a point midway in a scientific revolution that is unraveling the basic questions of how normal cells and tissues become diseased. "The revolution under way is going to be as important as the revolution in public health that was created by Pasteur's discovery of germ theory," contends biologist John Peterson Myers.

Environmental pediatrics creatively combines these new discoveries with the traditional tools of public health—animal studies, lab experiments, and epidemiology. And it continues working the way science has always worked, accreting knowledge one bit on top of the other.

Lab animals, chosen as close stand-ins for humans, have generated insights into the toxic causes of every one of the illnesses that beset children. Dr. Ana Soto, professor of anatomy and cellular biology at Tufts University, for example, using rats, has recently discovered the way bisphenol-A triggers precursors of carcinomas. In experiments with clam embryos, marine biologist Carol Reinisch found that a combination of three common water pollutants damages fetal nerve cell growth in ways that may be relevant to human illnesses such as autism and ADHD.

Epidemiology, another of the traditional tools of public health science, compares a group of people who are sick with a group who are healthy, to separate out what differences exist between the two and whether an external agent may have made the difference. "We can't do experiments on people, so we look around in the real world for something that's like an experiment, and that's epidemiology," explains Boston University epidemiologist David Ozonoff.

In traditional epidemiology investigations, researchers ask parents about their occupations or how far they live from a polluting factory or farm and from those answers extrapolate the child's possible exposures and how much of a toxicant might have entered his or her body. Though this approach has contributed useful information, it has serious weaknesses. The guesses are not always on the mark since people's memories are imperfect and the fact that a family lives downwind from a factory does not necessarily mean their bodies have absorbed pollutants. Furthermore, epidemiology cannot prove whether it's the pollutants that have triggered an illness. Traditional epidemiology is what government scientists did in Dickson, Tennessee, declaring an unusual rate of birth defects there, but declining to identify the toxic trigger.

Epidemiology can say, yes, that chemical is associated with this illness, but it stops far short of assigning cause and effect; animal studies can add further evidence to say, yes, that chemical causes that illness, and can describe a possible mode of action through which the substance wreaked its harm, but they cannot prove this is applicable to humans. These traditional tools take research only so far.

These limitations were narrowed substantially by the mapping of the genetic code and advances in molecular epidemiology. Molecular epidemiology fuses traditional epidemiology with molecular biology into a breakthrough methodology called "biomonitoring." Biomonitoring reveals definitively the number and types of chemicals that have entered, been absorbed into our bodies and our children's, and left behind some fingerprint, a biomarker. This methodology can even detect the chemicals that have invaded the unborn child in the womb. Gathering these data within population studies offers evidence both of mechanisms (how the disorder takes place) and of causality, a powerful tool for prevention.

Biomonitoring starts with drawing samples from the body, such as from blood, hair, urine, or saliva. The samples from a newborn might be his first stool, or a tiny sampling of cells from inside his cheek, or blood from his umbilical cord, an amazing reservoir of fetal history. These samples are full of biomarkers. One centimeter of hair tells the history of one month of exposure. Dr. Herbert Needleman pioneered this concept when he began examining baby teeth in the 1970s for traces of lead; his findings not only convinced the nation that lead had to be removed from gasoline, but it also set the baseline against which researchers could trace declines in lead poisoning once lead was banned.

Applying the new capacity of analytical chemistry, scientists can now measure levels of several hundred different chemical compounds in a single sample of blood, some down to parts per trillion, the equivalent of detecting a single grain of sand buried in the carpet of a three-thousand-square-foot house. These biomarkers taken together produce the measure of the total body burden from a person's, or a community's, total exposures from multiple routes of exposure.

This technique is sensitive enough to find a large array of chemicals in our children's bodies and ours that formerly escaped detection. It has begun, and will continue, to revolutionize the ability to link environmental toxins and chronic illnesses in the same way blood tests detecting the presence of antibodies have done for the study and control of infectious agents, concludes pediatrician Bruce Lanphear, director of the Cincinnati Children's Hospital Medical Center's Environmental Health Center. He adds that biomonitoring is one of the main reasons for the growth of the discipline of children's environmental health.

In the late 1990s, an arm of the CDC, the National Center for Environmental Health, under its then director, Richard Jackson, set out to catalog the national environmental chemical burden, creating what was lacking before—a baseline of biomonitoring information, expanding exponentially over the years.

Biomonitoring, then, is a successful recent technique to disclose the toxic substances in a child's body. That leaves the next question: So what? Did any harm occur? For this, environmental researchers turn to other new disciplines and technologies.

One new insight builds the foundation for another. The new study of hormone disruption rests on two decades of accruing data about the ability of environmental contaminants to interfere with hormones. This hybrid science incorporates the findings and methodologies of toxicology, endocrinology, molecular epidemiology, ecology, and behavioral biology.

Contemporary genetics has also transformed animal studies so they can produce more targeted outcomes. While mice have been bred for years as "mouse models," each with certain characteristics useful for experiments, now genetic engineering techniques have created the "knockout" mouse. In this lab mouse, researchers inactivate, or "knock out," an existing gene by replacing it or disrupting it with an artificial piece of DNA. Knocking out the activity of a gene provides valuable clues about what the gene normally does. One type of mouse has had its estrogen-signaling system "knocked out" so it can be used to evaluate the role of hormone-disrupting substances.

Patricia Rodier, professor of obstetrics and gynecology at the University of Rochester, is using a knockout mouse as a model to study the development of autism. Her work has already revealed the set of weeks in fetal life susceptible to damages that resemble some of autism's symptoms. Her field of study is developmental neurotoxicology, a branch of science that investigates the toxic effects of chemical exposures on the developing nervous system.

TOXICS + GENES = TOXICOGENOMICS

Very recently, out of the combination of the new toxicology with the study of genes, called "genomics," emerged a new tool kit, called "toxicogenomics." Its unparalleled capacity holds the potential to analyze interactions between a child's genetic inheritance and exposure to environmental toxins. Rather than studying one chemical and one effect at a time, it can look at multiple interacting pathways.

In the building that houses the National Institute of Environmental Health Sciences' toxicogenomics research program, most of the laboratories contain the familiar test tubes, microscopes, refrigerators, and other usual paraphernalia of scientific work. But in one small room, sitting inside what looks like a glass sarcophagus, is a device straight out of a *Star Trek* set. This is a microarray, a fairly small machine composed chiefly of thousands of little titanium pins. These pins take up molecules, sometimes hundreds of thousands of them, from the tissue of test animals or of humans, to which a researcher has attached fluorescent dye "labels"; these are then placed on slides for scanning by lasers. Under the scan, the molecules turn different colors, red or green, depending on which molecules are active. The scanner is hooked up to a powerful computer, which produces high-speed printouts of the fluorescent areas, enabling NIEHS scientists like molecular geneticist Richard S. Paules, director of the microarray group, to examine the genetic patterns evident in these printouts and find deviations from the norm. In the recent past, scientists were able to conduct genetic analyses on only a few genes at once.

The deviations serve as biomarkers, indicating that the genes have been damaged by an environmental contact. The damage changes the way the genes turn on or off—that is, the manner in which they do or do not perform their normal tasks. Dr. Paules explained that an abnormal DNA pattern might indicate cells that predispose children to cancer. The changes in gene expression also indicate whether and where a toxic insult created an effect.

In fact, Dr. Martyn Smith is using a variation of microarray technology in his lab to investigate the specific chromosomal biomarkers
that characterize childhood leukemia. His work is part of the University of California at Berkeley's study of childhood cancer, the largest in
the country. His microarray examines blood spots collected from newborns in several cooperating hospitals, looking for patterns that indicate
varying levels of proteins in cells. This study is known as "proteomics."
Higher than normal protein levels represent biomarkers for acute lymphocytic leukemia; in the printout, these high levels look like green
comets—the larger the comet, the more the damage. Through this
and other technological advances, Dr. Smith has made breakthrough
discoveries about the pattern of damages that can lead to cancer in
children.

These technologies will begin to bear fruit in about another
decade, according to Raymond Tennant, former director of the toxicogenomics center at the NIEHS, finally proving causation as well as
identifying the genetic variations that make some children more susceptible.

A specific strategy has already been created. At the EPA's request,
the National Research Council developed the concept for a fundamentally new way to test chemicals, avoiding the current need to rely on
animal studies. The new concept would look at a chemical's effects on
biological processes of human cells, applying the evolving understanding of how genes, proteins, and small molecules interact normally and
how toxics interfere with these normal pathways. These tests would
use automated experiments to test hundreds or thousands of chemicals
over a wide range of concentrations.

HEALING YOUNG MINDS

The building that houses the M.I.N.D. Institute, outside of Sacramento, exudes an aura of quiet and assurance. Soaring clerestory windows flood the pale wood walls and floors inside with natural light.
The rooms are uncluttered, yet cozy with soft chairs and sofas. Paint-
ings by children with developmental disorders and exceptional talent
decorate many walls. Every room, every hallway, gives off a sense of
space and unfussy orderliness. The glass of the fish tanks, built into the
dividers separating one family waiting area from the next, is so thick
that no child, no matter how enraged his tantrum, can break it.

This place is a frontier of combined treatment and research. Its focus is autism and other disorders of the childhood mind and body. Five sets of parents of children with autism built this building. That is, together they conceived the idea, donated the seed money, helped raise the rest, and still devote their talents to it. The building opened in 2003, the realized ideal where all the doctors that parents and their children need are in one place—no more dragging an upset child from one office to the next—and where all the clinicians and researchers from a vast spread of scientific disciplines have both their labs and one another in one place.

Pediatrician Randi Hagerman works with people with fragile X syndrome, the most common inherited cause of mental retardation and autism; her work is one way to investigate the genetics of mind-body disease. In another office, psychologist Sally Ozonoff is studying new babies born into families with one child with autism—such families are ten times more likely to have another child with the disorder—another path to explore the role of genetics. Paul Ashwood, an immunologist, researches the impact of diet and the immune system on the brain and body. Neuroscientist David G. Amaral is in charge of the nation's largest study of autism, following 1,800 children, half with autism, looking into genetic patterns, patterns of chemical exposure, and the interaction of those patterns. Psychiatrist Robert Hendren treats some of the most disturbed children and directs the institute with vast reserves of calm and wisdom.

The most modern equipment is right here, including microarray machines; from this work, the institute hopes to be able to identify subtypes of autism. Using biomonitoring, M.I.N.D. scientists have succeeded in finding four proteins that can serve as biomarkers of autism. Then, using proteomics, they were able to detect as least one hundred out of the thousands of proteins examined that were different in the autistic compared with the control group. Meanwhile, the effect of environmental factors such as PCBs, flame retardants, pesticides, and mercury as well as infectious and medical exposures specifically on autism is under scrutiny by epidemiologist Irva Hertz-Picciotto and toxicologist Isaac Pessah in less elegant quarters at a nearby campus of the University of California at Davis, of which the M.I.N.D Institute is a part.

Even technologies one would seldom associate with health studies have been turned into forensic tools. The Massachusetts Public Health

Department's Center for Environmental Health, in order to identify neighborhoods with higher than normal rates of illnesses, formerly had to look up every single residence, stick a pin into a map to represent that house, and then overlay maps of environmental pollution. But no more. Now their Geographic Information System does the analysis at the touch of a button. The University of Washington's Center for Child Environmental Health Risks equips farmworkers' children with Global Positioning System devices to track their daily exposure to pollutants. Computers organize enormous databases of biological data in a technology called "bioinformatics."

During a clinical exam of a child with certain symptoms, the doctor may use an MRI machine to see physical lesions in the brain, rotating the image to identify which vessels may have been destroyed, explains Dr. Martha Herbert. She has discovered that in autistic children with larger than normal brains, the growth is in the outer white matter, which is made of the insulated axons, or "wires," that connect the brain's neurons with each other. Her lab pioneered the technique to make these measurements, called "morphometric analysis," in which images of the brain from MRI scans of living people are divided into tiny parcels for measurement.

FORENSICS CONFRONT 9/11

When the Twin Towers melted to the ground, carrying with them thousands of human lives, it was one of the most horrible tragedies in the nation's history. But the tragedy was far from ended in those shattering moments. As the great buildings collapsed and burned, they emitted enormous plumes of smoke laden with metals, volatile organic compounds, hydrochloric acid, and many other extremely hazardous substances. The sky filled with dense clouds of pulverized cement, glass fibers, asbestos, lead from fifty thousand personal computers, flame retardants, radionuclides, polycyclic aromatic hydrocarbons, PCBs, and pesticides. The area around Ground Zero had the highest air concentrations of dioxins ever measured, probably from the incinerated plastics and other chlorinated materials in the buildings.

In the days that followed September 11, the administrator of the Environmental Protection Agency at the time, Christie Whitman, assured New Yorkers their air was safe to breathe and presented no health risk. She was very wrong.

For three months, Ground Zero's fires burned on and ash spread for miles over lower Manhattan and nearby parts of Brooklyn. During the recovery and cleanup, diesel exhaust from the trucks and machinery hauling out debris added to the pollution. For those months and more, this colossal miasma of chemicals infiltrated firefighters, police, and other first responders and the workers who came to clean up Ground Zero, putting them at high risk. Also at high risk, it is now apparent, were pregnant women exposed to that toxic cloud. And their unborn babies.

Boston University epidemiologist Dr. David Ozonoff has said that for the nation's children, the toxification of the environment is "a World Trade Center in slow motion." It is a trenchant metaphor. What, then, does 9/11 tell us about the connection between an environmental assault and injury? How were these women and their unborn children harmed?

Answers began to emerge soon afterward. Babies born to women who, during their first or second trimester of pregnancy, lived within a two-mile radius of the World Trade Center in the month after 9/11 were significantly smaller in birth weight and birth length. Further, regardless of the distance of their homes or workplaces from Ground Zero, women exposed in their first trimester on 9/11 had a significantly shorter pregnancy by an average of 3.6 days, so those babies were smaller. Their head circumference was smaller, too. The mothers seem unharmed.

Four ounces or one-third of an inch would not appear to be much to worry about, but in fact they point to potential trouble. Evidence shows that preterm and low-weight babies are more likely to suffer from lower IQ and learning problems, cerebral palsy, mental retardation, autism, asthma, and diabetes. Moreover, lower birth weight, even in the range of normal, is associated with more developmental and health problems in childhood.

These troubling birth outcomes among the 9/11 mothers were discovered by the Columbia Center for Children's Environmental Health. Right after the disaster, the center (and, separately, Mt. Sinai Hospital) set up a special project to determine whether the toxic fallout would affect the health of those exposed, especially vulnerable unborn babies. They were in a good position to do this because the center has long worked with and studied mothers and newborns in Washington

Heights, Harlem, and the South Bronx, whose exposures often mirror 9/11's. The center is housed in a warren's nest of small rooms in a complex of gray cement, unmistakably 1960s institutional buildings in Harlem. Dr. Frederica Perera runs the forty-person center, one of thirteen institutes originally funded by the federal government to research the environmental roots of chronic childhood diseases.

A short walk from the center takes you to the neighborhood of Washington Heights, which is home to many of these families. Block after block, the streets are lined with well-built brick apartment buildings attesting to a former history as home to middle-class families. Today it is a neighborhood in distress. An obstacle course of trash cans obstructs the sidewalks. Inside, the broken marble entry halls, scarred wood paneling, and apartments cut up helter-skelter into small units to cram in as many Latin American immigrant families as possible bear witness to hard times.

Working with families in this neighborhood, the center had discovered for the first time in the United States that air pollutants in soot can trigger impaired human fetal growth and development. The most dangerous components in soot are particles of polycyclic aromatic hydrocarbons (PAHs). High prenatal exposure to these tiny blades in soot were associated, especially among African American babies, with not only low birth weight, but also smaller head circumference. In Washington Heights, these particles spew from vehicles; around the fallen World Trade Center, they were released by the fires.

Another shared problem was found. In earlier years, while tracking the Harlem newborns, the center "for the first time had seen evidence that air pollutants can change chromosomes in utero," explained Dr. Perera. This damage increases a child's risk of cancer. The 9/11 babies display this defect, too: Babies born to mothers living within a mile of the World Trade Center showed a higher rate than normal of genetic damage.

This kind of analysis requires a combination of biomonitoring and epidemiology, a discipline Frederica Perera had invented, with another scientist, two decades earlier. Though her looks belie it, she is known as "the grandmother of molecular epidemiology," the founder of a new way of investigating the biomarkers among a group of people by examining changes in the molecules of their genes.

In three New York City hospitals, as each baby was born to a 9/11 mother, a small amount of his or her umbilical cord blood was taken and carried immediately to a darkened lab, where it was spun in a centrifuge to separate out the plasma and the red cells from the white, which contain the DNA. A small sample of maternal blood was also taken and processed. The hospitals took the babies' measurements. The biomarker Perera's team looked for in the umbilical cord blood was a DNA "adduct," a change that takes place if a small amount of a chemical attaches itself to and alters the baby's genome, its DNA. As a doctoral student, Perera had been the first to study DNA attachments and then link that damage to cancer.

The center has also followed the Washington Heights babies to see what effects exposure to pesticides might cause. This research began in 1998 when the center learned that the heaviest pesticide use in New York State was not in agricultural counties, but inside the homes of Manhattan and Brooklyn families. The families mostly used the then popular organophosphates Dursban (whose chemical name is chlorpyrifos, or CPF) or diazinon, to kill pests such as cockroaches. For decades, the EPA had approved these chemicals as safe.

Again, through an analysis of umbilical cord blood, the researchers found that every one of the hundreds of Harlem babies studied carried these chemicals in their bodies. The insecticides had crossed the placental barrier in strength. The result was serious: Babies prenatally exposed to the highest level of CPF or combined levels of the two insecticides weighed a half pound less on average, and Dominican American babies in the group were significantly shorter. Several years later, after CPF had been banned for residential use, the Columbia Center found that retarded fetal growth had disappeared: Infants weighed on average 6.6 ounces more than babies born before the ban, and their birth length also increased to normal.

Prenatal exposure to air polluted with soot and pesticides brings on another health problem as the babies grow: Asthma is rampant and unrelenting among the Washington Heights children. The neighborhood school has a clinic just for them. Here is another parallel between Harlem and 9/11. Asthmatic children living in the vicinity of the World Trade Center suffered more frequent and intense asthma attacks after the buildings' collapse showered down its brew of pollutants.

More bad news is likely for the 9/11 babies as the years pass, judging from the Columbia Center's tracking of Washington Heights babies. The newborns of Washington Heights who were exposed to high levels of either air pollutants or chlorpyrifos were not only smaller than normal, with more signs of DNA damage and asthma, but also developmentally impaired. When the center tested them at age three, these babies scored significantly lower on mental and psychomotor tests and showed double the risk of developmental delay than the norm. The more air pollutants the baby had absorbed in the womb, the lower his or her score as a three-year-old. The odds of developmental delay among the most "polluted" children was 2.9 times greater than that of children with lesser prenatal exposure. The same held true for those children with the highest prenatal exposure to CPF. Furthermore, not only did the most exposed children have significantly more developmental delays, there was an increase in early signs of behavioral problems, including ADHD.

These tidings are worrisome since lower test scores like these may foretell a future in which, without remediation, the children could fall further behind in language, reading, and math as they grow up, perhaps to live lives of diminished mental capacity and social disorder. And these tidings are worrisome because the 9/11 babies as well as others across the nation are exposed to the same levels of pollution—as Dr. Ozonoff says, World Trade Centers in slow motion.

PROOF

In sum, despite the barriers to federal testing of chemicals, the new breakthroughs in science have created a substantial body of knowledge. "We're at the level of sophistication of science where we don't need a dead body approach. We can put together a set of scientific information to identify a level of threat and take action to avoid that threat," declares Lynn Goldman, a pediatrician and epidemiologist and former EPA assistant administrator.

But to many minds, proof remains elusive, still around the corner. There are two reasons why. One is the scientific method itself. The other is the financial self-interest and ideology that purposely cultivate uncertainty.

The scientific method is a way of looking at and trying to understand the world. But, as we came to realize with some surprise during our research, uncertainty and controversy "flow through science like a river."

Science likes simplicity. But the world is infinitely varied. Science can succeed with one variable at a time, but it's almost impossible to study the various combinations of multiple chemicals that are today's reality. Furthermore, there's no group of unexposed children or adults in the world who could serve as controls. Science prefers straightforward responses, but many effects are subtle and hard to measure, such as behavioral outcomes. We expect answers now, but the effects of exposure may not appear until years later. The timing of exposure is so critical that a study of exposure at any one time may be different from a study that examined another window of exposure. It will take time to untangle gene-environment interactions.

The scientific method is currently geared to prove a negative of a question. Scientists will not say, "Rats exposed to the solvent TCE will result in birth defects," because, according to the rules of science, they would have to expose every rat at every possible dose level. So they turn the question on its head and say, "Treating rats with TCE will not result in birth defects." That's called a "null hypothesis." They can then disprove that question if a few rats develop birth defects after TCE exposure. But the experiment hasn't proved anything more than that.

Epidemiology studies, which examine human exposures and illnesses that have already happened in communities, also have an Achilles' heel. The study might show association—that is, TCE and birth defects were both found in Dickson—but stops short of asserting causation—that TCE caused the birth defects. Scientists are trained to prefer a false negative conclusion (for instance, a false assurance that chemical "x" won't harm anyone) than a false positive (that is, a false alarm that the chemical does harm).

So in science, finding no proof is rather easily accomplished, whereas the opposite, turning insignificance into significance, is extremely difficult. That is not, however, the optimal strategy for protecting human health.

These conservative traditions put pressure on researchers. Even scientists whose work has uncovered substantial links between exposure and chronic illness will pepper their verbs with conditionals such as "would," "might," "may," and "possibly," to satisfy the demands of those traditions. One researcher acknowledged, "We can't use powerful language because you need always to think about the critics." Yet, as environmental health scientists Richard Clapp and Polly Hoppin point out, scientists who equivocate impugn the research to which they refer.

Further, in everyday reality, public health officials who step forward to champion a cause-and-effect connection are liable to encounter the scenario described by David Ozonoff, who has observed and participated in many studies: "If you say some chemical is a problem, industry hires experts to say you're an idiot, the mayor says you're destroying property values, you can lose your job, or they cut your budget. On the other hand, if you say the chemical is fine, since no one knows whether you're wrong, the community will get discouraged and go away."

Those who have reason to support the status quo will continue to find proof always waiting around the next corner. They will argue: If human epidemiological studies indicate proof, that doesn't count; it shows association but not causality. If animal studies indicate proof, that doesn't count; animal assays are not applicable to humans. Nothing suffices for them but proof beyond a shadow of a doubt, because uncertainty equals inaction, and the status quo is desirable to those who profit from it.

As a result, in the tea leaves of science, one researcher will read certitude, another will read uncertainty, as these two conclusions show:

"Many of the new morbidities of childhood are linked with remarkably low-level exposures to environmental pollutants," says Dr. Bruce Lanphear, director of the Cincinnati Children's Environmental Health Center and an expert witness on behalf of families suing lead-based paint manufacturers in Rhode Island.

Contrast with: "There is no uniformity of opinion regarding the magnitude of impact that environmental toxicants have on the health of adults and children," says Dr. Robert I. Brent of the Alfred I. duPont Hospital for Children, Wilmington, Delaware, an expert retained by Dow in the lawsuit brought by families claiming the company's drug Bendectin caused birth defects in their children.

Without divine wisdom, absolute proof is rarely possible. After our years of research for this book, however, we strongly conclude that a sufficient weight of evidence is attainable and actionable "proof" is available. In general, if there's solid and consistent experimental evidence, that should be enough to say something does harm. "When it comes to the health of our children," says David Ozonoff, "most of us are willing to accept what we consider to be reasonable evidence. If I say, 'Get the solvent PCE out of the stream of commerce,' and in fifty years I turn out to be wrong, it really wasn't PCE causing cancer, I wouldn't feel bad about that. If you set the standard so high, if you instill enough doubt not to be able to act, then that's a social and political problem."

Remember that the tobacco industry argued for decades that there was no proof that cigarette smoking and secondhand smoke caused cancer, since no one had a full understanding of the mechanism by which the disease took hold, which is still the case. Eventually health officials moved ahead anyway, based on the weight of evidence. Remember that the lead manufacturer Ethyl Corporation argued there was no proof that their product would hurt either their workers or the public. Remember that no one understands how fibers of asbestos cause cancer, but there is no doubt they do. And recall that our country eradicated infectious diseases without fully understanding their mechanisms of action.

In the meantime, no golden rule of scientific endeavor has yet surpassed this, offered by Austin Bradford Hill, a noted English epidemiologist and statistician in the mid-1960s: "All scientific work is incomplete, whether it be observational or experimental. All scientific work is liable to be upset or modified by advancing knowledge. That does not confer upon us a freedom to ignore the knowledge we already have, or to postpone the action it appears to demand at a given time."

THE QUESTION OF CLUSTERS

Nothing illuminates the struggle to pin down proof as do clusters of a disease, such as birth defects in Dickson and cancer in Toms River. A cluster is "the occurrence of a greater than expected number of cases of a particular disease within a group of people, a geographic area, or a period of time." It would seem reasonable to expect that examining a grouping of one chronic illness in one place over one period of time would more easily reveal clues to cause and effect than looking at one case at a time. But that assumption has not been borne out. Out of the hundreds of studies of cancer clusters the CDC has made, they found only two that might represent cause and effect rather than chance (the CDC doesn't keep records on clusters of other diseases, we were told). The families of Fallon, Nevada, who have organized themselves into Families in Search of Truth, are so fed up with the standard cancer cluster investigations, which they feel are deficient, that they have secured money to hire and manage a new researcher (with University of Nevada at Reno and EPA funding). In this small town of 7,500 residents, seventeen children have been diagnosed with leukemia since 1997 (three have died). Among those children, the CDC identified a genetic variation in a gene that would otherwise help combat unsafe chemicals. But the agency found no evidence linking the cluster to any specific toxin, though it confirmed higher than normal levels of tungsten and arsenic in residents' blood.

Skeptics question whether clusters even exist. They say clusters are more likely a statistical fluke: If you shoot bullets into a target, many of them will by chance cluster themselves. Others say that clusters are connected only to occupational exposure or evil self-inflicted behaviors such as smoking or sunbathing.

Mostly it's agreed that clusters pose a challenge because some, if not most, are indeed simply random chance, while others may be real yet comprise too few children (what is called "adequate statistical power") for standard statistical analysis to decide they are or are not chance. But Suzanne Condon, director of the Massachusetts Public Health Department's Center for Environmental Health, argues, "You won't find clusters if you don't look for them." Her department, which runs from one hundred to two hundred investigations a year, has examined and positively substantiated many clusters of differing illnesses. The most famous was located in Woburn, a city near Boston, and portrayed in the book and movie *A Civil Action,* a story in which Condon appears offstage some years later.

Woburn offers a good place to mull over the essential issue of proof, so let us retell and then expand its story. As related in *A Civil Action,* the water coming out of the taps during the 1960s into some homes in the east section of the town of Woburn was poisoned by highly toxic chemicals that five different companies had dumped over many years in vast piles near the river that serves as the source of the town's water supply. Every time the town system turned on two specific well pumps, wells G and H, to meet water demands, the toxics were sucked out of the ground, into the river, and out through the taps. Among the families who drank this water, twenty-one children had died of cancer by 1986. Eight families sued the companies. The prosecuting lawyer couldn't come up with definitive evidence to link the dumping with the deaths, nor could the companies be exonerated. Without declaring guilt, the Grace Corporation settled with the eight families who had brought the suit.

The courtroom was, and is, not the place to find proof. The movie, too, omitted some of the book's facts. The weight of evidence, as it has emerged, is hefty and cohesive enough to confirm that in east Woburn, children exposed in utero to the contaminants in the water grew sick with leukemia (not every child, because genetic variability or other factors protected others). Two water studies showed the connection. One conducted in the 1980s, the Harvard Health Study, created a model to track the flow of water from the contaminated wells to homes and correlate that information with the cases of children with leukemia. Two decades later, two hydrogeologists from Ohio State University, Dr. E. Scott Bair, chair of the Geological Sciences Department, and Maura Metheny, one of his postgraduate students, came to Woburn and conducted another study. They bored deep holes in the riverbed to measure soil permeability, calibrated the rates at which the wells had pumped, and traced how contaminants would have flowed from the dump sites, under and out of the river, and from there to the suspect wells.

Then came the part that is new. They extrapolated the historical flow of the town's water back over the past twenty-six years, simulating many different variables—this pump off, the other on, changes in development near the river, varying rates of chemical seepage, and so on. They ended up with one hundred thousand equations with one hundred thousand variables, which they fed into their desktop computer. These computations would have been virtually impossible even for supercomputers twenty years ago. The result was a model that allowed them to pinpoint those homes where in the 1960s the women had been pregnant and to compare the risk of childhood leukemia relative to the amount of water delivered from the polluted wells.

The correlation was absolute. The children who died first had been born to mothers who while pregnant drank the most polluted water. The two children whose illness has gone into remission had lived in homes where the contamination was the lowest.

The capstone piece of evidence was the study comparing the children who had leukemia with a control group of other Woburn children without the disease, carried out by Suzanne Condon and her center. They found that "the risk of leukemia significantly increased as the amount of contaminated water from wells G and H delivered to the households increased," a conclusion completely in harmony with the earlier water flow studies. And facts since then have further vindicated these studies: In the years from 1986, when wells G and H were shut down, through 1994, no new childhood leukemia cases were reported in east Woburn and since 1995 less than half the number that would have been expected up through the most recent data gathering. In the aftermath of Woburn, the state of Massachusetts, with guidance from Dr. Richard Clapp, developed a registry to track cancer incidences that remains the best in the nation.

A coda to this tale: The major chemical contaminating east Woburn's water was TCE, the same contaminant as in Dickson's water. Though the movie focused only on the childhood leukemias, actually (as the book relates) children in Woburn also suffered from oral clefts, the same illnesses occurring in Dickson. Across the nation, over two thousand communities have reported chronic diseases from TCE.

FORENSICS TIES TOXICS TO AGGRESSION

It is worth considering the possibility—indeed, the probability—that our nation's struggle with early school dropouts, substance abuse, unemployment, crime, and incarceration is in part the result of the toxins pervading our children's early lives.

"I'm seeing younger and younger children with extreme behavior," states pediatric neurologist Martha Herbert. She tells of one four-year-old who had been treated with an antipsychotic drug usually prescribed for adult schizophrenics because the child had tried to kill a sibling. "I've had several cases like that," she adds. "It's scary because this kind of thing hardly ever used to happen."

Animal studies bear out the premise that toxicants can produce extreme behavior. For instance, mice reacted with increased aggressive behavior and reduced body weight when fed a diet of combined agricultural pesticides, even at low levels, according to a five-year study led by biologist Warren P. Porter at the University of Wisconsin at Madison. The chemical bisphenol-A even at very low doses also turns rats hyperactive and aggressive. Male mice exposed to DDT and other chemicals in utero were as adults more territorial than normal. It's possible that most if not all man-made hormone-disrupting chemicals increase aggression, especially if exposure occurs during fetal life.

Elevated levels of manganese (perhaps some from soy formula) have been found in the hair of youths incarcerated for felony crimes in four Southern California prisons; manganese in excess levels reduces the levels of dopamine in the fetus, which "may affect behavior during puberty when powerful stresses are unleashed on the dopamine neurons and altered behavioral patterns appear," a UC Irvine researcher opines. Children born to women exposed to an organophosphate pesticide at levels in use today displayed a significantly increased risk of poor social behavior by the time they reached four years of age, another study announces.

Yet most studies of human criminal behavior focus on aspects of social, economic, and child-rearing issues, slighting or ignoring the role of environmental harm. One of the exceptions was spearheaded by Dr. Herbert Needleman, who first linked lead to lower IQ (for which he was attacked by the lead industry). He organized a case-control study in 2001 in which he compared 194 youths aged twelve to eighteen arrested for violent crimes with a group of 146 nondelinquent controls from high schools in the city of Pittsburgh. The outcome: The delinquent youths were four times more likely to have high concentrations of lead in their bones than controls. This was true whether the teens were white or African American.

Neurologists with the New York University School of Medicine report four instances of unprovoked aggression, including two murders, by perpetrators who had been exposed to chemicals that disrupt neurotransmitters the way organophosphate pesticides such as Dursban do.

Recently, Dartmouth College professor of government Roger Masters, a student of Rousseau's theory of "human nature," delved into this question. He combined statistics, the science of brain chemistry, and social theory in a study that compared the rates of violent crime for all 3,141 U.S. counties to levels of industrial releases of heavy metals, as tracked by the EPA's Toxics Release Inventory. The rates of violent crime correlated exactly with the levels of heavy metals.

Is it possible that toxic chemicals contribute to the violence in schools such as Columbine and Virginia Tech? Many of the places where the students who committed these terrible crimes lived are near toxic waste dumps or, like Columbine, are toxic military bases.

Dr. Masters further theorizes that children with mental problems are at particular risk of turning to drinking and drugs. At first, he theorizes, a disturbed child who cannot clear neurotoxicants from his or her body might resort to alcohol and cocaine as a form of self-medication to reverse depressed mood and poor social performance. If alcohol or drugs do help, the user becomes increasingly dependent. But anyone already suffering from high levels of heavy metals is then in extra trouble, because lead and manganese interact synergistically with each other and with alcoholism. "It follows," Dr. Masters concludes, "that a community with any two or all three of these risk factors will have significantly higher rates of violent crime than those with only one factor." He further points out that educational policies such as the No Child Left Behind Act ignore the effects of environmental pollution on learning and "blame the victims, not only the children, but also teachers with classrooms with large numbers of hyperactive and learning-disabled students."

With the seismic shifts in environmental forensics, we can not only reach an understanding of the damaging effects of toxic substances, we can begin to focus on disease prevention. A high priority should be placed on identifying the pollutants that do the most harm and removing them from the environment.

Seven

PERPETRATORS

THE LATE MULLAH OF MARKET CAPITALISM, ECONOMIST MILTON
Friedman, proclaimed some years ago that business has no social responsibility other than to increase its profits. It has no responsibility to
pay a living wage, to help reduce unemployment, or to be a good citizen of communities in which it operates. And it has no duty to protect
the public from any harmful consequences of its profit-seeking operations, except as required by law. A corporate executive, he wrote, has
no responsibility "to make expenditures on reducing pollution beyond
the amount that is in the best interest of the corporation. . . ."

Unfortunately for our children, and for all of us, this is the prevailing—although by no means universal—ethos among corporate America at the beginning of the twenty-first century. In the name of profit,
industry continues to produce, deploy, and dispose of enormous quantities of chemicals and other hazardous substances that permeate our
environment and often cause illness and sometimes death. Companies
that do so are perpetrators of the crime against our children.

Corporate managers do not deliberately act to sicken or kill our
children. Most of them are no doubt decent human beings who would
not personally want to cause harm to anyone, much less to kids. But in
obedience to the supposed imperative of maximizing profit, many corporate executives pay little heed to the harm their operations cause or
convince themselves—and try to convince others—that there is no

harm. They claim, along with Friedman, that their only duty is to their shareholders.

Theoretically, government and its laws should be able to curb corporate environmental abuse. But corporations have become so powerful that they are in many ways uncontrollable and largely unaccountable. Armed with lobbyists, lawyers, public relations, advertising, and product defense firms, and using hired scientists, large corporations and their trade associations often can buy and shape how the laws and regulations are written, who enforces them, and how they are enforced. After decades of judicial appointments by conservative administrations, much of the federal bench shares the Friedman view of untrammeled corporate power. Victims of abuse by business and industry, including sick children and their families, are in an unequal contest in any effort to obtain justice from these megacorporations.

"In an ideal free market resting on private property," Friedman contended in support of his economic absolutism, "no individual can coerce any other, all cooperation is voluntary, all parties to such cooperation benefit or they need not participate."

But how voluntary was the participation of the children in Dickson, Tennessee, who were born with cleft palates, probably because their mothers unwittingly drank chemically contaminated water while pregnant? What benefits were received by the high school students in Port Neches, Texas, who developed leukemia, most likely because their school was next to a synthetic rubber plant? Is the fact that poor children in north Harlem have to live in the heavily trafficked and toxic-fume-ridden neighborhoods and so suffer from asthma not a form of coercion?

In this chapter, we will look at the records—some might call them rap sheets—of a number of companies responsible for putting toxic pollutants into our children's environment. These companies are not necessarily worse—or better—than others. We look at them because they are among the world's biggest and most powerful and produce and use enormous volumes of hazardous substances. Because of their size and their ubiquitous presence in our economy and our lives, they have been intensively scrutinized, and there is an extensive public record of their deeds and misdeeds.

It should be acknowledged at the outset that the chemical industry and all of the companies we discuss in this chapter have contributed a great deal to the ease, comfort, convenience, and sometimes health of our lives and to the prosperity of many communities and the nation as a whole. A number of companies have begun to develop and market products that are environmentally benign and pose little or no threat to children or adults and have taken other positive steps such as reducing their waste stream and performing internal environmental audits. That should not exempt them, however, from being held accountable for the harm and pain they have inflicted on our children.

A TOXIC LEGACY

Pittsfield, Massachusetts, was long a model of the hardworking, thriving New England industrial city. Set comfortably amid the soft Berkshire Hills, it is filled with handsome public buildings, sturdy frame houses, and an ethnically diverse population of Italian American, Irish American, French American, Scottish American, Hispanic American, African American, and Yankee families. The city is bisected by the Housatonic River and ringed by lakes, ponds, and streams. It was an archetypical small American city with its much loved Fourth of July parades, picnics on the lakeshores, England's department store, and cozy Wahconah Park, one of the nation's oldest baseball parks, where residents and vacationers spend lazy summer evenings half watching minor league games and half just eating hot dogs and ice cream, drinking beer, and enjoying the sunsets over the line of trees beyond the center field wall.

The farms that once blanketed the nearby countryside, many of which raised sheep for the area's now forgotten wool industry, are mostly gone, but for much of the twentieth century, the economic void they left behind was filled by an industrial colossus, the General Electric Company. At its peak, GE employed thirteen thousand residents of Pittsfield and Berkshire County, most of them in its transformer and aerospace divisions.

Although it was a hard-nosed employer, the company offered a decent wage that allowed its workers to enter the middle class, buy houses, and put their children through college. Jobs at the plants maintained Pittsfield as a prosperous, stable community.

Then, in the 1980s, under the direction of its CEO, Jack Welch, a hero of America's managerial class because of his harshly efficient leadership style, GE pulled its transformer and aerospace operations and later its plastics division out of Pittsfield. Fathers and sons who had worked in the company's plants for decades were left jobless. The city's tax revenues contracted, and its economy began a long downhill slide. England's closed, as did many of the upscale stores and movie houses on North Street, replaced by tattoo parlors, thrift shops, and social welfare agencies. The company's departure left behind a dispirited community faced with rising alcoholism, drug abuse, teenage pregnancy, and crime, some of it violent.

But GE left more than empty factory buildings, unemployment, and social decay, which the city is working hard to overcome. Its most lasting legacy is an environment badly polluted with chemicals and metals, including, most notoriously, polychlorinated biphenyls, or PCBs.

PCBs are a group of synthetic organic chemicals based on chlorine that are fire resistant and are good insulators. They were made in great quantity and widely used from the 1920s to the late 1970s, when their further manufacture was barred by the Environmental Protection Agency. The chemical is, however, highly persistent in the environment. Although it is no longer produced, it is still present in electrical equipment and other products throughout the United States and the rest of the world. The EPA classifies PCBs as a probable human carcinogen and a cause of damage to the liver, kidneys, thyroid, stomach, and skin. PCBs have also been found in a substantial number of studies to cause serious intellectual impairment in children.

The transformers and other products GE made in Pittsfield and at others of its facilities were filled with PCBs. When the company departed, it left a landscape inundated with hundreds of thousands of pounds of the chemical in the waters, land, and air in and around the city.

Tim Gray, who owns a nursery and garden shop in the nearby town of Lee, has for the past fifteen years directed the Housatonic River Initiative, a group of local activists formed to force the cleanup of the river, badly polluted by PCBs from the GE plant. The group has since extended its goal to a broader cleanup of Pittsfield's contaminated environment.

Tim, who has a degree in natural resource science from the University of Massachusetts, volunteered to take us on what he called his "toxic tour" of the city. A stocky man of medium height, with a Prince Valiant–style haircut, a brush mustache, and glasses, he ushered us into his van, which bore a license plate reading NO PCBS. As we drove off, he began an unchecked and seemingly uncheckable narration of what he saw as the harm that GE had inflicted on the city, his hands twitching with anger on the steering wheel as he spoke.

Pointing to the Housatonic River a few hundred yards from his nursery, Tim said that "the river from here to the GE plant is eight miles long and is the most contaminated in the nation. . . ." He added that a portion of the PCBs in the river end up in Long Island Sound and the Atlantic Ocean and contribute to the worldwide body burden of the chemicals in humans and animals. We drove into Pittsfield past some of the grand old buildings on East Street and then beyond the Dunkin' Donuts and the car dealerships. As we got closer to the GE plant complex, the area became somewhat seedy, with empty lots sur-rounded by chain-link fences, abandoned buildings, asphalt-covered yards, and small houses clustered together, several of them abandoned. At one site along the river, dredging operations were under way. The company, under an agreement with the EPA, is removing the PCBs from the river a couple of hundred yards at a time.

It is not just the river that is a threat to the community, however. "They dumped PCBs over so many years and so much leaked [into the soil] that underground there are moving plumes of PCBs, some of them a couple of feet in thickness and some of them acres and acres in size," Tim said. Moreover, in a portion of the city that had been peri-odically flooded, the Army Corps of Engineers came in the 1950s and removed the meanders in the Housatonic, creating new pieces of buildable land. Many in the community built homes and shops on this land, which was polluted with PCBs at levels far beyond the safety standards adopted by the EPA. For a while, GE gave away free soil from its grounds to Pittsfield residents as landfill, neglecting to tell them it was contaminated. Children played on the soil in their yards, and they and their families ate vegetables grown on them. PCB wastes were buried beneath a playing field under and next to a site on which an elementary school, the Allendale School, was subsequently built. The contaminated soil was later dug up, but a couple of hundred yards from the school there are two small hills containing high concentra-tions of PCBs and other toxic wastes.

"GE dumped within a one-hundred-mile radius," Tim said. "Their waste was hauled by tractor trailer wherever they were allowed. It's probably in every dump in Berkshire County."

"The thing that breaks my heart is the health part of it," he went on. "I have a yellow pad where I write down what people are telling me—families I've never been able to help. People are getting sick. At some point every day I think about it."

While General Electric, after years of negotiating with the EPA, agreed to clean up some properties as well as the river, the company has not acknowledged that there are serious health effects associated with its PCB wastes for children or anyone else. The company commissioned a study of Pittsfield by the Institute for Evaluating Health Risks, a private research group, which found "no clear and convincing evidence that PCB exposures were causally associated with adverse health effects. . . ." The study, conducted by Renate Kimbrough, a former toxicologist with the federal government, reported that "adverse neurobehavioral effects in infants and young children" had been reported in two studies but that the studies have "many uncertainties."

Asked about Kimbrough's conclusions, Dr. David Carpenter, director of the Institute for Health and the Environment at SUNY Albany, said, "That's just not true. There have been many studies that found health effects in children. Many other scientists see very strong evidence that PCBs pose serious threats to health, particularly cognitive effects on children but also a range of other illnesses.

"In Pittsfield, the whole city is a dangerous place to live," Dr. Carpenter asserted.

Internal company documents indicate that GE, despite its protestations, was aware for many years that its dumping of PCBs might cause health problems and liabilities for the company. In 1981, a manager in the transformer plant sent a memo to a superior noting that a PCB-based product called Pyranol as well as earth contaminated with the chemical "show up in unexpected places," and asked, "When will it show up in the drinking water?" He then went on to recommend that "a real thorough study should be made to at least identify where spent Fullers Earth and discarded Pyranol were dumped. If we do not do that now, our children and grand children [sic] will get bit by our neglect." Another memo, dated February 1992 but unsigned, presented a "List of Potential Environmental Liabilities" and pointed out that "debris from the General Electric Plant was dumped on private property in the city of Pittsfield in the 40's and 50's" and added that there were "underground storage tanks that could not be found."

As of this writing, there is no concrete evidence of neurological harm to Pittsfield's children. Of course, nobody has actually sought to find such damage. The state of Massachusetts had planned to study the records of schoolchildren in the city and correlate them with records of exposure to PCBs. But the plan was thwarted by the privacy law. "Federal law prohibits educational agencies from releasing to anyone any kind of record except with specific parental consent," explained Elaine Krueger, director of environmental toxicology for the Massachusetts Department of Public Health. "Massachusetts law," she added, "directs us to use this kind of data. We are working on how to remedy the conflict."

A compilation of the results of standardized tests in Massachusetts, reported by *The Boston Globe,* shows that Pittsfield children in grades three through ten performed under the state average in all subjects. But because of the lack of research, most of the evidence of harm to children remains largely anecdotal.

William Kowalczyk worked at GE in Pittsfield for twenty-seven years, nineteen of them in the transformer facility. He himself has liver cancer, but he worries most about the children who were born and grew up in the community.

"All the kids that grew up here are dying of cancer. In this neighborhood alone we had sixty-three people die of cancer. One family, all of them, husband and wife and six children, died of cancer. All of my grandchildren have ADHD. We live behind the hottest house in the county. The children all played in the brook and the dirt.

"I feel my life was shortened by a company that didn't care. We are condemned to die, but I don't want to see children have a life shortened by twenty years."

In nearby Lenox, Massachusetts, Dr. Siobhan McNally is a pediatrician who sees lots of children from Pittsfield. She noted that in 2003, the federal Agency for Toxic Substances and Disease Registry reported that two of three census tracts next to the GE plants had increases in bladder cancer, breast cancer, and non-Hodgkin's lymphoma but that the state Department of Public Health could not say that PCBs played a role in the diagnoses. "But here is what concerns me," she added.

"Developmental disabilities have increased all over the country, not just in hot spots. Why are we waiting for clusters? The big thing from my point of view is corporate responsibility. Before you dump stuff into the air, water, or soil, you have to do everything reasonable to make sure it is safe. To do anything less is nothing short of criminal."

Pittsfield is not the only community to suffer from General Electric's lack of environmental responsibility. The EPA, after long negotiation and litigation, required the company to spend millions to clean up PCBs it had dumped for years into the Hudson River in New York State. In 1998, the company agreed to pay a fine for violations involving toxic releases at its silicone-manufacturing plant in Watertown, New York, a site of repeated violations by the company. In 1995, it paid a fine and cleanup expenses at a former plant in Florida where it built transformers. In 1993, GE and others were told to pay damages for dumping chemicals in Riverside, California. The list of environmental misdeeds and penalties imposed on the company, compiled by Multinational Monitor, a nonprofit organization that tracks the transgressions of large corporations, goes on for pages.

We sent several e-mails to Gary Sheffer, the company's executive director for communications and public affairs, asking for comment on these issues. He finally replied by e-mail that another staff member of the public affairs office would contact us. No one ever did.

GE has been conducting a multimillion-dollar advertising campaign declaring that the new company philosophy is, "Ecoimagination." The coined word is supposed to suggest that the company will now engage in environmentally friendly projects such as solar and wind power—and projects declared eco-friendly by the company such as nuclear power and coal gasification. It ran television commercials showing trees growing out of smokestacks.

Perhaps GE is redefining itself. But some observers, at least, see the campaign as a transparent effort to improve its public image and promote public acceptance of nuclear power and other potentially dangerous technologies. *Flow,* a journal put out by the Department of Radio-Television-Film of the University of Texas at Austin, asserted, "For anyone who understands the scope of the GE corporate empire or who has an awareness of its long history of destructive environmental practices, the Ecoimagination campaign is risible."

If the Ecoimagination campaign means that GE is positioning itself to be more competitive in the green technology field by producing such products as high-efficiency turbines, it would be a welcome move. It would be supporting action to limit the pollution that is causing global warming. The company's green credentials might be more firmly established, however, if it would thoroughly clean up Pittsfield, Massachusetts, and other communities it has tainted at the expense of children's health and avoid, from here on, activities that degrade the environment.

THE SOURCE

In fact, General Electric did not actually make the PCBs it used in its electrical equipment and spread around so liberally in Pittsfield, the Hudson Valley, and elsewhere in the country. That task was performed by the Monsanto Company, another industrial giant, which produced the fluid using specifications set by GE and other manufacturers. They were produced in a town in Alabama called Anniston, in a section of the city populated largely by African Americans.

PCBs had been made in Anniston since 1929, first by the Swann Chemical Company, which was acquired by Monsanto in 1935. According to descriptions by residents, environmentalists, and scientists such as David Carpenter, contamination in that town is so severe that it makes Pittsfield and the Hudson Valley seem almost pristine by comparison. For almost forty years, the company "routinely" dumped millions of pounds of toxic wastes, including mercury and other hazardous substances but chiefly PCBs, into streams and open landfills in and around Anniston. Thousands of Anniston citizens filed damage claims against Monsanto on behalf of themselves and their children. Residents of the community had "exceptionally high" PCB levels in their bodies, according to Dr. Carpenter, who served as a consultant to plaintiffs in the lawsuits.

Monsanto spun off its chemical division, which produced the PCBs, into a firm called Solutia, Inc. Solutia agreed to spend over $600 million to settle the suits. The company also has been required to conduct cleanup activities to remove chemical contamination. But many, if not most, of the residents of the town are said to be dissatisfied. The money cannot remedy what has happened to them and their community.

Writing in the *Los Angles Times,* reporter Ellen Barry told the story of a boy named Manuel Washington whose family lived next to the Monsanto plant. As young boys will, he played in an open meadow and waded in a stream next to the plant. By age nine, Manuel was blacking out and having seizures. When he reached his late thirties, he was too weak to feed himself. And as his life reached an end, he was completely blind and suffered from psychosis, imagining himself attacked by monsters. He died of heart failure in a nursing home at age forty.

According to his sister, the autopsy report listed his death as "acute PCB intoxication in long standing" that damaged his pancreas, liver, kidneys, and brain. The $7,000 his family received as their share of the settlement with the company did not even pay for his funeral expenses, his sister said.

Neither Monsanto nor Solutia acknowledged any wrongdoing when they reached the settlement. As we have noted, making financial concessions without admitting that their activities had caused sickness or other harm is standard practice for corporations when the facts and the law catch up with them. The two companies still maintain that PCBs do not cause serious health problems.

Glynn Young, director of environmental communications for Monsanto, said when visiting Anniston in 2004 that "no data ever has confirmed a connection between PCBs and diseases," although he added that if it were his family, he would be upset.

Documents that emerged during the litigation against Monsanto, however, suggest that the company knew for decades that it was polluting Anniston with dangerous wastes and that it tried its best to hide that information from the community and government agencies. Journalist Michael Grunwald reported in *The Washington Post* that "thousands of pages of Monsanto documents—many emblazoned with warnings such as 'CONFIDENTIAL: Read and Destroy'—show that for decades, the corporate giant concealed what it did and what it knew."

As is the case with General Electric and many big industrial companies, Monsanto has left its toxic imprint on many communities and with a variety of products, some of which are of particular concern for children's health. Writing in the *Multinational Monitor,* published by the organization of that name, Russell Mokhiber, editor of the weekly *Corporate Crime Reporter,* described Monsanto as "a corporation whose financial health in the 1990s is built on products of dubious value that pose serious or unknown risks to consumers and the environment."

Among the products he listed were aspartame, the lower-calorie sweetener sold as NutraSweet; the herbicide Alachlor, sold as Lasso and one of the world's most widely used pesticides; and bovine growth hormone (BGH), which is injected into cows to increase milk production. Monsanto was also a major producer of 2,4,5-T, an herbicide known as Agent Orange in the Vietnam War, and the pesticide 2,4-D, the manufacture of which can leave a residue of highly toxic dioxin.

There is mixed evidence about the dangers of all these products, and Monsanto has defended them as safe. But there is a sufficient body of research to suggest that they all can present potent health risks, particularly to children.

Children are heavy consumers of soft drinks that contain aspartame. The sweetener has been approved for use by the federal government in the absence of substantial evidence of harm. The evidence about its safety is mixed. However, a large-scale bioassay in 2005 by the Cesare Maltoni Cancer Research Center in Italy found persuasive evidence that it was a potent cause of cancer, especially leukemias and lymphomas, in test animals. The report of the study concluded, "On the basis of these results, a reevaluation of the present guidelines on the use and consumption of APM (aspartame) is urgent and cannot be delayed." There is also evidence that the sweetener is addictive and can overstimulate and affect brain receptors. However, a major epidemiological study by the National Cancer Institute found in 2006 that the sweetener did not increase the risk of cancer in older people. The study did not include children.

Milk, of course, is basic to the diets of children, who drink it and eat it as ice cream, cheese, and yogurt in great quantities. Monsanto maintains, with the support of the Food and Drug Administration, that milk from cows injected with bovine growth hormone is no different from any other milk. The company joined with some dairy producers in a "Milk Is Milk" campaign to convince consumers of the safety of BGH. The company is so sensitive to suggestions that the hormone might be dangerous that it tried to intimidate the state of Maine into giving up its special seal of approval to milk without the hormone and, when that failed, sued a small organic dairy in Maine for advertising it raised its cows BGH-free. The case was settled out of court.

According to the Organic Consumers Association, a nonprofit group, there is a "wealth of scientific information" about the toxicity of

BGH and the differences between natural milk and milk from treated cows and that the differences suggest a higher risk of cancer. Treated cows apparently produce an increased level of the hormone insulin-like growth factor 1, whose harmful effects were described in a previous chapter. Consumers Union, after reviewing the literature on BGH, concluded that because children consume so much milk and other dairy products, and because of their rapid growth and development, "higher exposures to IGF-1 in childhood could set life-long processes in motion that determine later risks."

BGH already has been banned in several European countries, including Germany and Scandinavia.

The herbicide Lasso (alachlor) reaches children chiefly through drinking water. The EPA has determined that chronic exposure to high levels of alachlor in drinking water can cause damage to the liver, kidney, spleen, mucous membranes, and eyes and also has found evidence that it can cause cancer over a lifetime of exposure. While the health effects of Agent Orange and its ubiquitous contaminant, TCDD, a form of dioxin, are still considered controversial, it has been shown to cause birth defects such as cleft palate in test animals. It is also believed to cause several forms of cancer, to disrupt the endocrine system, and to cause reproductive problems, including miscarriages and spina bifida, a neural tube defect in fetuses that leads to gaps in the spine of newborn babies. There is also some suspicion that the effects of the chemical on DNA can be carried through several generations.

A KINGDOM BUILT ON CHLORINE

A number of other companies in addition to Monsanto manufactured Agent Orange. And that brings us to the Dow Chemical Company, one of those companies. Dow is one of the nation's older chemical fabricators and perhaps the world's biggest producer of chemicals, depending on how and when "biggest" is measured. The company was founded over a century ago by Herbert Dow, who built a chemical empire on the vast chlorine reserves contained in the briny waters that lay beneath the surface of central Michigan.

Chlorine-based chemicals have created a cornucopia of useful products that have been of great benefit to human society. But chlorine chemistry also produces many compounds that are among the most toxic and dangerous known to science. As the investigative journalist

Jack Doyle wrote in his book *Trespass Against Us: Dow Chemical and the Toxic Century,* over its history the company created jobs, prosperity, and economic growth that benefited the communities in which it operated and the larger society, both in the United States and in its foreign operations. "Dow Chemical, on this level, has made the world a better place. Yet still," Doyle added, "there is a substantial reality of chemical consequences that must be laid squarely on Dow's doorstep; an unpleasant reality of people and communities poisoned; of birth defects and ruined lives."

Doyle charged the company with leaving a "legacy of harm," involving the production and distribution of massive quantities of hazardous substances including asbestos, dioxins, 2,4,5-T and 2,4-D, perchloroethylene, bisphenol-A, vinyl chloride, and chlorpyrifos. It is also one of the biggest producers of trichloroethylene, the chemical that contaminated the water in Dickson, Tennessee; Woburn, Massachusetts; and many other communities. Dow not only produces a wide variety of potentially hazardous chemicals, but also makes the feedstocks used by other chemical companies to produce other toxic products.

Cheerios, made by General Mills, is a favorite breakfast cereal of American children. Our grandkids eat it all the time, sometimes just as a snack. But in 1994, General Mills had to destroy about fifty million boxes of Cheerios and another cereal contaminated with chlorpyrifos. The raw oats that are the main ingredient of the cereals had been sprayed with the pesticide, which had not been certified for use on the grain. The spraying had been done by an outside contractor without the company's knowledge. The company and the Food and Drug Administration said that the contamination posed no threat to the health of children.

But chlorpyrifos, made by a Dow subsidiary (now called Dow AgroSciences) under the trade names Dursban and Lorsban (and other trade names, including the resonant Killmaster), has been one of the most widely used pesticides in the world. As the "Evidence" chapter relates, it is a bad actor. Yet another chlorpyrifos study, this one in 2006 of mothers and children in California's Central Valley, one of the nation's great agricultural areas, found that Dursban and Lorsban, which are still used on crops, sharply reduces natural enzymes that protect newborns and children from toxic substances. The study, authored by Brenda Eskenazi, director of the Center for Children's Environmental

Health Research at the University of California at Berkeley, found that exposure to organophosphate pesticides such as chlorpyrifos can make children as much as fifty times more susceptible to illnesses from other toxic exposures.

After Congress passed the Food Quality Protection Act in 1996 to provide stronger safeguards for children, chlorpyrifos came under closer scrutiny, and in 2000, the EPA announced an agreement with Dow to eliminate the use of chlorpyrifos around schools, homes, and other places where it could come into contact with kids. Carol Browner, a former administrator of the EPA, said it was clear that exposure to the chemical presented an unacceptable risk to children. Dow continued to maintain, however, that the chemical posed no threat. "The rules have changed, but the safety of chlorpyrifos hasn't," asserted Elin Miller, a vice president of Dow AgroSciences. When DowElanco, predecessor of Dow AgroSciences, "voluntarily" withdrew chlorpyrifos from certain uses, it did so accompanied by a "veiled threat," according to scientists Nicholas Ashford and Claudia Miller. "In a letter to the EPA dated January 16, 1997, the president and CEO of DowElanco wrote, 'However, we must state unequivocally that our proposed initiatives are not promoted in any way by a conclusion that any current label uses create exposures capable of causing human injury, and any attempt to portray them in this light would only make difficult their timely and effective elimination.' " In other words, bad-mouth our product and we will make trouble about limiting domestic uses.

Others have challenged the company's see-no-evil view. While we were in the middle of writing this book, we received a telephone call from a lawyer in Indiana who had somehow heard about our project (we received a substantial number of such calls from parents, scientists, and health care and community activists who had stories to tell about sick kids). The lawyer, Roger Pardieck, was representing two families whose children had been exposed to products containing chlorpyrifos and had grown sick. One of them was the Ebling family, with a small boy named A.J. and a girl, Christina, who in 1994 moved into a new apartment complex that had been sprayed repeatedly with Dursban and another pesticide called Creal-O, a compound of diazinon and pyrethrin.

Both were healthy, normal children when the family moved in,

Mr. Pardieck said. But soon both began to be convulsed by seizures. The boy's seizures were eventually controlled by medication; the girl's proved intractable. The girl began to show severe neurological problems. She could not color between the lines in preschool. If she left the classroom to go to the bathroom, she could not find her way back. She became incontinent. "The little girl continued on a downhill slide cognitively and behaviorally," Mr. Pardieck said. "She is sixteen and has the capacity of about a three-year-old. The little boy will be twelve and has an IQ of 44."

Mr. Pardieck sent us videotapes of a local television station's reports about the children. They showed a home movie of Christina as a lively, spirited two-year-old playing happily in her home. A second segment showed her at eight years old, blue-eyed and pretty but obviously in her own mental fog. The last segment was heartbreaking. Christina at fifteen was slack-jawed and vacant-eyed, unrecognizable as the pretty girl of seven years earlier. She drools, and her speech is barely intelligible. She cannot even brush her own teeth. "She doesn't know what to do with the feelings she has," said her mother, Cindy. "She slaps and bites and pinches. She throws things at her father and pulls out his hair. The last two years have been hell as I watched my little girl progressively go downhill."

The tape shows A.J. at eleven, still a handsome boy, but sitting on the floor playing with a little toy car just like a five-year-old.

According to Mr. Pardieck, Dow's defense against his suit was that there is no proof that anything short of an acute dose of the chemical in excess of instructions on the label will cause anything but short-term injury. Used as required by the label, chlorpyrifos poses no danger to adults or children, the company contended. The local television program about the Eblings emphasized that the pesticides were applied by untrained technicians, not that the chemicals might be inherently dangerous.

Charles Lewis, then executive director of the Center for Public Integrity, had another story to tell about Dursban when he introduced his organization's study of "the politics of pesticides," several years ago. He told of Joshua Herb, "who came into the world a healthy, happy baby but now is a ten-year-old paraplegic, confined to his home with twenty-four-hour nursing care, an oxygen system to breathe, and health care bills of about $30,000 a month. As an infant, Joshua was exposed to Dursban, a particularly potent pesticide." Joshua's parents sued Dow and won an out-of-court settlement.

In 1995, the EPA fined Dow $876,000 for hiding from the agency for over ten years the 327 lawsuits and other claims against the company of adverse effects resulting from exposure to chlorpyrifos.

Polluting companies and their defenders invariably insist that anecdotal stories about sick children exposed to their products prove nothing about cause and effect—and they are correct. That does not mean, however, that there is no cause and effect, just that it is not proven.

TROUBLE IN TOMS RIVER

Dow became an even bigger company in the 1990s when it acquired another major chemical concern, the Union Carbide Company of Bhopal notoriety. Some twenty thousand people died and well over a hundred thousand sickened when the company's plant released a cloud of lethal methyl isocyanate gas in Bhopal, India, in 1984. But Dow inherited other environmental health issues when it acquired Union Carbide. One of them is in Toms River, New Jersey, the scene of one of the largest known clusters of childhood cancers in America.

The town of Toms River, a community within Dover Township, lies about halfway down the coast of New Jersey, where Toms River enters the Atlantic Ocean. Spreading out over the coastal flatland from a neat town center is a suburban landscape of well-kept single-family houses, strip malls, and a few sprawling shopping centers. It is a reasonably affluent community, with median family incomes and home prices above the state average. While vacationers are now a major contributor to the town's economy, Toms River has also been home to a variety of industrial plants, including Union Carbide and Ciba-Geigy chemical facilities.

For many years, the two companies disposed of a witches' brew of chemical wastes from a number of sites into the Dover County water supply. Thousands of fifty-five-gallon drums from the Union Carbide facilities containing chemical solvents, including trichloroethylene, were illegally buried by a contractor on a farm inside the township. Also discovered in the soil and water were phenols and a chemical byproduct of styrene-acrylonitrile production and other compounds, many hazardous. Wastes from the Ciba-Geigy plant found in the groundwater and in the river contained volatile organic chemicals, including trichloroethylene and chloroform, as well as polycyclic aromatic hydrocarbons, dyes, and heavy metals, including arsenic, cadmium, chromium, mercury, and lead. The air in the township was also found to contain elevated levels of volatile organic chemicals.

Between 1979 and 2005, 122 children in Toms River–Dover Township became victims of cancer, chiefly leukemia and brain cancer—and the count is not yet complete. That is a significantly higher number of cancers than would be expected, according to Dr. Jerald A. Fagliano, a program manager for New Jersey's Department of Health and Senior Services. (Why *any* childhood cancers should be "expected" is another issue.) The excess number of cancers is especially high among female children. The rates were highest in areas of the town served with contaminated water. There are also elevated rates of autism, heart defects, and other chronic childhood illnesses in the area.

The department found that after the contaminated water had been removed from the town's drinking water system, the number of new cancers among children under the age of four went down to zero, which fact struck us as a rather strong clue to the causes of the children's cancer.

Nevertheless, both the state and federal government investigators decided there was insufficient evidence to prove a cause-and-effect relationship with the chemical contamination, said Dr. Fagliano. There was, however, enough of an "association" between the water and the sick children to take action, Dr. Fagliano added, and noted that the wells containing contaminated water had been removed from the town's water supply.

Parents of cancer-stricken kids in Toms River, however, have little doubt about the cause of their kids' suffering. And if they had not organized and demanded action by government, it is doubtful any cleanup would have been demanded. Linda Gillick, a tranquil-looking but tough-talking mother of a boy with cancer, helped found an organization to confront the companies and government, to offer support to parents of sick kids, and to just generally raise hell. We met with her in the group's offices, a small, white frame house with rooms painted cheerful play school colors and decorated with pictures of children at play, stuffed animals, and flowers.

When the companies and government ignored the epidemic of

cancers, she said, "it took fighting and education for the town to come over to our side." She added with a laugh that "our township officials have been much more supportive over the past few years. At the beginning you would have sworn I put the contamination into the water."

Ms. Gillick's son Michael was diagnosed with neuroblastoma, a malignant tumor that starts in nerve cells and assails young children, when he was three months old. He developed tumors all over his body, and Linda and her husband, Rusty, were told by his doctors that he would not live to see his first birthday. That was in 1979. In 2008, as this is being written, Michael is twenty-nine years old. We did not meet him face-to-face, but he must be a young man of extraordinary courage. He is only thirty-nine inches tall and, photos indicate, his face and body bear harrowing marks of his ordeal. He has been in almost constant pain throughout his life and has had to take morphine for most of it. He has had last rites said over him on two occasions. But he never gave up fighting or, as his mother told us, living life as fully and normally as he could manage. He has somehow also managed to have a sense of humor about himself. When he graduated from high school, he told the oncologist who had treated him since he was first diagnosed and who had said he would not live to be a year old, "Uncle Joe, I love making a liar out of you."

The companies reached a financial settlement with the families of the sick children. Under its terms, the families are prohibited from giving any details. This is a common practice when companies settle environmental suits. Usually, the corporation will also deny it did anything wrong. This practice can limit adverse publicity for the company, but it also limits public scrutiny of the offense and is an impediment to public health protection.

The companies paid a fine, and the township is still trying to get Ciba-Geigy to clean up its site, Ms. Gillick said. She is not satisfied with what she considers the light treatment of the polluting companies.

"They need jail time," she said. "When you fine them, it's pocket change. Millions of dollars mean nothing to them."

Joseph Kotran, whose daughter Lauren has also battled cancer for most of her life and who sat in on our conversation with Ms. Gillick, also thinks that polluting industries have too much power. "It's the companies who oversee the testing and the cleanup. They have total control of the data and can package it the way they want to and then present it to the Environmental Protection Agency. At meetings, you will see the EPA deferring to the companies."

As we were wrapping up our conversation, Gillick took a phone call and we could hear her softly asking the parent on the other end: "Have you taken her for a second opinion?" She asked what hospital the child was in and offered advice about what to do next. Before she hung up, she said, "I love you, honey."

FROM GUNPOWDER TO TEFLON

Founded at the beginning of the nineteenth century, E. I. DuPont de Nemours & Company is the oldest and, for much of its existence, the biggest chemical manufacturer in the United States. Starting out as a manufacturer of gunpowder, the company steadily expanded its development and sales of chemical products and by the early part of the twentieth century was one of the world's great industrial powers, producing thousands of innovative and useful products such as synthetic fibers, pesticides, paints, dyes, plastics, insulation, and pharmaceuticals. It also has been a major source of pollution, pouring billions of pounds of dangerous substances into the environment.

Beginning in 1924, DuPont began producing tetraethyl lead for the formulation of "no knock" gasoline, despite knowledge that lead was poisonous, causing madness and death to some workers exposed to it. Although evidence of the baneful effects on the minds and development of children continued to pile up, leaded gasoline continued to be used through the 1970s, emerging in clouds from the tailpipes of cars and trucks. According to one leading neurotoxicologist, "The addition of lead to gasoline is one of the greatest public health failures of the twentieth century." DuPont was among the lead, gasoline, and auto manufacturers who opposed phasing it out or urged a much extended phase-out. "But," said a physician quoted by Gerard Colby in his book *Du Pont Dynasty,* "why make all this fuss about tetra lead? The Du Ponts make other poisons there in even greater quantity which kills a man like that. . . . And those plants are still going full blast."

DuPont has been one of the chief contributors to the explosion of synthetic organic chemicals introduced into the marketplace—and the environment—since World War II. By 1980, the company was included on the Environmental Protection Agency's list of the nation's top twenty emitters of cancer-causing substances. It also was the foremost producer of chlorofluorocarbons, the chemical that was found to be destroying the planet's protective ozone layer.

One of DuPont's most popular and widely used products is Teflon, which it has been making for over fifty years. Teflon is everywhere in our lives, from nonstick frying pans, computer chips, and laboratory equipment to popcorn bags, stain-free clothing, fast-food wrappers, shampoos, and dozens of other applications. Teflon contains a chemical called perfluorooctanoic acid, shortened to PFOA or sometimes C8, which has now been found to be widely dispersed in the environment and in human bodies. A study of newborn babies at Johns Hopkins Hospital in Baltimore found that nearly all of them had detectable levels of PFOA in their blood. Moreover, the chemical is highly persistent in the environment and in the human body.

DuPont failed to disclose for decades that PFOA from its plants was contaminating water supplies and getting into the air and soil. It was not until the Environmental Working Group, a Washington-based research organization, investigated and publicized the problem that the public and the government paid attention. "For many years, people didn't have a clue they were being exposed to these chemicals," said Lynn Goldman, a former assistant administrator of the EPA and one of the researchers at Johns Hopkins. Citizens in communities on both sides of the Ohio River between West Virginia and Ohio were dismayed to learn a few years ago that there were significant levels of PFOA from DuPont's Parkersburg, West Virginia, plant in their drinking water. Other communities around the nation are now finding similar contamination.

Even after the 3M Company voluntarily gave up production of PFOA, beginning in 2000, because of what its scientists were finding out about its impacts on the environment and human health, DuPont continued to produce and distribute great volumes of the compound. An e-mail message by one of DuPont's lawyers presented during litigation spoke volumes about the company's protestations: "Our story is not a good one, [sic] we continued to increase our emissions into the river in spite of internal commitments to reduce or eliminate the release of this chemical into the community and the environment."

In December 2005, DuPont reached a settlement with the Environmental Protection Agency under which it agreed to pay a penalty of $10.25 million plus another $6.25 million in environmental projects, which the EPA called the largest civil administrative penalty in its history, for withholding information from the agency. The agreement resolved allegations the company failed to report "substantial risk (to human health) information about chemicals they manufacture, process or distribute in commerce." Under the agreement, DuPont admitted no liability for its conduct.

The company also claimed, and still does, that PFOA in the environment poses no serious risks to human health. But a body of research has found that at high doses it causes birth defects, developmental problems, low birth weights, depressed immune system, and liver cancer in laboratory animals. The effects of chronic exposure at lower levels have not yet been determined as of this writing. Early in 2006, however, the EPA quietly published a notice in the *Federal Register* stating that the agency "can no longer conclude that these polymers will not present an unreasonable risk to human health or the environment." The agency said that the compound was a likely cause of cancer. In response, DuPont issued a statement saying it believes its products are safe.

Debra Cochran, who lives in Pageville, Ohio, across the river from the DuPont plant, does not believe that PFOA is safe. Her daughter Kimberly was born without enamel on her teeth. Ms. Cochran is sure that PFOA, a compound in the fluoride family, was in the water she drank while pregnant and is the reason. All of Kimberly's teeth had to be surgically removed when she was two years old and replaced by a plate of dentures. She is facing more oral surgery and then plastic surgery when she is older.

Of course, much of the nation's drinking water supply is deliberately infused with fluorides as a dental health measure. A 2006 report by a panel of the National Academy of Sciences National Research Council concluded that the federal government is permitting excessively high levels of fluoride in drinking water, putting children at risk of severe tooth enamel damage. But Ms. Cochran, who has been active in investigating what is happening to children in her area, is convinced that the contaminated water she drank while Kimberly was in her womb is the reason for her daughter's problem. There is a high incidence of dental problems among children along the river, she said.

"There are jokes about people in West Virginia not having all their teeth. Well, this may be a reason."

There are other problems with children's health in the area, Cochran added. At a hospital in Columbus, Ohio, where she takes Kimberly for treatment, a doctor said that most of the patients he saw are from that part of the state, and most of the children come for cancer treatment. DuPont agreed to pay for an independent monitoring program to track illnesses in people exposed to PFOA but sharply curtailed the number of those it originally agreed to monitor, Ms. Cochran said.

Early in 2006, the EPA announced it was asking DuPont and seven other companies that sell products containing PFOA to eliminate public exposure to the chemical. "The science on PFOA is still coming in, but the concerns are there," said Susan Hazan, EPA acting assistant administrator. "Acting now to minimize it is the right thing to do for our health and the environment." DuPont and the other companies indicated they would comply.

DuPont, along with a number of other chemical companies, has made gestures in recent years toward adopting a more sensitive approach to its impact on the environment and public health. Several years ago, it agreed to participate in an EPA-sponsored program to voluntarily screen some twenty commonly used chemicals for the risks they presented to the health of children. DuPont is also among the few companies joining the effort to reduce the amount of carbon released by industry into the atmosphere and contributing substantially to global warming. It also played a maverick role by supporting creation of the Superfund for cleaning up toxic waste, contrary to the wishes of much of the industry. But the company, like others, has been saying appropriate things about the environment for many years and still continues to produce high volumes of chemicals that potentially put our children at risk.

THE COMMON DENOMINATOR

GE, Monsanto, Dow, and DuPont are not, of course, the only companies whose operations produce hazardous products. The list of perpetrators could include many that have placed high volumes of toxics into the environment, including BASF, Velsicol, American Cyanamid, Hoechst, and others. We could not examine the operations of all of them in a book of this length. We did, however, try to talk to executives of the American Chemistry Council (formerly the Chemical Manufacturers Association), the industry's chief trade association, to ask for their views about chemicals in the environment and their effect on children's health. We hoped to offer the industry group an opportunity to present its version of the information we were uncovering in our research and to obtain industry views on the issues taken up in this book. Despite our effort, we were unable to do so.

The council lobbies Congress and the executive branch for laws and regulations favorable to the industry. Between 1998 and 2004, the council spent $27 million on lobbying, according to the Center for Public Integrity, a nonprofit group in Washington. Sometimes it itself writes or has written the laws and rules that are then adopted by politicians to whom the industry has given campaign contributions and wined, dined, and entertained. It also conducts public relations campaigns and helps set up "citizens' groups," which purport to represent consumers on the issues it seeks to influence. It also arranges for scientific research that supports the industry and seeks to place allies in government posts and advisory boards. In 2001, the council entered into an agreement with the National Institute of Environmental Health Sciences to fund research "for understanding the possible effects of environmental factors and chemicals on human reproduction and fetal and childhood development" and a similar agreement with the EPA. The Government Accountability Office found fault with the partnerships' lack of public oversight.

One of the council's more intensive lobbying campaigns was to reduce the reporting requirements under the government's Toxics Release Inventory.

The council even has input into United States diplomacy. According to a report released by a House committee in 2004, the chemical industry has worked with the Bush administration "to undermine a European plan that would require all manufacturers to test industrial chemicals before they were sold in Europe."

The council did send us in advance of our aborted meeting with them a document that proved to be quite instructive about the industry's attitudes toward some of the issues discussed in this book. It was the council's comments on a draft report about toxic chemicals and children's health prepared by the North American Commission on Environmental Cooperation, a group created under the North American Free Trade Agreement. The council's comments were, in essence, an attempt to deny that toxic chemicals are a threat to children's health or to minimize their danger.

The comments dismiss as "fundamentally flawed" the report's finding of direct links between emissions on the government's Toxics Release Inventory and health effects. Well, it may be difficult to prove beyond a shadow of a doubt that particular releases of particular toxics in a particular community caused particular illnesses in particular children, but the weight of evidence shows again and again that that is precisely what happens. The record on children's health in areas where there are high releases of mercury or lead or polycyclic aromatic hydrocarbons, for example, is well documented. But it is a common practice by polluters to demand absolute scientific certainty that their emissions cause harm, knowing full well that such certainty is almost always illusory.

The council cited a federal report on national indicators of well-being that claimed that 83 percent of children under eighteen in the United States are in excellent health. We are not sure how "excellent health" is defined, but as reported earlier in this book, almost twenty-one million out of seventy-three million kids in this country suffer from cancer, asthma, debilitating birth defects, or neurological disorders ranging from ADHD to autism. And those numbers do not include the millions of additional children who suffer from other chronic illnesses such as diabetes and obesity.

Disputing that chemicals in commerce are largely unregulated and that children face significant exposure to chemicals, the council points out that of the more than eighty thousand chemicals on the inventory of the Toxic Substances Control Act, three thousand are high-volume-production chemicals with releases of a million or more pounds a year. The implication seems to be why worry about exposing our kids to only at least a million pounds each of three thousand chemicals! The paper goes on to say that industry is voluntarily assessing the "hazard effects" of these high-volume chemicals. One might ask, what has taken industry so long to examine the impact its products have on children? And why should we assume that self-testing by industry will produce honest results if those results would affect corporate profits?

The council even challenges the overwhelming evidence that children are particularly susceptible to toxic substances in their environment. It states that children "may be more or less susceptible . . . to

certain environmental agents." That is a correct but misleading statement. While some adults may be more at risk to some chemicals under
certain circumstances, as we have reported in our "Victims" chapter,
study after study has demonstrated that fetuses and young children are
far more at risk from most of the toxic hazards in the environment.

One of the seemingly positive initiatives by the council is its Responsible Care program, a voluntary effort by the chemical industry to
reduce its emissions of pollutants. According to a council newsletter,
the program has produced a 75 percent reduction in industry emissions
since 1988. According to the Environmental Working Group, however, the program was chiefly a response to growing public distrust of
the chemical industry after the chemical disaster in Bhopal. According
to the Environmental Working Group, "Responsible Care was, and remains, primarily a public relations effort to polish the image of a notorious industry that has consistently fought against meaningful health
and safety precautions for workers or communities. During the 1990s,
CMA [the council's forerunner] spent $1 million to $2 million a year on
implementing Responsible Care at its member companies, but more
than $10 million a year on advertisements about the program." The
Environmental Working Group report said that the industry program
had accomplished little more than what was required by government
regulations.

OTHER INDUSTRIES, OTHER HAZARDS

We have focused on chemical companies in this chapter because of the
range and volume of toxic substances they put into the environment
that can harm our kids. But these companies are by no means the only
perpetrators of the crime against our children. The lead, mercury, and
other metals emitted by the smelting industry wreaks physical and
neurological havoc on children who live within range of their miasma.
The spoils from mining operations leach toxics into water supplies.
The U.S. automobile industry fought every step of the way against pro-
posals to increase safety, save fuel, and lower polluting emissions from
their vehicles and lost much of their market to foreign competitors in
the process. Pesticide use and wastes from hog farms and other factory
farms make agribusiness a significant contributor to the assault on children. Pharmaceutical companies sometimes carelessly market drugs
that can harm the bodies and minds of or even kill children. The mili-

tary is a notorious polluter, placing highly toxic materials such as rocket fuel into the environment, where it can reach children through water and air. As we shall note later, the nuclear industry also appears to be a contributor to illness among children.

Then there is the fossil fuel energy industry, companies that extract, refine, and use oil and coal, which when burned pour billions of pounds of pollution into the air and water each year. These pollutants include mercury, lead, benzene, oxides of sulfur and nitrogen, solid particulates, and other substances that take a heavy toll on kids. They are also the major emitters of the carbon dioxide that is saturating the earth's atmosphere and causing the planet to warm rapidly, which undoubtedly will bring another environmental assault on children, as we have indicated earlier.

Yet most of the energy industry, companies such as ExxonMobil, which earned a world record $40.6 billion in profits in 2007, continues to deny that its emissions contribute significantly to global warming or even that there is a threat from global warming. Exxon and others in the industry use their formidable resources to block efforts that would require them to reduce its pollution, including carbon. Eric Schaeffer, who resigned as director of enforcement of the EPA, charging that the George W. Bush administration was not enforcing the laws, said that Exxon was particularly notable for its "hard-nosed resistance to settling claims brought by the government and cleaning their plants up." Nor have they made much of a contribution thus far to developing alternative, nonpolluting sources of energy. In their intransigence, they were strongly supported by the second Bush administration, which has been described as a government of, for, and by the oil industry.

MOTIVES

Why do companies do what they do? Why do they pollute and poison our children's environment? The prime motives are no doubt company profit and personal wealth for managers. But that does not fully explain corporate behavior. In many (if not most) cases, manufacturers could find safer alternatives for their pollutants and still turn a profit, as DuPont did with chlorofluorocarbons and, no doubt, will do with Teflon. Many of the products they now make are not essential to the well-being of consumers. Corporate representatives claim that they make what the public wants. But if that is so, why does industry spend billions of dollars a year on advertising and marketing to persuade people to want their products? Elizabeth Sword, former executive director of the nonprofit Healthy Child Healthy World coalition, believes that "Madison Avenue is one of the biggest problems. They convinced the public that clean has a smell. Why does status mean the size of a car and home and the greenness of your lawn?" As the rapidly growing markets for organic foods, green buildings, and hybrid cars demonstrate, there is money to be made on cleaner, environmentally benign, and less dangerous products.

Corporations are also subject to the tyranny of the immediate. They must show progress in their quarterly reports and take care that the stock market keeps increasing the value of company shares. Managers therefore tend to take a short-term view of their responsibilities, discounting long-term future benefits that would accrue from making safer products that would protect public health.

Eric Schaeffer, who founded the Environmental Integrity Project when he left government, is convinced that "for some companies it's just a racket—how to game the system in ways that avoid having to spend money on compliance or to minimize what you have to do. You are trained in business schools to cut your costs." He added, however, "There is a level of industry response to environmental regulations which is emotional. It is not a pocketbook issue, but a response to the idea of being told what to do. They want to make their own decisions—and now they have the power to do so."

Just before he stepped down as director of the National Institute of Environmental Health Sciences, Dr. Kenneth Olden commented to us, "The chemical industry didn't create pesticides or PCBs or chlorofluorocarbons to poison us. They created them to improve the quality of our lives. Now, you have to make sure there are firewalls."

Firewalls to protect our children from potentially harmful products made by industry are erected by government in the form of laws, regulations, enforcement actions, and penalties. In recent years, however, government has not been building those walls and, in fact, has frequently collaborated with industry to tear breaches into those walls. That can be changed by an informed electorate.

Eight

CO-CONSPIRATORS

FREE MARKET CAPITALISM DOES SOME THINGS VERY WELL. IT IS AN efficient engine for producing and distributing goods and services and for creating wealth. It encourages and rewards creativity and innovation and provides ample opportunity for economic and social mobility. It can buttress national power.

There are some things, however, that the market does not, cannot, or will not do. The system does not distribute wealth equitably, nor does it necessarily meet real social needs and priorities. The market is a voracious consumer of natural resources and generates enormous quantities of polluting waste. And the market for the most part places little or no value on the environment or public health, regarding them as "externalities" with no relationship to the economy. Those "externalities" are the instruments of the environmental assault on our children.

Only governments can control corporate power and correct or mitigate the flaws of the free market system and its practitioners. As Patrick Parenteau, director of the Environmental and Natural Resources Law Clinic of the Vermont Law School, observed, "Corporations do what they do for a very clear and understandable and even legal set of necessities. They have to be forced to do the right thing through legislation, litigation, and the electoral process, so they don't have any other choice."

There have been periods in U.S. history when governments at all levels made conscientious efforts to restrain excessive corporate power and correct abuses of the system. Toward the end of the nineteenth century and beginning of the twentieth, the federal government moved to limit economic monopolies, to improve conditions in the workplace, to conserve publicly owned natural resources, and to impose rules for sanitation and the safety of its citizens. In the 1880s, New York City's mortality rate began to decline when government built the Croton Reservoir, delivering cleaner water. Over the following decades, public health measures eradicated the infectious diseases that historically claimed many children's lives. During the New Deal era, initiated in the 1930s under President Franklin D. Roosevelt and extending in spirit through the administration of Lyndon B. Johnson, governments acted to improve the condition of workers and to create a social safety net for those buffeted or cast aside by the prevailing economy.

After the first Earth Day in 1970, the United States government embarked on an ambitious effort to curb the industrial assault on the environment and human health. Congress passed clean air and water acts to regulate dangerous substances in commerce and hazards already disposed of in the environment. The executive branch created the Environmental Protection Agency to give form to the laws and to enforce them. The courts punished offenders and required them to reform their operations in conformance with the new laws and regulations.

Government intervention to protect health and the environment was not based on ideology or partisanship. It was a pragmatic political response to the growing recognition by science and the public that with our technology, our consumption, and our waste, we were fouling our own habitat and harming ourselves and our children. Nor was environmental activism in Washington and around the country a politically partisan initiative. Republicans and Democrats in Congress joined to pass the landmark environmental laws. President Richard Nixon, a staunch conservative—or what was considered to be conservative in that era—signed the legislation, created the EPA, and assigned able public servants to carry out the new laws.

That was then.

THE BACKLASH

In the 1980s, President Ronald Reagan came to office vowing to get government "off the backs" of the American people. The pace of environmental regulation slowed, and enforcement of the law plummeted. The Reagan administration was restrained by a Democratic Congress, but much of the momentum for environmental protection that began with the first Earth Day was lost. With Newt Gingrich's "Contract with America" in 1994, another massive attempt was made to roll back environmental rules and restraints on harmful corporate activities. As part of the "Gingrich revolution," the Office of Technology Assessment, an independent agency that helped provide nonpartisan science to the federal government, was abolished. The excesses of the "Contract" agenda were tempered while the White House was occupied by President Bill Clinton and Vice President Al Gore and because easing environmental protection was unpopular with the electorate, even in conservative areas of the country.

No administration since the early 1970s, Democrat or Republican, has had a particularly admirable environmental record. When George W. Bush ascended to the presidency, however, his administration methodically began to dismantle the firewalls that had been painstakingly erected over the years to shield the American people from the dangers of industrial activity and abuses of corporate power. It was not a political climate salubrious for children.

In carrying out his programs, Mr. Bush could count on the support of Congress, both houses of which were controlled by the Republican Party for most of his term in office. The chairman of the Senate Environment and Public Works Committee, Senator James Inhofe of Oklahoma, had a near perfect record of voting against legislation to protect the environment and called warnings of man-made global warming "the greatest hoax ever perpetrated on the American people."

At the end of his first term in 2004, the Children's Environmental Health Network, a national, nonprofit organization of physicians, nurses, public health workers, scientists, environmentalists, parents, teachers, labor union representatives, and others, looked closely at the Bush environmental record and gave it a grade of F, failing it in almost every activity related to protecting children's health from pollution. The network's report said that its evaluation "makes clear that, in general, this Administration's track record is toxic to our children, lessening protections for children and missing opportunities to keep toxicants out of children's environment."

In other words, the United States government had become co-conspirators in the crime against our children.

DIRTY SKIES

In 2000, members of the Bush administration, including Vice President Dick Cheney, held a series of meetings with energy industry executives. Both Bush and Cheney had come out of the oil industry. While much of what went on at those meetings has been kept secret, it is clear by now that the agenda was an opportunity for industry to present its wish list of favors desired from the government. The companies did not get everything they wanted, but they did get most of it, including a program called Clear Skies.

Among other things, the administration relaxed legal requirements that power plants install pollution control equipment if they make substantial improvements to their facilities. The program also ignored a campaign promise by Bush to lower carbon dioxide emissions, the chief greenhouse gas. Perhaps the most egregious element of the program was an EPA ruling, dictated by the White House, that drastically weakens regulations drafted in the Clinton administration to reduce the emissions of mercury from coal-fired power plant smokestacks. Eric Pianin reported in *The Washington Post,* "The regulatory turnabout was engineered by Jeffrey R. Holmstead, the EPA's senior air quality official and a former industry lawyer." Mr. Holmstead, a Bush political appointee, had worked with a libertarian group that advocated market solutions to environmental problems and also as a partner with Latham & Watkins, which represented electric power companies and other industries.

Coal-fired plants are by far the biggest source of airborne mercury, a potent neurotoxin that can cause brain damage and interfere with reproduction. The EPA's own scientists estimated that as many as 630,000 babies a year are exposed to high levels of mercury while in their mother's uterus, as described earlier in this book.

The Clear Skies program would put off until 2018 the goal of reducing mercury emissions from the plants by 70 percent. It would also adopt a market-based process of capping and trading mercury emissions to reach that goal. The cap-and-trade approach, which allows a facility that can meet the goals cheaply to sell pollution "credits" to a plant that would prefer not to invest in pollution controls, has been used successfully to reduce the amount of sulfur and nitrogen oxides spewed from power plant stacks and carried by winds long distances from the source. But critics say it is an inappropriate way to control mercury pollution, which concentrates close to the polluting facility. If a polluter pays another power plant for credits to emit mercury, the purchasing company's stacks can continue to spew mercury, which will continue to build up in the area and accumulate in the bodies of mothers and children.

"This is going to be a bad rule, one that allows mercury levels to rise, particularly in the West," said Fred Krupp, head of Environmental Defense, a national environmental group that pioneered the concept of cap-and-trade programs to curb other airborne pollutants.

In adopting the weaker mercury rule, political appointees ignored the work and recommendations of professional staff at the EPA and a federal scientific advisory panel. EPA officials further omitted the results of a Harvard study, funded by the EPA and coauthored by an EPA scientist, that showed that the costs of mercury pollution and the benefits of strong regulations were higher than previous estimates. Bruce C. Buckheit, who retired as director of the EPA's Air Enforcement Division in 2003, said that the mercury rule making involved "a degree of politicization of the work of the Environmental Protection Agency that goes beyond anything I have seen in my career in government." The EPA's own inspector general charged that senior management had instructed staff members to arrive at a predetermined conclusion favoring industry in preparing the mercury rule.

The electric utility industry welcomed the new rule, which would cost millions of dollars a year less to implement than what had been proposed during the Clinton administration. When the rule was published, Martha Keating, a former EPA scientist, thought she had seen it before. Much of it, she said during an interview on the PBS program *NOW* in January 2005, was an almost word-for-word replication of a proposal given to the Bush administration in 2003 by Latham & Watkins, the law firm where Jeffrey Holmstead worked before he was appointed to the EPA's senior air quality job.

POLITICIZING SCIENCE

The mercury rule is only one example of perhaps the most disturbing facet of the Bush administration's dereliction of its duty to protect our children—its manipulation and perversion of science to meet political and ideological goals. Federal agencies such as the National Institutes of Health, the EPA, the CDC, and the FDA have long conducted and supported scientific research for government policies to protect the health of American citizens. But in recent years, many in the scientific community have strenuously questioned the integrity of government science.

By 2006, over nine thousand top U.S. scientists, including Nobel laureates, medical experts, university science professors and chairpersons, and former heads of federal agencies, signed a statement expressing distress about what they saw as "the misuse of science by the Bush Administration." The statement, distributed by the Union of Concerned Scientists, charged that "when scientific knowledge has been found to be in conflict with its political goals, the administration has often manipulated the process through which science enters into its decisions."

To avoid acting to curb pollution that is causing climate change, the statement said, "the administration has consistently misrepresented the findings of the National Academy of Sciences, government scientists and the expert community at large. To avoid issuing a scientifically indefensible report, EPA officials eviscerated the discussion of climate change and its consequences." It pointed to censorship and political oversight of government scientists and to the fact that highly qualified scientists have been dropped from federal advisory committees and replaced with individuals working for or associated with industries that would be directly affected by the output of those committees.

Stacking science advisory panels was one of the Bush administration's tactics for assuring it would be provided with the results it wanted. In 2002, for example, the administration named a group of business-friendly scientists to the advisory committee of the National Center for Environmental Health, an arm of the Centers for Disease Control and Prevention, without consulting with the director of the center, Dr. Richard Jackson, a strong advocate for public health. Dr. Jackson departed shortly thereafter. The appointees included Roger McClellan, former director of the Chemical Industry Institute of Toxicology; Becky Norton Dunlop, a vice president of the conservative Heritage Foundation, which opposes government regulation; and Dennis Paustenbach, a toxicologist who is frequently retained by industry to do risk assessments (as described in the next chapter). They replaced scientists such as Dr. Ellen Silbergeld, a toxicologist at Johns Hopkins University, much of whose research has focused on toxics in the environment that harm children.

In 2002, the CDC's Advisory Committee on Childhood Lead Poisoning Prevention was considering the need to redefine to a lower level the amount of lead in kids' bodies that constitutes poisoning. That summer, the Department of Health and Human Services dropped three internationally recognized experts on lead poisoning, Dr. Bruce Lanphear, Dr. Michael Weitzman, and Dr. Susan Klitzman, from the advisory committee. To replace them, it nominated several people with ties to the lead industry. Some saw the nominations as an attempt to protect the industry from lawsuits.

Donald Kennedy, editor in chief of the journal *Science,* a former president of Stanford University, and himself a distinguished scientist who once headed the Food and Drug Administration, wrote in the journal about a professor of psychiatry who was called by the White House about his nomination to serve on an advisory panel. The professor said he had been questioned to "determine whether he held any views that might be embarrassing to the President." One of the questions was whether he had voted for Bush. When the professor answered that he hadn't, the caller asked, "Why didn't you support the President?"

"This stuff would be prime material for a Robin Williams comedy shtick, but it really isn't funny," Dr. Kennedy wrote. "The purpose of advisory committees is to provide balanced, thoughtful advice to the policy process—it is better not to put the policy up front."

The Bush administration often turned its scientific research over to industry. In 2005, records obtained by Public Employees for Environmental Responsibility showed a sharp increase in the number of cooperative research and development agreements with companies and trade associations under the administration. The group's report noted that the American Chemistry Council had become the EPA's "leading research partner." One of the joint projects with the chemistry council was a study that would have given parents $970 and a camcorder in return for continuing to spray pesticides in rooms used by their babies and infants, to observe their health effects. The study was canceled following expressions of public outrage at the prospect of using kids as test animals.

"Under its current leadership, the EPA is becoming an arm of corporate R&D," asserted Rebecca Roose, program director of the public employees' group.

In 2004, the EPA dismissed new studies of health risks from the use of formaldehyde, including leukemia and other cancers, and accepted an air pollution rule that could save the wood products industry many millions of dollars. The rule it accepted was based on a cancer risk model presented by the Chemical Industry Institute of Toxicology. Its risk assessment for formaldehyde is about ten thousand times less stringent than the level used by the EPA in setting general standards for formaldehyde. "This rule making veers radically from standard scientific and regulatory practices," said David Michaels, a professor at George Washington University and an assistant secretary for environment, safety, and health at the Department of Energy during the Clinton administration.

In organizing a review of studies to bisphenol-A and eighteen other chemicals, the understaffed (two-person) center within the National Institute of Environmental Health Sciences charged with the essential task of evaluating chemicals for their possible harm to human fertility or fetal development similarly turned to a for-profit company for assistance. In this case, the center entered into an eight-year, $5 million contract with Sciences International to assist in this review. When it became public that Sciences International's other clients included Dow Chemicals and BASF, both of which manufacture bisphenol-A, the company was fired. The ensuing report, though it did express some concern for pregnant women and fetuses, otherwise found either negligible or minimal concern; leading bisphenol-A researchers took strong exception to that conclusion, pointing out that Sciences International had excluded from its review a vast array of studies that found harm.

For thirty years, the jobs and responsibilities of government employees have been outsourced, contracted out to private companies. Just as former military roles, such as building bridges, patrolling the streets, and guarding prisoners, have been outsourced during the war in Iraq, so too have the roles in science been taken from public servants and awarded to the private sector. There is little evidence that outsourcing saves money; in fact, it's likely to cost our nation more. Even more important is the fact that employees of the government, whose mission is to serve the public, have been replaced by employees of private companies, whose mission is to make money.

THE REVOLVING DOOR

Another administration tactic is to place political appointees from or friendly to business in key positions to carry out—or not carry out—the laws. The appointment of Jeffrey Holmstead to the senior air quality post at the EPA is only one of dozens of examples of industry foxes being named to guard the government's regulatory henhouses, many at senior levels of the administration. The top White House policy adviser on the environment, as chairman of the Council on Environmental Quality, was James Connaughton, who came out of a law firm that represented giant companies such as General Electric and ASARCO. This is not, of course, a new practice, but in recent years the process has escalated. A nonprofit coalition of public interest groups created to monitor this practice issued a report that stated, "The Industry-To-Government revolving door . . . establishes a pro-business bias in policy formulation and regulatory enforcement." The report also warned of the equally rapidly spinning revolving doors that propel government officials—and members of Congress and congressional staffers—from government to industry or government to lobbyist.

"The revolving door," the report concluded, "casts grave doubts on the integrity of official actions and legislation."

SLIDING AROUND THE LAW

When it wasn't blocking environment and health regulations, the administration weakened them or delayed implementing them. In 2005, for example, the EPA decided to delay rules to protect children and construction workers from exposure to lead-based paint. Instead, the agency said it is considering voluntary standards for paint companies and construction firms. An EPA spokeswoman said that the agency was exploring alternatives that might be more effective as well as less costly to industry and the public.

The nation made progress in cleaning up its water supplies in the last third of the twentieth century, but in recent years that progress has been slowed or reversed. The U.S. Public Interest Research Group, a consumer advocacy organization, reported in 2004 that more than 60 percent of the nation's industrial and waste treatment plants that discharge wastewaters were putting more pollution into the water than allowed in their permits. "The numbers point out that enforcement is not a priority for this administration, and clearly little to nothing is being done to deter polluters from breaking the law," said Richard Caplan, who authored the report.

The number of lawsuits filed by the EPA against alleged polluters plummeted after 2001, according to Eric Schaeffer, former head of the agency's enforcement office, who resigned in protest from the agency and now runs the Environmental Integrity Project, a nonprofit organization that advocates effective enforcement of environmental laws. "Government isn't protecting our children very well," said Schaeffer. "There are lots of really good people grinding away at the problem, but the political level puts limits on what you can do. Even in the best political environment you operate under constraints—there is always a battle with industry, which is always well organized and has its own scientists. But now industry has the ear of the White House."

Schaeffer noted that the OMB estimated that a ton of sulfur dioxide pollution in the air produces $7,000 in quantifiable public health costs, whereas the cost of removing a ton of sulfur from power plant emissions costs less than $500 a ton. "It is a public policy failure when you have something that costs the economy $7,000 and could be cleaned up for $500 and it isn't done. What an indictment of our political system when you can't make this happen!"

In 1986, Congress gave the EPA the mandate to create a Toxics Release Inventory, which required industry to report their emissions of harmful substances into the environment. The requirement was widely hailed as a democratization of environmental protection, providing the public with information about potentially harmful pollution, and put pressure on industry to reduce its emissions. But in 2006, twenty years after it became law, the EPA removed some of the law's teeth. Instead of having to report releases of five hundred pounds of the most toxic substances, polluters would have to report only releases over five thousand pounds, a tenfold decrease in reporting requirements. And they would have to report only every other year instead of every year.

In an op-ed article in *The New York Times,* Senator James M. Jeffords of Vermont and Julie Fox Gorte, a vice president of the Calvert Group, an investor in socially responsible companies, called the move "just one more example of the Bush Administration's efforts to quietly undermine our nation's environmental protections. Washington should be working to expand corporate disclosure and accountability rather than moving to allow polluters to conceal their toxic releases."

In 2001, the EPA announced it was withdrawing a Clinton-era rule to reduce the amount of arsenic in the nation's water. President Bush asserted that the rule was being dropped because it had been rushed through the rule-making process and was not based on "sound science." In fact, the EPA had been working on the rule for a decade, and the National Academy of Sciences had concluded that the old standard for arsenic was inadequate. The real reason for the decision to kill the rule (which was reversed after public outcry) was political. According to David Frum, a former Bush speechwriter, Karl Rove, the White House political guru, had pressed dumping the rule as a way to win votes in New Mexico, where high levels of arsenic in the water would require substantial expenditures to reduce those levels.

FOLLOWING THE BUSINESS AGENDA

The Bush administration also made it difficult for states to sue companies that may have caused harm, environmental or otherwise. In a process that has been described as "silent tort reform," federal agencies waged what *New York Times* reporter Stephen Labaton called "a powerful counterattack against active state prosecutors and trial lawyers. At the urging of industry groups, the federal agencies have inserted clauses in new rules that block trial lawyers and state attorneys general from applying both higher standards in state laws and those in state court precedents." So not only was the federal government failing to act to protect our children, it was preventing state governments from doing so.

The people's representatives on Capitol Hill usually rubber-stamped whatever the administration did to weaken environmental laws and turned a blind eye when it failed to carry those laws. Members of Congress, particularly conservative Republicans who dominated the legislature, frequently proved themselves quite capable of taking the initiative to lend a helping hand to their friends and campaign donors in industry.

In 2005, Representative Joe Barton, Republican of Texas, introduced legislation to enable oil refineries to expand capacity and coal-fired power plants to upgrade their facilities without taking steps to curb air pollution. The bill would weaken the Clean Air Act ostensibly as an emergency measure to boost energy production in the wake of Hurricane Katrina. The bill was supported by most Republicans, while Democrats said it was a gift to industry using the rapid rise of gas prices as an excuse. "The entire bill is written for ExxonMobil," said Representative Edward Markey, Democrat of Massachusetts, who noted that refinery profits had doubled from the previous year.

Individual members of Congress frequently step forward to help business. Bill Frist, the Republican majority leader in the Senate until 2007, introduced legislation that would protect Eli Lilly and other pharmaceutical companies from lawsuits claiming injury to children from the mercury-based preservative thimerosal in vaccines. The provision was withdrawn after widespread public objections. But then it reappeared as a midnight rider to a national security bill and was rammed through Congress following the Republican victory in the 2002 midterm elections. No one claimed responsibility for the mysterious amendment, but Senator Frist, a physician whose family owns the biggest for-profit health care companies in the nation, once said that a candidate to head the Food and Drug Administration did not get the job because within industry "there was a great deal of concern that he put too much emphasis in safety." And, as we reported earlier in this book, Frist ignored a plea for help from the families of children with cleft palates in Dickson, Tennessee.

The Democrats regained narrow control of Congress in the 2006 election and were in a position to block further erosion of environmental safeguards. The Democratic leadership have at least voiced support for movement on such issues as global warming.

CO-CONSPIRING BY THE STATES

State governments, meanwhile, are not always on the side of the environment and children.

The state of Texas, for example, is one of the more polluted in the nation. The southeastern portion of the state is carpeted with refineries, petrochemical plants, chemical storage warehouses, and other polluting facilities. The air in cities such as Houston, Port Arthur, Beaumont, Texas City, and Port Neches is often acrid with hydrocarbons and toxic chemicals such as butadiene, and children living there are at higher risk. But because the oil and chemical industries wield strong economic and political power, and because Texas has been dominated by conservative politics in recent years, the state's environmental and safety rules and enforcement are relatively lax.

The Texas Commission on Environmental Quality has insisted that its regulations for protecting citizens from toxic chemicals in the air are "ultrasafe." But in a series of articles on air pollution in 2005, *Houston Chronicle* reporter Dina Cappiello found that safe exposure levels set by the state for nearly two thousand chemicals identified by the federal government as causing cancer or other serious health problems were "hundreds of times higher" than those imposed by other states. Cappiello's series provoked substantial comment and public concern in Texas, and shortly after it appeared, the Texas State Legislature took up a bill to impose tougher rules as well as financial penalties on companies that produce hazardous air emissions. It was defeated. The chairman of the Texas House's Environmental Regulation Committee claimed that antipollution rules would drive out industry. "The sign would say 'Texas: Closed for Business,' " he warned, though it seems unlikely that the oil, chemical, and petrochemical industries would abandon their massive infrastructures in Texas because of strengthened environmental requirements. These industries make heavy political donations to local elections.

The late Dr. Marvin Legator, professor of environmental toxicology at the University of Texas Medical Branch in Galveston, concluded in a study of Texas's environmental guidelines that "the method the state of Texas utilizes to protect its citizens . . . is one of convenience, not scientific principle."

Lufkin, Texas, is an hour or so north of Houston. Its surroundings are somewhat more rural than the Gulf Coast industrial corridor, and sources of pollution are not as obvious. But the area has its problems. They come, according to Dr. Bill Shelton, a pediatrician and radiation oncologist in Lufkin, from a big paper mill with a chlorine-based bleaching process that sends dioxin into the area's waterways. The mill, which has shut down but may reopen, also put tons of manganese into the local environment, he said. Several large power plants in the area spew large amounts of mercury into the air—and then into the land and water. There is also a pipeline, now nearly a hundred years old, that is rusted, pitted, and patched yet continues in service, carrying hydrocarbons right through the city.

Shelton, a jovial, bearded bear of a man in his seventies and known as "Doc" to all, is an avid bass fisherman. He became concerned about pollution in the area following massive fish kills in nearby Lake Sam Rayburn, one of the state's biggest bodies of water. When tested, he said, the fish were found to contain high levels of dioxin and mercury. He began to notice a very high incidence of cancer and neurological disease among residents of a low-income community called Rivercrest at one end of the lake. People in the community depend for some of their subsistence on the fish they catch from the lake and the Angelina River, which empties into it, he noted.

"From practicing medicine all these years, I know that almost every single adult from Rivercrest has either cancer or a neurological problem and every child is in special education in the Lufkin school system," he said. He added that he knew about cancer in the area because he ran a cancer clinic. His clinic, however, did not treat children with cancer, who were sent to a hospital in Houston.

Shelton, along with Lufkin residents Walter West, a retired space engineer with the National Aeronautic and Space Administration (NASA), and Diane Avrilett, who runs a catering service and restaurant called the Lunch Box, has spent years trying to get industry and government to clean up their acts. Together, they have had little success.

"We have spent thousands of hours dealing with the Texas Commission on Environmental Quality and with the EPA regional office in Dallas," Shelton said. "What stands out is that both of those agencies yield to and are controlled by pressure from industry. It comes right down to the apathy of the public and the fact that our mayor, our city council, our state representatives, our senator, Kay Bailey Hutchinson—every single one of them is exactly like the Bush administration: Politics first."

When the paper mill applied for a new permit that would allow it to continue to dispose of waste into the lake, the city of Lufkin hired three buses to carry residents to the hearing to support the application, Shelton said. He added that the sheriff's office kept citizens opposed to the permit off the buses and then escorted the buses to Austin for the hearing.

Texas is only one of many states with permissive attitudes toward pollution. Virtually all of the southern states, for example, have banded to block legislation and rules that would lower pollution from within their borders that drifts to the Northeast and taints the air in the eastern states, states that have been trying to achieve cleaner air.

Some states, however, have been active in trying to protect the environment and children's health and have generated much of the forward motion there has been in recent years. New York, especially when Eliot Spitzer was its attorney general, and California, which has moved vigorously to address air pollution and toxic contamination, have filled some of the gaps left by the federal government. So, too, have most of the New England states and states in the Northwest and the rust belt.

MILITARY-INDUSTRIAL CO-CONSPIRATORS

There are arms of the federal government that are themselves major polluters whose activities and wastes harm the environment and children. The U.S. military is one, and so is NASA. The chemicals and other materials that are used in weapons and other devices employed by the armed forces are often exceedingly dangerous. The Pentagon has estimated that it would take seventy years and $20 billion just to clean up *former* military facilities contaminated with toxic, radioactive, and other hazardous substances. Many active facilities are contaminated with these materials.

Members of the military and their families, including their children, are exposed to these pollutants. So are those who live in the neighborhood of the bases and depots. If the toxics get into the water and the air and even into the food supply, as they often do, they can cause harm over wide areas. But in recent years, the Defense Department and the industries that supply it, supported by the White House, have resisted broad cleanup efforts or even acknowledging that their wastes can harm human health. In the name of national security, the administration sought to ease environmental rules for the military or exempt it from some regulations.

One example is a long-simmering dispute over perchlorate, a chemical used by the armed forces and by NASA in rocket propellants. Because it has a limited shelf life, unused stocks have to be replaced regularly, and as a result, large amounts must be disposed of as waste. And since it is relatively inert, when it gets into the environment it stays there a long time. Perchlorate may harm human health and the environment, at low concentrations. It has been found to interfere even in tiny amounts with a pregnant or nursing woman's uptake of iodide into the thyroid gland and thereby blocks the secretion of hormones that are vital for the growth and brain development of fetuses and infants as well as for normal metabolism and mental function in adults. This chemical has been found in water supplies of at least thirty-five states, from New England to California, including the Colorado River, which provides drinking water to over fifteen million people. It has been found in lettuce, milk, and other foods.

Despite its ubiquity and dangers, there are, as of this writing, no federal standards setting safe levels of perchlorates in drinking water or food. In 2001, the EPA proposed a level of no more than one part of perchlorate in a billion parts of water as a safe level for fetuses and babies. It based its proposal on the findings of a study conducted under a joint agreement between the agency and the Pentagon, testing perchlorate on laboratory animals to identify a level at which perchlorate did not impair the animals' thyroid function. The study was paid for by the Perchlorate Study Group, a group of companies that make or use perchlorate, including Kerr-McGee, Lockheed Martin, and Aerojet.

The Defense Department and the companies were dismayed with the EPA proposal, which the Pentagon feared would cost billions in cleanup costs—and possibly lawsuits—and interfere with weapons development. Colonel Daniel Rogers, chief of environmental law and litigation at the Pentagon, called the EPA risk assessment "biased, unrealistic and scientifically imbalanced." The Perchlorate Study Group then took the highly unusual step of hiring an outside consulting firm to critique and challenge its own multimillion-dollar study! The firm, Exponent, Inc., which frequently takes on technical assignments for industry, found that perchlorate was safe at much higher levels and that the nation's drinking water posed no health threat from perchlorate contamination. The Pentagon then proposed a safety level of 200 ppb.

The White House sent the controversy over perchlorate to the National Academy of Sciences, which set up its own study panel. The academy came up with a safety standard of 24 ppb. The study was funded by three of the companies of the Perchlorate Study Group, and two members of the panel had done work for Lockheed Martin. An investigative report from the *Press-Enterprise* of Riverside, California, found that data showing that perchlorate at low doses had inhibited the thyroid function of the subjects studied had been withheld from the material given to the National Academy of Sciences.

Early in 2006, the EPA issued a "preliminary remediation goal" of 24 ppb for groundwater at Superfund sites. The agency's own Children's Health Protection Advisory Committee wrote to the EPA's administrator Stephen Johnson, objecting that the proposed standard did not protect newly born children. But that standard did not satisfy the Pentagon, the defense contractors, or the White House, which believed it would still be too costly to enforce. As a result, more delay followed without imposing a limit on perchlorate contamination that would protect children, with no end in sight.

Some states, including Massachusetts and California, have moved to lower the health risks from perchlorate. But a national effort will be required.

The perchlorate story is by no means the only case history of how the military has been both a major polluter and a major roadblock to cleaning up the pollution. Jon R. Luoma, writing in *Mother Jones,* noted, "From Cape Cod in Massachusetts to McClellan Air Force Base in California, the Pentagon is facing mounting criticism for failing to clean up military sites contaminated with everything from old munitions to radioactive materials and residues from biological-weapons research. Now, citing the demands of the war on terrorism and working with sympathetic officials in the administration and Congress, the department has stepped up efforts to remove substantial parts of its operations from environmental oversight." It has succeeded.

The Department of Defense owns more than 1,400 facilities contaminated with TCE, the same chemical found in Dickson, Tennessee, and Woburn, Massachusetts, among other towns, and one of the most prevalent contaminants of water across the nation. PCE, a related chemical, also pollutes military compounds.

One of the military bases heavily polluted with TCE is Camp Lejeune, a Marine Corps base in North Carolina. The Agency for Toxic Substances and Disease Registry found that at least one hundred babies exposed in utero to the TCE-contaminated water on the base between 1950 and 1980 suffer from leukemia and birth defects, including spina bifida and cleft palates . . . just as the children in Dickson, Tennessee, do. It took an amendment attached to the 2007 Defense Authorization by Senators Elizabeth Dole (R-NC) and James M. Jeffords (I-VT) to force a study of the link between exposure to TCE and PCE and these disorders at the camp.

In 2001, an EPA risk assessment found that TCE was five to sixty-five times more toxic than an estimate made nearly two decades earlier. The findings suggested that TCE was, at least in part, responsible for thousands of cancers and birth defects every year. The assessment was to be the basis of stringent new rules to reduce public exposure to the chemical. But after the Pentagon challenged the science underlying the EPA study, the risk assessment was jettisoned.

Ralph Vartabedian, a writer for the *Los Angeles Times,* concluded from this episode, "What happened with TCE is a stark illustration of a power shift that has badly damaged the EPA's ability to carry out one of its essential missions: assessing the health risks of toxic chemicals. The agency's authority and its scientific stature have been eroded under a withering attack on its technical staff by the military and its contractors." He added that the Bush administration's political appointees at the EPA sided with the military.

Writing in the *International Journal of Occupational and Environmental Health,* Dr. Jennifer Sass, a senior scientist with the Natural Resources Defense Council, stated that "the willingness to set aside science and public health in favor of protecting the polluters—the Pentagon and defense contractors—is appalling. The situation also illustrates the troubling, cozy relationship between the Pentagon and industry that tests the limits of ethics, conflict of interest rules, and in some cases, federal law."

The National Academy of Sciences subsequently took on a study of TCE, which concluded that the evidence is stronger than ever that the chemical causes cancer. If the EPA does resume its work, a new drinking water standard will not be set until 2008, if ever.

Nearly half a century ago, President Dwight D. Eisenhower, who had spent most of his life in the army, warned the nation as he was leaving office to beware the growing power of the military-industrial complex. His warning was prophetic and, it seems, unheeded.

Government and its policies can change, especially when citizens are paying attention, understand what is happening, do not like what they see, and demand change. Recently, there has been a dramatic change in the public perspective on global warming, and as a result, public policy has begun to shift. The shift is primarily in Congress,

where voter dissatisfaction with a range of Bush administration policies, especially the war in Iraq but others including the environment,
brought a Democratic majority to Capitol Hill in 2006. If there can be
movement on climate change, there can also be movement to address
the toxic plague that is harming our children. It can be hoped that the
next administration, whatever its party, will get the message.

Nine

WITNESSES FOR THE DEFENSE

KNOWLEDGE, IN SCIENCE, IS ALWAYS INCOMPLETE. BUT MANUFAC-tured uncertainty is different. It is disinformation, biased and mislead-ing. Its goal is to paralyze the regulatory and legal systems and to sway public opinion. This switch allows the manufacturer to mischaracter-ize fundamental policy conflicts over protecting health as, instead, dis-agreements over science.

Tobacco was the first industry to manufacture scientific uncer-tainty. "Doubt is our product since it is the best means of competing with the 'body of fact' that exists in the mind of the general public. It is also the means of establishing a controversy," a tobacco executive wrote in a 1969 memo to his colleagues.

By the mid-1980s, Big Tobacco had developed a full-scale strategy built on creating doubt. Nicknamed the White Coat Project, it was commissioned by Philip Morris, conceived by the large public relations firm Hill & Knowlton, and carried out by the Weinberg Group, a com-pany whose services include "product defense." The project centered around hiring physicians and other scientists willing to produce stud-ies, give public testimonials, and appear in advertising—in this case, upholding the alleged innocuous nature of secondhand smoke. The scientists were often paid through a law firm so that their studies could be classified as attorney work products that could not be subpoenaed.

Studies produced by the scientists-for-hire were praised as "sound science," while opposing research was denigrated as "junk science," terms created purposely for the tobacco industry.

This strategy of doubt and others in the tobacco wars served as fully staged dress rehearsals for the struggles over environmental pollutants that followed. Many of the people and companies subsequently offering defense for toxic chemicals (and drugs) learned and honed their skills on assignments from the tobacco industry. These are the witnesses for the defense. They are the scientists willing to present an argument on behalf of an industry, backed by the lawyers, public relations firms, lobbyists, and the legislators amenable to the influence of campaign contributions. These witnesses for the defense now make up their own profitable industry.

The interests of the tobacco and chemical industries converged and merged in the early 1990s, precipitated by the EPA's designating tobacco smoke and secondhand smoke as causes of cancer. The tobacco companies, worried not only about sales, but also that states, cities, and businesses would adopt smoking bans, quickly organized a blockbuster counteroffensive. "Our overwhelming objective is to discredit the EPA report," a Philip Morris executive wrote. To do so, the company created a front group to carry out a full-scale "sound science" work plan in defense of tobacco and widened its focus in order to conceal its true mission as a shield protecting tobacco. Issues such as the regulation of food additives and auto emissions were added to its agenda, and the chemical industry signed on as a sponsor. The goal evolved: to create controversy over evidence that environmental toxins cause disease.

The two industries then launched a collaborative campaign in 1996 to obtain federally funded but previously confidential research data so that the scientists they hired could reanalyze the data to reflect industry positions. With this intention, they drafted language for a new law that Congress enacted in 1999. This Data Access Act allows anyone to obtain scientific data produced under federally funded research studies. The next year, the companion Data Quality Act was slipped into a thick federal appropriations bill that neither legislators nor their staff had time to read and was passed without debate. Control of the way all federal agencies make use of these laws was given to the Office of Management and Budget, the powerful surrogate for the White House, staffed at that time by antiregulatory and antienvironmental appointees inclined and, with these laws, equipped to hamstring environmental regulation.

The Data Quality Act seems harmless, even beneficent. It allows any company subject to regulation to contest the "sound science" behind any and every piece of raw data that federal regulatory agencies might use for setting rules. In fact, it can be used to produce paralysis by analysis. It offers an established procedure for stalling, killing, or altering government documents and decisions. For Philip Morris, the goal was to reopen the debate about the effect of secondhand smoke on lung cancer, as an internal company memo explains. That goal was attained; in 2005, the OMB inserted data quality requirements into the EPA's cancer guidelines, an opening for industry to challenge the guidelines and to weaken and delay their being put in place, and the National Institutes of Health downgraded warnings about smokeless tobacco.

Jim Tozzi, former deputy director of the OMB and a longtime Washington insider, was the brains and strategist behind both acts. After serving in key positions in five different administrations, he set up shop as a point man in the tobacco wars, with $2.5 million from Philip Morris, allowing the tobacco industry to remain in the background, as Philip Morris stipulated. Tozzi proved his creativity and adeptness at regulatory chess. He successfully kept secondhand smoke from being assigned a medical code to identify it; without such a code, doctors cannot attribute illness to secondhand smoke, and as a result, secondhand smoke cannot be counted as part of our national health costs.

It was Tozzi who wrote the two short paragraphs that make up the Data Access Act and the two paragraphs that make up the Data Quality Act, as well as the voluminous guidelines for the OMB's use of the act, and brought the trade associations for the chemical and other industries into planning this strategy.

Disarming in his voluble, engaging honesty, Jim Tozzi is a whiz at his job. "I've spent my life setting up a process that regulates the regulators," he explains. Tozzi pleasantly affirms his satisfaction in delaying the removal of lead from gasoline for three years. Tozzi's expansive offices, bearing his different company names on different doors, take up an entire floor in a well-kept building in the heart of Washington, D.C.'s lobbying district. An ashtray filled with cigarette butts graces the middle of a cozy round conference table in his private office suite. "At one Cardozo Law School lecture I gave, a student asked, 'Are those good government statutes or no government statutes?' " He gives one of his hearty, cheerful belly laughs and adds, "That's one good question."

CREATING DOUBT ON BEHALF OF A PESTICIDE

Now comes the part of the story where polluting chemicals move center stage.

On the wall just behind Tozzi's conference table hangs an elegantly matted and framed tear-out of a front-page article from *The Washington Post* of August 2004. It tells the story of how he brought the first challenge under the new Data Quality Act to protect the continued use of an herbicide called atrazine, whose main producer is the Swiss company Syngenta (formerly Novartis), now the world's largest agrochemical manufacturer. Atrazine is cheap, and it works. It is the most widely used weed killer in the corn belt of America, sprayed over tens of millions of U.S. acres. Some of it evaporates and comes down in rainfall, contaminating streams in at least twenty-three states. Atrazine is found in over 90 percent of water samples in farming communities. Its use was never allowed in Switzerland.

Atrazine's cancer-causing effect in lab rats and the high rates of prostate cancer among atrazine factory workers brought the chemical under EPA scrutiny. The agency accepted Syngenta's argument that the mechanisms that cause cancer from atrazine in rats do not exist in humans and that high rates show up among their workers only because the company is so vigilant in screening for cancer, though recent studies seem to contradict Syngenta and the EPA.

In 2003, in the same month that the European Union banned atrazine, the EPA, at the end of a nine-year-long review, decided to keep it on the U.S. market with no new restrictions. Before arriving at that decision, the EPA had held about fifty private meetings with Syngenta and consulted two advisory committees composed of only Syngenta and EPA representatives.

Nonetheless, when evidence surfaced that atrazine was turning male frogs into hermaphrodites, the EPA asked for more studies. That led Syngenta, through its crop protection division, to award a contract of about $2 million to Ecorisk, a company that had regularly worked to defend Syngenta products. In 1998, Ecorisk contracted biologist Tyrone Hayes to work on this task. He had patented a technique using amphibians to detect hormone disruption, because the hormones and mechanisms of sexual development in frogs are similar to those in humans. Hayes, whose interest in frogs dates back to his childhood in the marshy backyard of his home in South Carolina, was at age thirty the youngest tenured full professor in the history of the University of California at Berkeley.

His Ecorisk contract contained a clause prohibiting him from publishing his findings without their approval, a clause common among product defense companies. Two years later, his studies showed that exposure to atrazine during their fetal stage at levels thirty times lower than currently permitted in water converted the frogs' male hormones to female, causing the male testes to produce eggs rather than sperm. Hayes says, "We showed that these animals are chemically castrated" and feminized. Syngenta cut off funding and interfered with publication of his data. At that point, in 2000, Hayes quit Ecorisk, concerned that his findings would not be published "in a timely manner"—that is, in time for the EPA to consider them in regulating the product. Hayes is now using his own patented technique in his own lab to study the combined hormonal effects on frogs of chemical mixtures, including atrazine, and finding harmful effects at even lower levels of exposure.

After Hayes quit, a campaign to discredit his findings and question his credibility and professionalism followed, including a broadside critique from the Kansas Corn Growers Association.

At the same time, Ecorisk with Syngenta funding hired a new set of scientists who conducted studies that found no risk from atrazine at the EPA-allowed levels. With these studies, they changed the weight of evidence and created doubt. Syngenta now asserts, "We have a database of nearly 200 studies conducted on atrazine during the past 10 years," and complains that it is "egregious to fixate on the view of one scientist." The company avows its "commitment to sound science."

Still worried about the hermaphrodite frog problem, Syngenta turned to Jim Tozzi for help in influencing the way the EPA would regulate atrazine. Tozzi's weapon of choice was the very Data Quality

Act he helped create. He filed the first challenge under the act, questioning the validity of Hayes's studies. In 2006, reacting to Tozzi's challenge, the EPA said it could not validate Hayes's studies because it did not yet have an accepted test for measuring hormone disruption. And thus the agency reaffirmed its earlier decision.

Tozzi's most recent challenge struck at the National Toxicology Program, which evaluates the toxicity of some of the chemicals in use today and has consistently been one of the most thorough and impartial science arms of the federal government. Under the Data Quality Act, Tozzi sought to bar this program from continuing to evaluate the cancer-causing potential of atrazine, claiming it would be a waste of resources since it had already been reviewed and given a pass by the EPA. His efforts delayed the review for a substantial period of time, though it eventually went forward.

WHAT SCIENTIFIC CONTROVERSY ACCOMPLISHES

The defense of a chemical, such as atrazine, by researchers working on behalf of the chemical maker has become common practice. "Traditional covert influence of industry has turned brazenly overt in the last several years," warns James Huff, associate director for the study of cancer and chemicals at the National Institute of Environmental Health Sciences.

Made-to-order science serves many goals: It arms companies with the research and expert testimony to delay or derail regulation, to overwhelm families who bring lawsuits against them and influence verdicts, and to mollify the fears of communities faced with pollution and of consumers faced with toxic products.

"Our company [has] one goal in mind—creating the outcome our client desires," pledges the Weinberg Group, the company that had proved its talents running tobacco's White Coat Project. The firm boasts that it "successfully guided clients" manufacturing Agent Orange out of regulatory trouble; it has defended asbestos and chlorine-based compounds; and it has campaigned on behalf of the phthalate family of chemicals and bisphenol-A (both used to make plastics) against evidence that they can disrupt human hormones. DuPont hired the firm to help cope with the cases of birth defects and cancer in the West Virginia town where its Teflon production facility is located and to counteract the science presented for the EPA's pending assessment of the Teflon chemical PFOA (perfluorooctanoic acid). In a letter soliciting DuPont's business, the Weinberg Group offered to "reshape the debate . . . to establish not only that PFOA is safe . . . but it offers real health benefits."

Industry funding does not in itself necessarily undermine science. It would be an error to categorize researchers simply by where they work or whom they associate with. There are scientists who work in partnership with industry for the advancement of science and to uncover new knowledge wherever they find it. There are conscientious scientists who think that the risks associated with many chemicals are relatively low or even exaggerated. There are scientists who say that money has no influence on their work. One such scientist is epidemiologist Kenneth J. Rothman, who says of his work consulting for industry, "You're not biased if you're correct."

Scientists working for manufacturers do not, however, on the whole bite the hand that feeds them. "Studies of potential occupational and environmental health hazards that are funded either directly or indirectly by industry are likely to have negative results"—that is, they report no harm, concludes Lorenzo Tomatis from his years of experience as longtime director (now deceased) of the International Agency for Research on Cancer, which studies and evaluates chemicals suspected of causing cancer. Various reports support his conclusion. In one, journalists Dan Fagin and Marianne Lavelle looked at the correlation between funding sources and the conclusions of all the studies to date of four particularly egregious chemicals—atrazine, alachlor, formaldehyde, and perchloroethylene. They found that only 6 industry-financed scientific articles mentioned potential harm, while 32 gave favorable opinions of the chemicals; in contrast, among the articles financed by independent sources, 71 studies uncovered potential harm, while only 27 indicated that the chemicals were benign.

In another study of industry influence on science, endocrinologist Frederick vom Saal, the nation's foremost expert on the plastic ingredient bisphenol-A, reviewed the 115 published studies of the effects of low levels of the molecule on lab animals. He found that 90 percent of the 104 independent research studies concluded that bisphenol-A altered brain chemistry or had some other harmful effect; in contrast, all 11 studies paid for by plastics or chemical companies absolved the chemical of any harm. Similarly, when the American Plastics Council gave $600,000 to the Harvard Center for Risk Analysis to review 19 studies of bisphenol-A, the center reported that none of them had found any unusual risk, though three of the reviewers refused to sign the report.

WAYS TO SKEW STUDIES

"There is sometimes fraud in science. . . . But I am convinced that deliberate distortion of science without lying is far more common, and far more serious in its impact and implications for our country, our posterity, and ourselves," writes John C. Bailar, a scholar at the National Academy of Sciences, in a journal article, "How to Distort the Scientific Record Without Actually Lying."

Though science seems driven by facts, there are, as Dr. Bailar warns, many ways to distort the outcome. Studies can focus only on adult male workers, without considering the probable increased sensitivity of children and women. Studies can fail to look for the genetic variants that make an often substantial portion of people vulnerable, though others will not be. They can select sampling sites, such as the least polluted among a set of wells, to minimize finding traces of the chemical under study. Researchers can mix exposed people in the supposed uncontaminated control group or can mix both exposed and unexposed community residents or workers in the test group. The scientist hired by GE to report on possible illnesses among workers in their PCB-manufacturing plant in Pittsfield, Massachusetts, added a group of clerical workers who worked isolated from the chemicals into the supposedly exposed group; the report gave the plant a clean bill of health.

Hired scientists can do sloppy work, as did the consulting firm Ecorisk in studies of the effect of atrazine on frogs, according to the EPA. In at least two of the studies, the "unexposed" group of frogs had actually been exposed to atrazine—it seems the frogs hopped between treated and untreated tanks of water. In another case, 80 to 90 percent of the frogs died, apparently as a result of poor care, so these tests failed to reach any conclusions.

Studies can choose to look at lab animals at a stage of their life when harm is less likely to occur. In an industry study of phthalates, researchers reported that no effect occurred to marmosets exposed and then examined when they were already in puberty. But with these chemicals "it's the developing organism that's affected," Dr. Ted Schettler explains. Or scientists can literally bury the evidence, as Dr. Schettler found when, after obtaining the raw data under the Freedom of Information Act, he discovered that the animals that showed any impacts had been discarded.

Researchers can select the lab animal sure to prove their point. A study of bisphenol-A commissioned by the American Chemistry Council, for example, selected a type of rat that had been bred over hundreds of generations to be extraordinarily insensitive to estrogen hormones, exactly the hormone disruption that bisphenol-A seems to cause. And researchers can misrepresent. Gilbert Ross, medical director of the American Council on Science and Health, in blasting the studies of bisphenol-A by vom Saal and others, claimed that "almost all of these scares are based on tests of rats . . . fed huge doses . . . then extrapolated to humans and used by activist groups and their public relations machines as a basis for scare campaigns." To the contrary, the bisphenol-A studies are "scary" exactly because it is at extremely low doses of the levels to which most Americans are exposed that adverse effects occur in lab animals.

With a disease such as cancer, a study can be conducted over so short a time that illness doesn't have the chance to show itself. Or cancer studies can count only deaths, not incidences, since almost half the people with cancer live for at least five years after treatment. Or select a certain study time frame. Dr. Kenneth Rothman, hired by the butadiene manufacturers during a lawsuit brought by families of students with cancer in Port Neches, Texas, who attended the high school next door to the plant, ran a study that counted only students who had died, not those surviving, and also excluded students who died of cancer but had attended the school before the years he chose for the study.

These studies have one thing in common: The chemical in question stays on the market. As Bernard Goldstein, former dean of the University of Pittsburgh's School of Public Health, ruefully quips, "Industry and government have a button on their laptops that says 'More research needed.' "

Corporate science has other, equally powerful weapons that go beyond skewed studies. To influence public opinion and regulatory decision making, it is seeking to redefine risk itself, calling this new strategy "good epidemiological practice." After writing that lung cancer "was very likely caused by large, relatively uncontrolled sources of community air pollution, particularly arsenic-emitting smelters," the American Council on Science and Health concluded, "However, the risks for community exposure are likely to be quite small, between 1.5 and 2.0 [times the normal rate]."

In the context that smokers face a ten to twenty times higher risk of lung cancer than nonsmokers, industry is trying to create a mind-set that a doubled risk isn't so bad. If 1.5 or 2.0 times the normal rate becomes unexceptional, then when chemicals exuding from a waste dump are found to raise the likelihood of childhood leukemia by that magnitude or less, corporations can discredit that rate of increased risk as unimportant.

It has proved difficult to remedy the problems of science-for-hire. There are no procedures for disbarring scientists. It took over four decades, until 2006, before a federal court found the major U.S. tobacco companies guilty of committing fraud, conspiracy, and racketeering through their efforts to create doubt in the public's mind about health harms from tobacco smoke. The court ruling focused on the fact that tobacco industry funding of research at prestigious medical schools and universities was essential to its illegal enterprise. The same might be said when it comes to other environmental pollutants.

PRODUCT DEFENSE IN ACTION

The march of scientists into the arms of industry is driven by dwindling grant support from the government, by researchers' dependence on industry for information since it's the corporations that generate the data and hold them secret, and by the fact that industry is turning increasingly to outside researchers. Even corporations with their own in-house laboratories are following the trend toward outsourcing, explains Dr. Carol Burns, an epidemiologist on the staff at Dow Chemical's Midland headquarters. She and only two other epidemiologists and a statistician serve Dow in-house worldwide.

There are some scientists and consulting companies with particular experience on behalf of corporations. Exponent is one such company. It is traded on NASDAQ and has a staff of approximately eight hundred covering seventy scientific and engineering disciplines, located in eighteen offices throughout the nation and three overseas. Many of its staff are unimpeachable scientists, concludes David Michaels, a former high official in the U.S. Department of Energy who has researched the question of "manufactured doubt" in science. Some may be attracted by earnings far above the norm in the nonprofit and public sectors; top Exponent staff members earn over $1 million a year. The company's early life was nurtured by contracts from Philip Morris and from auto companies fighting accusations of auto safety failures. Though Exponent will not divulge the names of its clients, they are easy to find throughout the scientific literature. They include over fifty top corporations and trade groups such as Dow Chemical, DuPont, DaimlerChrysler, Exxon, Ford Motor Company, GE, Honeywell, the American Chemistry Council, the National Mining Association, and the American Petroleum Institute, as well as law firms that represent corporate defendants faced with toxic lawsuits. Exponent's government clients include the U.S. Army and the Department of Energy; this is the company those departments hired to fight against a more stringent standard for the water-pollutant perchlorate.

Exponent often provides expert witnesses for companies faced with lawsuits over pollution. Pacific Gas & Electric (PG&E) hired it recently to fight the lawsuits brought by hundreds of residents of two California counties who claimed that the chemical chromium 6, leaching into groundwater from the company's nearby power plants, caused their families' cancers and other illnesses.

Dr. Dennis Paustenbach, then an Exponent vice president, headed the team. He was well versed in chromium 6, having consulted for PG&E in an earlier, eerily parallel lawsuit involving a different California community, the lawsuit portrayed in the film *Erin Brockovich*.

Paustenbach's strategy was the usual product defense: to manufacture doubt that the evidence was adequate to prove definitively that chromium 6 in water could cause such harm. Appointed simultaneously as a member of a panel that the state of California had set up to look into chromium 6's toxicity and to recommend possible changes in the state's regulation of the chemical, he wrote most of the panel's report, including a chapter exonerating chromium taken, almost verbatim, from material he had written for another chromium-industry client. He also used his position to recommend scientists who had backgrounds working for industry on chromium as advisers to the panel. Plaintiffs charged that, through these maneuvers, PG&E manipulated the panel and concealed key evidence, leading to a loosening of the California standards.

One of the scientists Paustenbach recruited as a paid expert for PG&E was Dr. Philip Cole, who heads the Epidemiology Department at the University of Alabama at Birmingham's School of Public Health; he and his department members are frequently hired by chemical companies. Among his duties for PG&E, Cole was paid $600 to watch the *Erin Brockovich* film. Cole's viewpoint is clear: He has written that cancer rates in the United States have declined over the past fifty years, except for lung cancer.

Paustenbach and his group also conceived the idea of retaining a Chinese scientist (for $2,000) who had studied and published about an area in China where chromium 6–polluted drinking water was associated with high cancer rates, and having this by-then-ten-year-old study ghost-*re*written, with PG&E funding. This new version, however, reversed the scientist's original conclusion. Paustenbach then had this version presented to the California panel and had it published in a scientific journal under the Chinese scientist's name, without any indication of PG&E involvement and secret funding. The journal retracted the ghostwritten article once the improprieties were discovered years later.

Paustenbach, who had earned over $7 million consulting for various corporations on chromium over twenty years, was eventually voted off the panel for his conflict of interest. He resigned simultaneously, so he would not need to disclose the panel's decision. Asked during his deposition whether he thought the money he had earned was a conflict of interest, he responded, "Of course not. Frankly, in the world of conflict of interest in the scientific community, the amount of money is irrelevant." Exponent continued to lobby for industry with PG&E money even after Paustenbach's departure. Both the Erin Brockovich and the more recent lawsuits were settled out of court, the former for $333 million and the second one, in 2006, for $295 million.

INFLUENCING REGULATION

Product defense also involves influencing regulation. Paustenbach and his sometime employer Exponent have exercised this role, too, for example, during the EPA's long deliberation about how to regulate dioxin. In 1985, the EPA had rated dioxin one of the most dangerous carcinogens . . . until challenged by industry to reexamine its assessment. The agency did so twice; their latest analysis judged the risk as ten times greater than its previous assessment. But before it could complete its work, the EPA appointed an advisory subcommittee of twenty-one outside experts, including Paustenbach. Among the twenty-one, six had financial ties to companies whose products generate dioxins. These scientists consistently tried to subvert the panel's work, according to panel member epidemiologist Richard Clapp, of Boston University's School of Public Health.

Paustenbach repeated the strategy he had employed on behalf of PG&E and inserted material into the draft EPA subcommittee report denying dioxin's ability to cause cancer at low doses. This material came from a paper he helped write for Exponent, paid for by the U.S. Chlorine Chemistry Council (as mentioned before, dioxin is a by-product of chlorine).

The occupants of academic ivory towers sometimes behave like those in the offices and labs of consulting firms. "The last quarter of the twentieth century has seen the commercialization of university science become more aggressive and more pervasive," concluded physicist and philosopher Sheldon Krimsky from his investigation of academic conflicts of interest.

An occupant of one of those ivory towers is the Harvard Center for Risk Analysis. The center, asserting that the Chernobyl nuclear reactor meltdown in 1986 resulted in 2,000 deaths from thyroid cancer in Russia, compares this number to the 7,800 skin cancer deaths a year from sun exposure in the United States (a questionable figure); that assertion overlooks the tenfold increase in thyroid cancer among Ukrainians exposed as children and adolescents—fifty thousand new cases of thyroid cancer are projected—and the vast number of deaths in Europe from the fallout. The center's funders are many of the nation's Fortune 500 companies. Its founder, John Graham, was selected by the George W. Bush administration to head the department within the federal Office of Management and Budget that controls the way federal agencies apply the principles of regulation and risk assessment, including the Data Access and Data Quality acts.

Early on, the center provided its gold seal of academic prestige to the tobacco industry. Graham took the position that secondhand smoke was relatively benign, at the same time soliciting money from Philip Morris. When the company sent a contribution, Graham returned the check, asking instead that "an arrangement be made with Kraft General Foods," a Philip Morris subsidiary, and explaining the center's interest in "the challenges in food safety and pesticide regulation." Kraft did make a gift. The center's newsletter has downplayed the effects on children of exposure to pesticides and ingredients in plastics, while taking funds from the makers of those chemicals. With about $500,000 from the industry's Styrene Information and Research Center, it studied styrene, an ingredient in the manufacture of rubber polymers, and concluded that concern was unwarranted, though more research was needed. In contrast, the EPA and the international agency that assesses potentially harmful products rate styrene as a possible human carcinogen.

When the Food Quality Protection Act was under congressional consideration, the American Farm Bureau Federation, a trade organization lobbying against the act, contracted with the center for an economic analysis of its potential effects. The center's resulting report asserted that the cost of compliance would drive food costs so high that poor people would have to cut back their purchases, resulting in one thousand added deaths from malnutrition a year.

The University of Alabama is another academic magnet for corporations seeking scientists to participate in research and lawsuits. Dr. Elizabeth Delzell, an epidemiologist who holds a professorship in Philip Cole's department at the university, has a history of corporate service. Over a period of twenty-one years, she and one or another collaborator were hired by Ciba-Geigy and Shell Oil for different studies of worker health in their manufacturing facilities in Toms River, New Jersey. A finding of elevated patterns of cancer among workers, attributable to the plants, would have strengthened the connection between the cancers among the town's children and the toxics used in the plants. Delzell never found any such evidence. Her studies concluded that any elevated rates of various cancers were probably related to the workers' former employment in a plant in a different city, and/or overall there were no elevated cancer rates, and/or cigarette smoking appeared to be the major cause of whatever rates were elevated, and/or there was no clear evidence of a causal relationship between the chemical used in the plant and any specific type of cancer, and/or "the small study size necessitates interpreting the findings cautiously."

IBM also hired Delzell and Exponent for assistance in lawsuits brought by over 250 employees at plants in San Jose, California, that make microprocessor chips in special rooms. Similar lawsuits are active in IBM plants in East Fishkill, New York, and Essex Junction, Vermont. Her job was to do epidemiological research; Exponent's was to give expert opinion testimony. The workers claimed the "clean rooms" are clean only for chips but deadly for employees who spend every day inhaling the vapors of the cleaning solvents into which they dip the wafers to process them. They say that IBM knew these dangers for decades. Clean room veterans seemed to suffer from high rates of certain cancers, while other IBM families gave birth to children with terrible birth defects—a skull too small for the child's brain, blindness, fused ribs, missing kneecaps. These solvents belong to the same group of organic solvents as the TCE found in Dickson, Tennessee, and Woburn, Massachusetts.

The San Jose IBMers' suit claimed their chemical exposure caused their cancer. It was the first toxic tort case involving the microelectronics industry to go to trial anywhere in the nation. Richard Clapp, the Boston University epidemiologist, was hired on behalf of the workers to analyze the Corporate Mortality File records tracking employee deaths that IBM had secretly kept since 1969. When he cross-referenced these files with worker histories, he found the proportion of deaths from brain and kidney cancers and non-Hodgkin's lymphoma among manufacturing workers was more than double the national averages, and they were far younger than the norm when they contracted cancer. Court orders sought by IBM blocked the publication of Clapp's study and its use in the San Jose trial and from any public view.

Delzell studied the cancer rates, too. But she concluded that cancer among IBM employees was 16 percent lower than average, and the overall death rate was 35 percent lower. Critics of her study point out it compares IBM workers with the populations of California and New York as the control group, which, since this group includes everyone, also includes the elderly and people too sick to work. This is a well-known way, "the healthy worker effect," to bias a study's outcome.

IBM also hired an Exponent vice president to critique Clapp's study. He did this vigorously, repeatedly labeling Clapp's study as "junk science," the term popularized by the tobacco industry to denigrate critics. With Clapp's data kept from the jury, the test plaintiffs in the first California suit lost their cases, in March 2004. Since then, IBM has settled out of court with a number of the other California and New York families of children with birth defects and of dead employees. Even with the litigation behind them, IBM continued to fight to keep Clapp's study from public view, though ultimately he won the right to publish.

RESTRAINING TOXIC LAWSUITS

Families say that they bring lawsuits against corporations as the last resort. But a recent Supreme Court ruling will make it harder for them to win. The ruling arose from a lawsuit that William Daubert, on behalf of his wife, and Betty Mekdeci brought against Merrell Dow Pharmaceuticals, Inc. They claimed that Bendectin, the company's drug to control pregnancy's morning sickness, caused their babies to be born with birth defects. Though the jury found that the company had acted fraudulently and negligently, and awarded $19.2 million to the plaintiffs, Dow appealed the case to the state court and finally to the Supreme Court, which reversed the decision, and the Daubert precedent was born. After this dispiriting court experience, Betty Mekdeci turned to citizen action for redress and founded a birth defects research center.

Daubert requires the judge to examine the underlying research that the plaintiffs and defense propose to present and admit only evidence he or she deems scientifically reliable and relevant to the case. This process effectively usurps the jury's role in assessing testimony, plaintiffs say. Often the only evidence that remains allowable are epidemiological studies, which are usually funded or strongly influenced by industry, according to Mekdeci. Since Daubert, the number of all tort trials has decreased steadily and summary judgments that dismiss a case have risen.

The lower court judge in the Bendectin case wrote, "President Eisenhower warned the nation of a military industrial complex. This case clearly demonstrates what a medical industrial complex can accomplish. The testimony exposed scientific literature created for purposes of legal defense. . . . The testimony revealed a sycophantic relationship between 'scientists' and their funding source—one of the main scientific witnesses, [Dr. Robert Brent] was a retained expert for Merrell Dow for 18 years which clearly affected his objectivity. . . . [The testimony revealed] the use of litigation funds for scientific research manipulation and the factual editing of supposedly scientific research literature by the very lawyers defending in litigation."

Daubert may prove to be one more tool the medical industrial complex can use in claiming "sound science," to exclude incriminating evidence from a court proceeding and to delay or halt cases against corporations. Because the burden of proof is on the plaintiff to prove the defendant's fault, the judge can disallow a plaintiff's evidence more readily. Daubert also raises court costs by hundreds of thousand of dollars for plaintiffs with limited funds, while corporate defendants can hire teams of lawyers and scientific experts and charge them off their taxes as business expenses. "These requirements favor the powerful in our society over the weak and vulnerable," concludes David Michaels.

The notion that tort cases are out of control and in dire need of reform springs from an opinion-molding campaign of the corporate sector and its lobbying arm, the Chamber of Commerce, which spent $24.5 million of its $150 million annual budget in 2004 trying to curb class action lawsuits. But defenders of tort law assert it is a critical instrument to bring equality and honesty to a free enterprise system.

W. R. Grace cast blame on tort trial lawyers for "clogging up the legal system" during a lawsuit brought by the families of Libby, Montana, who had become ill with lung diseases or died after decades of inhaling asbestos dust from the company's vermiculite mine. When they sued the company, they found that Grace had moved $4 billon of its assets "out the back door" over a period of twenty years into other W. R. Grace companies formed for just this purpose, then had re-formed a new W. R. Grace with the scant remaining funds. The new W. R. Grace then filed for Chapter 11 bankruptcy protection. The victims collected a few pennies on the millions due from the settlement. In Dickson, Tennessee, the families suing Scovill-Schrader for contaminating their water and causing their children's birth defects found that the company took the same path, stripping away its assets to leave the liability with a bankrupt company.

Tort litigation is also one of the few ways to bring corporate documents into the public domain. As Boston University epidemiologist

David Ozonoff sifted through letters and documents that came to light during court cases against the manufacturers of asbestos, he found evidence not only that corporate executives had known for decades that asbestos was dangerous, but also that they had outlined and put into practice a defense strategy to protect their product and company profits if and when the information came to light. Asbestos danger is not limited to workers; the fibers are carried home on their clothing, to seed an especially horrible kind of cancer among their children. The asbestos companies thought nobody would ever see the documents, Ozonoff explains. Similarly, the materials with their damning revelations of the tobacco industry's tactics are open to the public and researchers only as an outcome of lawsuits.

Most toxic tort lawsuits end with a settlement, as, for example, in Toms River, Anniston, and San Jose. The harm cannot be undone, and no actions take place to prevent similar harm from occurring to other children. The company pays the families a usually inadequate amount of money, while continuing to aver that the chemical in question is harmless and requiring the families to pledge never to discuss the details of the settlement. All the documents remain secret.

REVOLVING DOORS

Witnesses for the defense—scientists, lawyers, and administrators—move often and easily from jobs in government to jobs in for-profit consulting companies and the law firms and trade associations that serve industry, then back again. Like Jim Tozzi, they gather the insider knowledge, experience, and personal contacts in government that make them valuable on behalf of corporate interests.

Exponent recently added Elizabeth L. Anderson to their staff as a vice president. She had worked for the EPA for more than ten years, directing the agency's risk assessment programs and its first use of risk assessment in regulating toxic chemicals. When she left to head the consulting company Sciences International, she was hired by the tobacco industry to debate her former colleagues at the EPA over the "sound science" of the risk analysis that set the amount of the pesticide phosphine that tobacco companies would be allowed to use to fumigate stored tobacco. She was also editor in chief of the journal *Risk Analysis* at the time (and remains so still). On her own initiative, she offered to use her editorship there to "expedite the peer review process" at the journal of a study she and two coauthors had written that minimized the risk of that fumigating pesticide. Her coauthors were R. J. Reynolds Tobacco Company employees, and funding came from a tobacco-pesticide coalition. The purpose, as she wrote, was to show the EPA the paper to influence regulations. Sciences International is the very company that the National Institute for Environmental Health Sciences hired to evaluate the risks of bisphenol-A and then fired when the company's conflicts of interests surfaced.

This is an unexceptional example of the way the revolving door works. A study sponsored by the Center for Public Integrity was able to trace where half of the forty EPA officials who left top-level jobs in toxics and pesticides regulation during the past fifteen years went to work. Of those, eighteen went to work for chemical companies, their trade associations, or their lobbying firms; two went to work for environmental organizations.

THE COURT OF PUBLIC OPINION

Science is marshaled not only to influence regulators and judges, but also to sway public opinion, which in a democratic society is just as important. Science-for-hire is folded into the advertising and policy statements that make their way into public consciousness. Fabulous sums are spent on ads directed at consumers, with great effect, such as the $185 million outlay between 1952 to 1996 by the trade group the Chemical Manufacturers Association, now renamed the American Chemistry Council. Working their influence more subtly are the dozens of industry-funded nonprofit policy organizations that exist to offer "sound science" opinions on demand.

Notable among them is the American Council on Science and Health, which, though strongly antitobacco, consistently defends controversial chemicals. Was there ever any real health problem at Love Canal? asked the council's president rhetorically; then she answered herself, saying that the problem was the media-induced stress placed on residents. Similarly, she has asserted that there is no link between cancer and the PCBs now contaminating the Hudson River. The council wrote and published a book dedicated to debunking the notion that children are more vulnerable than adults.

The council was founded with and continues to receive a substantial portion of its funding from corporations; almost all the remainder comes from corporate-financed foundations. Its science director,

Gilbert Ross, was convicted of medical fraud, imprisoned, and labeled a "highly untrustworthy individual" by the judge in his criminal trial.

Add to the legions of nonprofit policy institutes that influence public perceptions the legions of front groups, which insiders call "grass-tops" or "Astroturf," to distinguish them from the genuine grassroots.

Burson-Marsteller, one of the world's largest public relations firm, staffs groups with names such as the Alliance for Fire Safety, which may sound like an advocate for consumers but is in reality a collection of brominated flame retardant manufacturers fighting a possible ban on their products because it's been found that they may interfere with human reproduction. Organized as the Council on Water Quality, a group of companies (which includes Lockheed Martin) is attempting to influence the EPA standard for perchlorate in water, as the previous chapter described.

The Center for Consumer Freedom is a group founded by a Washington lobbyist whose seed money came from Philip Morris. It now carries on (against the organic food movement, among other targets) with financing from meat, fast-food, restaurant, and beverage companies. It has campaigned against the possible infringement of the right of Americans to have high fructose sugar in their soft drinks.

The language that groups like these employ is often hysterical and venomous. As *New York Times* editorial writer Verlyn Klinkenborg observed, such groups "blur the distinction between corporate interests and the individual and collective rights of humans. Their inflamed rhetoric assumes that your interests—as a free consumer—are synonymous with the interests of corporations. . . . In fact, the language is as Orwellian as it is possible to get. . . . It hides from people exactly how their interests are being abused."

INTIMIDATION AS A DEFENSE STRATEGY

Those who stand up for the right of children and society as a whole to live without unseen and unintended pollution are demonized, labeled as fearmongers, Luddites, "we know what's best for you" nannies, food (or such) police, extremists, environmental Chicken Littles, anti-science, and—that dirty word—activists. They are charged with hidden motives: advancing a political agenda or seeking political power, raking in a fortune in representing the injured, returning the country to a more primitive form of living, "triggering endless mischief," and "terrifying Americans, distracting us from the very real everyday risks around us."

Nasty epithets are minor nuisances, compared with the treatment given to activists, scientists, and writers who are too outspoken and too effective. They can become the targets of intimidation and reprisal. When one of the two authors of a recent impeccably researched exposé of the way manufacturers suppressed knowledge of the harm vinyl chloride caused workers testified as an expert witness, the defense lawyers subjected him to a five-day deposition. The lawyers also subpoenaed his coauthor, the experts who had reviewed their book for accuracy, and the publishers' documents about the peer review process. One of the subpoenaed experts described all these actions as unique in her experience. In a practice that began with the publication of Rachel Carson's *Silent Spring,* industry has conducted well-financed public relations campaigns to savage books and articles critical of their products and operations. A reviewer who signed his piece simply with his credentials as an MD and MPH wrote a scathing critique in the *New England Journal of Medicine* of a book by biologist Sandra Steingraber about the environmental roots of cancer. He was later unveiled as the director of toxicology for W. R. Grace, the company sued for polluting the water in Woburn, Massachusetts. And an article Steingraber wrote about mercury in fish, at the request of the magazine *Childbirth Forum,* was pulled just before publication. The editors informed her that "it cannot be used at this time based on a directive from the magazine's sponsor, Procter & Gamble, [because] it is too controversial." Procter & Gamble has been active in opposing mercury regulations, Steingraber explains.

The manufacturer of the insecticide lindane brought a libel suit against the small Ecology Center of Ann Arbor, Michigan, for writing that it made no sense that the EPA had withdrawn its approval of the chemical for agricultural and military uses yet the FDA allowed its continued use on children as a treatment for head lice. Lindane has been banned in fifty-two other countries.

Efforts to intimidate and silence critics through so-called SLAPP suits—strategic lawsuits against public participation—like the suit against the Ecology Center, have been used by corporations against scientists, nonprofit groups, journalists, and others without the resources to defend expensive lawsuits. A number of states have reacted to this practice by adopting "SLAPP back" legislation.

University of Pittsburgh researcher and physician Herbert Needleman was the first, but not the last, to face an industry harassment strategy that demands the researcher's raw data—in his case, data on the toxicity of lead. Just recently, paint manufacturer Sherwin-Williams, fighting a lawsuit brought by the state of Rhode Island for the harm their lead products had caused, used the same strategy, subpoenaing raw and even unpublished data sets from several researchers. (They lost the subpoena and lawsuit.) Jim Tozzi tried it on behalf of Philip Morris and secondhand smoke, calling for a scientist to release her data so that his firm could subject what he claimed were her "various analytical and statistical errors and bias" to "objective scientific analysis."

DOWNPLAYING HARM IN THE ARENA OF PUBLIC OPINION

Minimizing and denigrating environmental risk through the repeated set of specific assertions also molds opinion. The first of these allegations is that the dangers of toxics are relatively minor compared with the greater dangers of smoking, obesity, sunburn, and other lifestyle self-inflicted illnesses.

Dr. Joseph K. McLaughlin, president of the International Epidemiology Institute, a biomedical research center in Rockville, Maryland, gave this response when *The New York Times* asked his reaction to an *American Journal of Epidemiology* study discussing a possible link between hair dye and cancer: "If the results were true, and that's a big if, it would mean that, in the grand scheme of life, using hair dye may present a remote risk to your health, but it would still be less risky than crossing the street, driving a car, not wearing a seat belt, or drunk driving."

It's true that children face hazards such as those in their lives, and Dr. McLaughlin's statement is accurate. But this argument is often disingenuously championed to keep the public's mind off the hazards of environmental pollutants by those who deny environmental toxins can pose any harm. Dr. C. Everett Koop, former U.S. surgeon general, often states that "the enemy is not tiny amounts of chemicals that have proved safe over many years." He chastised both a division director of the National Institute for Occupational Safety and Health in 1997 and a committee of the House of Representatives in 1999 for their "borderline hysteria" in proposing to warn health workers that latex gloves could cause serious, even life-threatening, allergic reactions. Today, as a visit to any hospital will demonstrate, latex has indeed been identified as a potentially harmful substance. Subsequently, it was revealed that Dr. Koop had a $1 million contract with a leading manufacturer of latex gloves and that the CDC study he had cited as the alleged basis for his testimony had never been done.

This denial of potential harm fails to recognize the compounded and often synergistic risks of exposure to multiple chemicals. Yes, having your hair dyed may indeed present a minuscule risk. How about if you add that thimbleful of dye to another tiny amount of, let's say, chemical fragments from a disintegrating plastic water bottle and mix in a fraction of pesticide residues from some apples, and phthalates in your lipstick, and a couple of grains of dioxin from your steak? How about if you are pregnant? How about if you are one of those people with a genetic variation that leaves you more vulnerable and any damage done can be passed on to your children? How about if your six-year-old ingests, inhales, and absorbs some of those tiny yet cumulative amounts during the course of his or her normal day? Nor should it be forgotten that the majority of chemicals are in use without being proved safe.

Another commonly heard assertion is that the withdrawal of a chemical will undermine the American way of life. The urban myth that grew up around "the Alar scare" belongs to this category. As perpetuators of such myths tell the story, "media hysteria" spread "misinformation" about Alar, a chemical once routinely sprayed on apples to ensure their uniform growth for ease of mechanical harvesting. Apples make up a large part of children's diet, so the issue took on special significance. "It was misinformation," the myth alleges, to call Alar a carcinogen, alleging that "studies later confirmed that Alar was safe." The myth further alleges, "It took years for the apple industry to undo the damage." Growers foretold the ruination of their industry. The truth is that the EPA did classify and continues to classify Alar as a carcinogen and never confirmed it was safe. The truth is that publicity about Alar's health risks did cause a sharp drop in apple sales for two years, but then the public resumed eating as many apples as before; and the truth is that the industry is just as profitable as pre-Alar, according to its own trade association.

A third assertion is that similar products occur in nature; therefore, man-made, synthetic chemicals cannot be harmful. According to this thesis, all living matter is made of "chemicals." Dioxin, for example, spews forth from volcano eruptions, from forest fires. Potatoes naturally contain arsenic, our bodies produce some bits of formaldehyde. Carbon dating of chemicals in whales allegedly found some that were identical to synthesized flame retardants, a science writer reports; he concludes, therefore, that these are natural compounds, possibly originating in marine sponges, and if so, man-made flame retardants must be harmless.

The fallacies in this assertion are many. Our bodies have had aeons to adjust to naturally occurring toxics. We (and other creatures) produce enzymes that can dismantle natural toxicants; synthetic toxic chemicals accumulate, while naturally occurring toxins can be metabolized and excreted. And as pointed out by John Wargo, Yale University professor, it makes no sense to add man-made harm to harm that may occur from nature.

MONEY AS THE ULTIMATE DEFENSE TOOL

When all is said and done, money remains the most powerful witness for the defense of possible pollutants. For this reason, petrochemical corporations pour hundreds of millions of dollars, both individually and collaboratively, into the coffers of politicians at all levels of government every year, much of it through their trade associations. The makers of chemical, biological, and bioengineered weed-and-seed products for agribusiness, for example, make up CropLife America, which distributed nearly $9 million to congressional campaigns between 1987 and 1996. The top contributors were Dow at $2.5 million, followed by FMC at $1.5 million.

The manufacturers of pesticides for home and garden have their own lobbying group with the comforting name of the Responsible Industry for a Sound Environment (RISE), whose thirty-six members include Monsanto, DuPont, and Dow, among others. RISE and its member firms spent more than $15 million in 1996, the year the new Food Quality Protection Act was under debate, to employ 219 lobbyists, including 69 former congressmen and staffers and federal officials. The cosmetics trade association spent $3.5 million over two years to defeat safe cosmetics bills in California.

The fate of just one chemical can inspire an industry lobby. When the EPA was reviewing the best-selling weed-killing herbicide 2,4-D, used in products such as Dow's Weed B Gon, to decide whether or not to allow the chemical to remain in use, its manufacturers formed the Industry Task Force on 2,4-D Research Data. This coalition spent at least $34 million on more than three hundred studies and surveys to present to the EPA to influence the outcome. The group is still intact, and the chemical, suspected of causing non-Hodgkin's lymphoma and one of the toxins highlighted in *Silent Spring* forty years ago, remains in use today.

Corporate contributions clearly, and not surprisingly, reward legislators who support corporate interests. The contributions are proportionate to the legislator's power on specific issues and his continued adherence to industry's position. For example, corporations generally favored the Clear Skies Act, which, contrary to its name, loosened their obligations to reduce air pollutants. In late 2005, in a Senate committee vote, senators who voted for the act received an average of $381,369 in contributions that year from companies that would benefit from the legislation, compared with $126,839 in contributions from such companies to senators voting against it. The two sponsors of the Clear Skies Act in the Senate, James Inhofe (R-OK) and George Voinovich (R-OH), netted an average of $614,986, almost twice as much as other, less powerful senators voting for the act and almost five times as much as those voting against.

Similarly, oil industry contributions correlated with how members of the House of Representatives voted on the proposed Gasoline for America's Security Act of 2005, which offered incentives for the construction of new refineries. In the year before the bill came up for vote, the representatives later voting for the bill received a total of $2,231,520 from companies that would benefit from such legislation, in contrast with the $411,462 (or 18 percent) contributed to those voting against the bill.

The public good is overwhelmed by private gain when money speaks with such a loud voice.

By manipulating public opinion, influencing regulation and legislation, and overpowering the legal system, while controlling the manufacture of chemicals and the only real knowledge of their structure and safety, the petrochemical industry has striven to make Americans passively overlook or accept a degraded environment.

But the power of a parent's concern, of a scientist's integrity, of the transformation that comes from knowledge, breaks these unseen chains. And this is what is happening across the nation. Environmental health is David to the industry's Goliath. Don't forget that David prevailed.

Ten
_____

POSSE COMITATUS

ELISE BLAXILL AND HER DAUGHTER MICHAELA PLAY A GAME TO-gether. Elise, a slender, auburn-haired woman, places a couple of pillows just inside the door of their home, an expansive, beautifully furnished house off Brattle Street in the most elegant section of Cambridge, Massachusetts. She sits on the pillows, her back against the door, and tells her nine-year-old daughter, also fair-haired and pale-skinned, to stand at the other end of the long entrance hall. She instructs Michaela to walk toward her when she smiles but to stop walking when she is not smiling.

Michaela stands expectantly at the far end of the hall. When her mother smiles she moves forward, then stops immediately when the smile disappears. When she finally makes it to Elise, she throws herself into her mother's arms, hugging and kissing and giggling. The hair of mother and child mingle in a glowing auburn cloud.

It is a simple game. But it represents remarkable progress on what has been for the Blaxill family a long, arduous, and often anguished journey.

For the first year and a half of her life, Michaela was a bright, happy child. Then things began to change. She started losing words she had already learned. She no longer would look at her parents. She exhibited unusual behavior patterns such as compulsively lining up objects. She did not look around when her name was called. "It was clear," said Elise, "that things were going terribly awry."

Elise and her husband, Mark, took Michaela to Children's Hospital in Boston for a full workup. "The doctor gave us the bottom line," said Mark. "Michaela was diagnosed with PPDA"—pervasive developmental disorder, also called autism spectrum disorder. "The doctor said, 'Go get a blood test and come back in six months.' He gave us some pamphlets to read."

That was it, according to the Blaxills. No explanation about possible reasons for the autism, no treatment, no course of action was suggested, no hope offered.

It got much worse. Michaela no longer recognized or communicated with her father. She began biting herself, banging her head against the hardest objects she could find. She would spend hours rocking herself back and forth; sometimes she would stand on her head. Some of this behavior, the Blaxills would learn, was because Michaela was trying to relieve the intense gastrointestinal pain that was a symptom of her autism. But her parents did not understand at first. No one told them.

"It was a nightmare," said Elise. "I would wake up every day and want to throw up. It was like having a child that was growing up naturally and then the Wicked Witch came and put a spell on her. You know she is still there and you have to find all the incantations and all the wizardry to get her out. She is behind twenty doors and you have to unlock them all."

Without much help from the medical or scientific communities, the Blaxills began to look for their own keys. Both Elise and Mark are highly intelligent and well educated. Elise is an art consultant, Mark a senior vice president of a Boston-based consulting firm, with degrees from Princeton and Harvard Business School. They are fortunate enough to have the resources to mount an intense effort to rescue their child.

The first thing Mark did was to get on the Internet, where he found thousands of American parents of autistic children searching for answers, looking for and offering support, demanding help.

"We found this subculture out there," Mark said. "It was not the world that Children's Hospital was sending us to, but it was kooky things like alternative medicines, Listservs, and websites. It is a messy, hurly-burly world." Mark calls this world "the bazaar," an unruly marketplace of information and ideas. He distinguishes it from the world of orthodox science and medicine, which he describes as "the cathedral."

When it comes to autism, he said, "The cathedral scientists have nothing to do. They say, 'It's genetic; live with it.' Or they offer the most miserable form of punishment—applied behavioral analysis, as if nothing in the brain is working except the most primitive signal and response parts. . . . In the bazaar you get vitamins, gastrointestinal things such as ways to deal with the child's diarrhea. It addresses their whole world of dysfunction, the larger systemic context of the child's illness, her yeast infections—the stuff that comes from the environment."

Mark turned a room of their house, painted a deep red, into what he calls his war room. Its shelves are lined with textbooks on biology, chemistry, and other relevant sciences. The floor is piled with monographs. (On a shelf is also a copy of John le Carre's novel *The Constant Gardener,* which is about alleged crimes of the pharmaceutical industry.) He has educated himself deeply enough to publish several papers on the epidemiology of autism in peer-reviewed scientific journals. He is also a director of a nonprofit organization called SafeMinds, which seeks to influence policy on mercury in the environment, which many parents, including the Blaxills, are convinced plays a role in their children's autism. There now are a number of volunteer organizations focusing on autism, including Defeat Autism Now! (DAN!), founded by the late Dr. Bernard Rimland, which is pressing for cures for autistic children as well as finding ways to prevent it. We sat in on a meeting of the New England chapter of DAN! attended by more than 1,500 parents, physicians, and scientists.

Michaela had been inoculated with the usual childhood vaccines to which mercury had been added as a preservative. Mark first heard about thimerosal, the mercury-based preservative used in vaccines, from his online chat rooms. Initially, he was skeptical about claims that it was a cause of autism. Then he met Lyn Redwood, a nurse with an autistic child who did some dose calculations and found, he said, that on the day the child is vaccinated, he or she receives ninety times the permissible limit set by the Food and Drug Administration. The Centers for Disease Control and Prevention, companies that make or use thimerosal, and the Institute of Medicine have issued reports saying there is no basis for suspecting the preservative plays a role in autism. There also have been studies that indicate that exposure to mercury causes significant damage to brain function, particularly exposure in fetuses and children.

Autism is clearly marked by an inflammatory process going on in the brain, Mark said. "We know that methylmercury triggers inflammatory processes in the brain. . . . If I had to bet money today, I'd bet that mercury in vaccines is the responsible agent." And some of the parents wonder why, if mercury in vaccines is harmless, members of Congress surreptitiously slipped into a national security bill provisions exempting from legal liability companies that make or use thimerosal.

Using what they learned in the bazaar and from the literature, the Blaxills began a long, slow process of healing their daughter. "Michaela is like a big onion, and you have got to peel away the layers one at a time," said Elise. "The last five years have been a gentle, rigorous, unrelenting peeling away of those layers. And that is really how we have done it. Nothing has been a silver bullet, the one thing that has made this child whole."

The first layer of the onion was Michaela's diet. They found that her gastrointestinal pains were relieved by feeding her a casein- and gluten-free diet. Elise gives Michaela special foods. She makes her waffles out of almond flour and carrots, gives her honey instead of sugar, and has her drink only pear juice. Michaela was given therapy for her hearing, for her speech, to help her integrate her sensory reception; she has therapy to help her work out the knots that form in her stomach. She was chelated—given medication to remove the mercury from her system. Elise "fought like crazy" to get Michaela into a school in Lexington, Massachusetts, that specializes in children with neurological problems. The caregiving is all-consuming. Elise herself takes fourteen vitamins a day because "I cannot get sick." For Mark, dealing with Michaela's autism is "a second career."

The Blaxills estimate it takes $100,000 a year to take care of their daughter. The city of Cambridge pays $50,000 a year for Michaela's special school. The money was well invested. Michaela is not completely healed, but it is evident that she has emerged from the dark, sealed chamber of pain and incomprehension into which she was cast by her autism.

While we were talking to her parents, Michaela returned with her counselor from one of her therapy sessions. When she first saw us, she covered her eyes with her hands. But when Elise introduced us, she gave us a sweet smile and said, "Hello, Alice . . . hello, Philip."

The great distance Michaela has traveled looks like a miracle. But it isn't. It is a result of the dedication, enormous labor, intelligence, and expense of her parents. It is also a result of their willingness to question and challenge the conventional wisdom about their daughter's illness, to regard current science and governmental policy based upon it with skepticism, and to keep an open mind about what could be done. Left without answers and without help, they took matters into their own hands.

ACTING FOR THE CHILDREN

The Blaxills are not alone in their activism. A growing multitude of Americans—unhappy with the indifference and arrogance of corporations, dissatisfied with the inaction and disgusted with the pandering of government, frustrated by the answers they are getting from the mainstream medical and scientific communities—are acting to find ways to heal our children and to protect them from the consequences of a degrading environment. Many, of course, are the families of sick kids. The parents and grandparents of children in Dickson, Tennessee; Woburn, Massachusetts; Toms River, New Jersey; and dozens of other communities around the country who believe that industrial poisons caused illness, pain, and suffering among their kids are the most dedicated of these activists. Like the Blaxills, they believe that the private and public institutions that are supposed to look out for the welfare of children are not doing so and are not paying attention to their complaints. Chloe Silverman of Cornell University, who examines the sociology of medicine, said that the current model of medical practice tends to marginalize the expertise and participation of parents in the treatment of their children. In part because of the commercialization of medicine, she noted, alternatives to the standard treatment of illness are often suppressed.

The activists also include doctors, public health officials, scientists, environmentalists, community organizers, academics, journalists, lawyers, labor union officials, some government civil servants brave enough to stand up to their political overseers, a few maverick politicians, and a very few members of the business community.

We think of these Americans as citizens on a mission to obtain justice for the children—a kind of *posse comitatus*. Of course, they are not like the posses of movie westerns. They usually have no official standing; no sheriff deputizes them. And unlike the movie posses, their membership is likely to include at least as many women as men. Their weapons are not six-guns and rifles, but research, information, organizing, political action, demonstrating, moral suasion, and searches for alternatives to the current economic, industrial, scientific, and medical practices that are harming our children. But the Latin roots of the term derive from the words for "power" and "county" or "community." This posse is composed of citizens of a democratic society acting as individuals and communities, often banded in volunteer organizations to correct a grave injustice afflicting our children.

When traveling around the country gathering stories for this book, we encountered and interviewed many members of this posse and learned from many more. And those we encountered are only a fraction of the small army of intelligent, intensely committed, sometimes angry citizens who are working to change a system that is harming our children. We tell only a few of their stories here, but the fact that there are so many of these bright, active Americans out there working to help their kids and ours is cause for hope in the future.

A SENSE OF DUTY

It is often the parents and family members of sick kids who join the posse. But, as we noted, there are many others. Some come just because they are called by their conscience, their sense of duty, their unwillingness to accept injustice.

One such is Hilton Kelley, a tall, muscular forty-four-year-old with closely cropped gray hair and beard who looks like an African American version of Russell Crowe in the movie *Gladiator*. Kelley grew up in Port Arthur, Texas, but moved to Hollywood, where he was having a successful career as a stuntman and actor. He worked in movies with Eddie Murphy, Danny DeVito, Ted Danson, and Robin Williams. While he was working on the *Nash Bridges* television series with Don Johnson, he took a trip back to his hometown for the annual Mardi Gras. He was depressed by what he saw. "Port Arthur was going downhill. All those companies were emitting tons of chemicals into the air."

Among the emitting companies were refineries and petrochemical companies run by Chevron Phillips, ExxonMobil, Precor, Motiva, Huntsman, and BASF. They were on the western side of the city, whose population is almost completely African American. The air was foul, the health of the populace poor, particularly among children.

On the plane back to California, Kelley experienced an epiphany. "It was a spiritual encounter so real that I had to come back and do something about it," he said. He dreamed he was standing on line on top of a cliff, and in the front of the line was Jesus Christ. "When I got to the front, he said something. I thought I would lift into the air; instead I started to descend. I grabbed the edge of the cliff and pulled myself up. I realized that this was not Jesus. He was trying to pull me away from God's work. I believe God wants the community to be saved, to be uplifted. I believe he wanted me to go back."

Kelley returned to Port Arthur and began a new career organizing the community, to reverse the decline of its environment. It is a formidable job. "Sulfur dioxide, benzene, butadiene, volatile organic chemicals, particulate matter, we have it all. Sometimes it's on your car, you can smell it in the air, a chemical odor looming over the community. You are inhaling those fumes all day long and constantly clearing your throat, coughing up mucus. That's money you smell, my mother used to say. A lot of folks here who said that died of cancer." The children are most vulnerable to the pollution, he said. One of every five families has a child who requires nightly treatment for asthma. There is a high rate of brain tumors among the city's children.

Kelley has been trying to organize the community to force industry in the city to clean up its act. "There is power within each individual to make change, but they have to have the will within themselves to do something." He formed a group called Community In-Power and Development Association, Inc. (CIDA). It keeps residents of the area informed about pollution levels, educates them about what the pollution means for their health, and tries to get the industry to reduce its emissions of toxic pollutants. Kelley has also gone to Washington to lobby Congress to do something about the pollution in Port Arthur. He strongly opposed President Bush's Clear Skies legislation, which, he said, would increase rather than reduce pollution.

So far, CIDA has slightly over one hundred members, Kelley told us, about twenty-five of whom are active. "A lot of people fear retaliation if they speak out. Industry is a big provider of resources here, although those are not landing in the hands of our people. Jobs are few and far between, and most of them are going to foreign workers. A city councilman works for ExxonMobil." The day we interviewed Kelley, he had just met with a local judge to see if there were legal options for reducing pollution. But the judge, he said, "is basically for industry and believes they are doing the right thing." He said that the National Association for the Advancement of Colored People (NAACP) chapter in Port Arthur has never "stood up against the refineries" because one of the companies donates money for the chapter's annual dinner every year. The churches also do not speak out because the companies "give them a pittance, penny ante things." One of the churches told him not to pass out materials about industrial pollution.

"God has no soldier in this town; all have sold out," Kelley said. "But I am fighting for the kids who are as yet unborn, who cannot speak for themselves, for the most susceptible—for the pregnant mothers and for the women who want to get pregnant but never can because of the fibroid tumors that means their uterus has to come out."

One of Kelley's projects was to start a "bucket brigade." A simple plastic bucket, containing a plastic bag and an air pump, costing about $75, has turned out to be one of the more effective weapons for posses—or brigades—in many communities. It is used to collect air samples near polluting facilities. The samples are sent to a laboratory for analysis on a gas chromatograph mass spectrometer, which identifies the "fingerprints" of toxic gases in the air. With the buckets, citizens can perform their own analysis of air pollution when industries are recalcitrant about giving them information or reducing pollution and government is negligent in its duty to protect them.

The bucket brigades were initiated by the lawyer Ed Masry, whose firm hired Erin Brockovich and fought against the power company PG&E. The brigades then were translated into a national and, increasingly, international movement by Denny Larson, director of the Global Community Monitor in San Francisco.

One of the most active and successful of the groups is the Louisiana Bucket Brigade, led by Anne Rolfes, which monitors the air in that state's "Cancer Alley," a dense warren of chemical and refinery facilities. The brigade's website tells a story about Dorothy Jenkins, president of Concerned Citizens of New Sarpy (Louisiana), who used to call the operators of a refinery near her home to complain about the constant odors and plume of black smoke and was told there was nothing to worry about. Then she got a bucket and had the air tested. Now, when the refinery managers and government regulators tell her there is nothing to worry about, she answers, "Why, then, was there a benzene reading of 14 in my air sample, a reading that violates the state standards?" The bucket, the Web article concluded, "gives community members power to hold institutions accountable to provide a safe and healthy environment."

Belonging to a bucket brigade—or, indeed, joining the activist posse in any capacity—at the very least gives people the sense that they are doing something for the children and themselves, that they are not helpless in the face of an indifferent, unreachable industry and inattentive government. Ann Tillery, the redheaded grandmother whose grandson Justin died of brain cancer over five years ago, is a member of a bucket brigade in the Port Neches–Beaumont area of Texas.

"I am an environmentalist because I lost my grandson," she said. "This has become my life's work. I believe this is what God wants me to do." Just before we visited her in 2005, she had tested the air outside of a refinery using a monitoring bucket borrowed from Hilton Kelley. She found what she said were high levels of butadiene and benzene in the air. "Then the security man came out and made us leave. He said it was because of national security." She also clips obituaries of children who die of cancer and keeps a running list. She and other activists had to "raise hell" to get three employees for the Texas Cancer Registry. The registry lists cancers only from towns of more than five hundred people. It lists only two leukemias from Port Neches, but Tillery and Dale Hanks, the lawyer who represented the families of the kids from "Leukemia High," say they know of dozens more in that city.

Residents of the area make fun of her for her activism, she said, because "they really don't know what is happening and they really don't care."

Dale Hanks and many of the lawyers like him who try to obtain justice and compensation for the families of sick and dead kids, and to hold the polluters accountable, are members of the posse in good standing. Industry has assailed lawyers making tort law claims for injured citizens as ambulance chasers seeking only to enrich themselves. The business community has lobbied the federal government to make it more difficult to sue for damages and has been supported in its effort by the conservative governments of recent years. But Hanks, who sits in his cramped little office on a side street in Beaumont, Texas, and takes on multibillion-dollar corporations and their expensive lawyers, scientists, and consultants, is clearly not in it for the money. He has immersed himself deeply in the science of chemical-related cancers and keeps careful track of the latest developments in science and law. When he speaks of the sick children and dead young people he has represented, he does so with empathy and regret.

If not for lawyers like Hanks, industry might be even more careless about putting dangerous substances into the environment. It is not so much the money that the lawyers win in damages that helps rein in corporate activities. Such payments are usually little more than pinpricks to big companies and their enormous resources. What the companies do fear is the adverse publicity such lawsuits bring. That is why they so often settle without going to court—and without acknowledging that their products cause harm.

ACTIVIST INSTITUTIONS

On occasion, a national organization has been born out of a parent's indignation. When Betty Mekdeci was pregnant, she was given a widely prescribed drug called Bendectin to combat morning sickness. ("It's a clever name—it sounds like a blessing," she said.) Her son David was born with birth defects: an underdeveloped right hand and arm, chest wall defects, and a malformed heart valve. She consulted several doctors, all of whom asked her what drugs she was taking during pregnancy, so she wrote to the Food and Drug Administration to ask about the safety of Bendectin and other medications that were prescribed during her pregnancy. Finally, contacts at the FDA encouraged her to pursue Bendectin as a cause of her son's defect.

Mekdeci, who had worked at writing promotional materials for hospitals and the medical equipment industry, then entered a long legal battle with Merrell Dow Pharmaceuticals, which had acquired the company that originally made the drug. Although the jury found for the plaintiffs and awarded David damages, the award was overturned by the trial judge, who ordered a second trial. The Mekdecis lost the second trial and appeal (ending in the Supreme Court's

Daubert ruling, as the chapter before this describes). But as a result of the attention focused on the legal action by Mekdeci and other mothers who had taken the drug and given birth to children with defects, Bendectin was pulled from the market.

Although a number of studies have found no link between Bendectin and birth defects, others have found links between Bendectin and limb defects, heart defects, diaphragmatic hernia, and oral clefts, and it remains off the market.

Betty Mekdeci, her mother, and her husband, Mike, spent years doing research on possible links between Bendectin and birth defects. While doing so, they found that parents of kids born with defects had relatively little access to information such as sources of statistics about birth defects and research into causes. So they founded a nonprofit organization called Birth Defects Research for Children, Inc., based in Orlando, Florida.

One of its projects was to create a National Birth Defect Registry, a program that looks at potential connections between birth defects and exposures to environmental hazards including radiation, medications, alcohol, smoking, pesticides and other chemicals, lead, mercury, dioxin, and other pollutants. It was to Mekdeci that a sympathetic worker in the Tennessee Department of Public Health pointed the mothers of children in Dickson born with cleft palates. And it was Mekdeci who pointed us to Dickson for this book.

As word got out about Mekdeci's program, she was contacted by organizations and lawyers for Vietnam War veterans who had been exposed to Agent Orange. She set up another registry, and as she collected data on the vets, she found a pattern of birth defects, including learning, attention, endocrine system, and immune system problems, in the vets' children. The registry also produced some evidence that grandchildren of exposed veterans suffered birth defects, although, she acknowledges, that case is much harder to make because of all the additional variables in the next generation. Some of the registry's early funding came from the Washington, D.C.–based National Coalition Against the Misuse of Pesticides (now Beyond Pesticides), directed for many years by Jay Feldman, who is one of the tough guys of the posse. Mekdeci is convinced that the dioxin, which is a contaminant of Agent Orange, is responsible for the birth defects. Her organization's research, she said, has uncovered important outcomes to exposure to dioxin, including damage to the endocrine and immune systems, and has contributed to the growing understanding of the fetal origins of many adult diseases.

Mekdeci's son is now in his early thirties. He graduated from university and works in her organization.

Lois Gibbs, another angry mother, founded the Center for Health, Environment & Justice. Gibbs is the housewife who organized the residents of Love Canal, New York, where homes were built over a chemical waste dump. State and national governments did nothing to help its residents until the pugnacious and persistent Gibbs made their plight a national issue. After persuading the government to buy out the contaminated homes, Gibbs moved to the Washington, D.C., area and set up her center.

Now operating for over a quarter of a century, the center helps communities around the country address problems such as leaking landfills, polluted drinking water, incinerators, and hazardous waste sites. Gibbs follows a "bottoms up" approach, training activists in local communities, providing technical and organizing assistance, and encouraging parents and their supports to become active in politics. Many of the center's programs are devoted specifically to protecting children, including a "safe schools" initiative that issues a periodic "Poisoned Schools Report."

Environmentally induced childhood illness, she said, "is a serious public health crisis. It's a serious crisis for the next generation."

Gibbs's organization occupies a modest suite of rooms in a small office building located several miles outside of Washington in Falls Church, Virginia. Luxurious quarters, such as those occupied by the American Chemistry Council and most other industry trade and lobbying organizations, are generally nonexistent for the volunteer groups formed to address children's health issues. When we first visited Greater Boston Physicians for Social Responsibility, their headquarters occupied a labyrinthine basement that it shared with a homeless shelter in a Cambridge church. The group later moved to an office shared with another nonprofit group and operates out of a few document-littered cubbies. The new quarters at least have windows that let in sunshine.

Healthy Child Healthy World, a group of parents, physicians (including Philip Landrigan), and celebrities (including Meryl Streep), has the long-range goals of eliminating the exposure of kids to toxic substances and educating parents on what they can do to protect their children. We visited its then director, Elizabeth Sword, at the group's headquarters, at the time in Princeton, New Jersey, where it occupied a few rooms in a walkout basement of a frame house. We reached Sword's office through a narrow corridor lined with cartons of brochures. On the wall of Sword's small office was a poster for a "Chlorine-Free Environment" and on a table an open can of organic cocoa, a container of Organic Valley milk, and a box of environmentally correct antibacterial disinfectant wipes. There was also a placard that read "PERSEVERANCE."

The offices of West Harlem Environmental Action in New York City take up a few unadorned rooms in an aging building on Harlem's bustling 125th Street. The organization was started to demand health safeguards and jobs for local residents when a sewage plant was being built at the edge of the community on the Hudson River. But Peggy Shepard, the handsome, elegant woman who runs the organization, said that it remains in operation to address broader issues of environmental and social justice for the African American and Hispanic American residents of northern Manhattan. West Harlem Environmental Action is the organization that cooperates with the Columbia Center for Children's Environmental Health in its research with community residents. The group now speaks out on a wide range of problems that affect children's health, including lead and other toxic contamination, water quality issues, high asthma rates, insect pests in houses, and the need for open space, and links these issues to community economic development. "Our community is a dumping ground, like so many people of color communities," said Shepard. "It is not just one dry-cleaner but twenty dry-cleaners and all the other things. Nobody is looking at the vulnerability of children to the multiple sources of pollution in the community."

Shepard believes the mainstream national environmental organizations do not focus sufficiently on the public health problems of communities, particularly the problem of children's health. But the national groups, including their scientists, lawyers, lobbyists, and communications staffs, do play an important role in the effort to quell the toxic assault on our kids. They keep a careful eye on what the corporations and the government are doing and keep the media and the public informed of their deeds and misdeeds. They press for legislation to protect the environment and public health and for effective enforcement of the laws. If government does not act, some of the groups will take polluters to court. The environmentalists track emerging science affecting environmental protection, and some of them do their own advanced research. They also act as intermediaries between the scientific community and the public, spreading the word about what is new and important. Through its political arm, the League of Conservation Voters, and the Sierra Club's political action committee, the environmental community seeks to elect politicians who will act to preserve the environment and public health. In this latter endeavor, they were not notably successful in the early years of this century.

Several of the national groups, including Lois Gibbs's, have been particularly active in the cause of children's environmental health. One of the most active and effective in recent years has been the Washington-based Environmental Working Group, which focuses heavily on scientific research affecting public health and on bringing the results of such research to the attention of the media and to the public. It also uses its work to harry lawmakers and law enforcers into acting to protect the public.

The group started with a report on pesticides in children's food, which Ken Cook, its president, said played a role in persuading Congress to pass the Food Quality Protection Act. The group has done some forty reports that focus on children and the environment; one of them, on the load of toxic chemicals in the umbilical cords of newborns, captured wide national attention. It has done extensive investigation of possible links between autism and the exposure of fetuses and children to mercury. It called wide attention to the fact that Teflon from DuPont plants was contaminating the environment.

"You can't test every kid," Cook said. "What you try to do is look for patterns and try to be protective through policy. You can't tell parents to shop their way out of these risks, which is crazy. If the medical director of DuPont says they can't tell us how Teflon got into kids, then responsibility ought to be not the shopping cart–level decisions, but a policy one."

Another group that has been particularly active on behalf of children is the Natural Resources Defense Council, which has been around since the 1970s—with a remarkable continuity of leadership—and is a bedrock of contemporary American environmentalism. The group, headquartered in New York City, with offices around the country, has been successfully conducting research, promoting environmental legislation, litigating, and raising public awareness about a wide range of issues ranging from the protection of wildlife and public lands to air and water pollution and the impact of toxic substances on human health. When we visited their Washington office in 2005, the group was in litigation with the EPA, charging that the agency had failed to impose the tenfold margin of safety on exposure to a group of organophosphate chemicals mandated by the Food Quality Protection Act.

Jennifer Sass, a PhD scientist with the group, is worried that policy deals with only one toxic pollutant at a time, while in our industrialized civilization we carry multiple carcinogens and mutagens in our bodies. "Risks are not distributed evenly," she said. "If you say, 'On average we are all living longer,' that doesn't describe how children, the sick, the elderly, and certain communities will pay the price and others won't."

Still another environmental group that works to reduce the toxic threat to children is Environmental Defense. Some environmentalists look a bit askance at the group because it often is willing to cooperate with industry to achieve its goals. But it produces excellent science and policy goals. As for its ties with industry—sometimes they work to improve the environment.

In addition to the big national organizations, there are dozens of smaller groups, many of them linked to a specific disease or a particular source of pollution, that are working to reduce the toxic load on children, to find treatments and cures, and to alter economic and political conditions that permit the crime against our children to continue.

WHITE HATS

As we noted, there is no sheriff leading this posse. If one were to be named, however, there would be no more likely candidate than Dr. Philip Landrigan, chair of the Department of Community and Preventive Medicine and director of occupational and environmental medicine at the Mt. Sinai School of Medicine in New York City. Dr. Landrigan, a pediatrician and public health physician, has been called by some "the father of environmental pediatrics." His fingerprints can be found on a remarkable number of initiatives around the country to safeguard children from environmental hazards. As senior adviser on children's health protection to the administrator of the EPA in 1997, for example, he was responsible for establishing the Office of Children's Health Protection at the agency. He was chair of the National Academy of Sciences committee that produced the report *Pesticides in the Diets of Infants and Children,* which led to passage of the Food Quality Protection Act. He founded and is director of the Center for Children's Health and the Environment, a national organization head-quartered at Mt. Sinai dedicated to protecting kids against environmental threats to their health. He is a board member or adviser to many organizations dedicated to protecting children, including the Children's Environmental Health Network, the Healthy Child Healthy World coalition, the Institute for Children's Environmental Health, and the Collaborative on Health and the Environment. He is also an adviser to the authors of this book. Together with Dr. Herbert Needleman, he has written books about environmental threats to children's health and how to protect children from them.

Landrigan is a slender, gray-haired man with a minimalist mustache and a quiet, genial manner that belies his deep dedication to children's health. He began his career as what he described as a "medical detective" gathering intelligence on epidemics for the Centers for Disease Control and Prevention in the 1960s, at first in Central America. Then he was sent with another pediatrician to El Paso, Texas, to investigate reports of lead poisoning among children there. On the plane, he read a report by a committee of the National Academy of Sciences, which, he said, had been handpicked by the lead industry, that concluded lead emissions from smelters did not constitute a health problem.

The two doctors took blood samples from children in a nursery school right in the middle of exhaust plumes from a nearby ASARCO lead smelter. All of the kids had elevated lead levels. Then they drew concentric circles around the smelter and found that the farther out they went, the lower the rate of lead poisoning among children. "The smelter's smoke-stack was the bull's-eye," Landrigan said.

While engaged in his work, an executive from the company that owned the smelter invited him to lunch at the Cattlemen's Club in a glass-enclosed penthouse atop one of El Paso's tallest buildings. "After we finished our steaks, he took me to the window and said, 'Doctor, we are very proud of this town. We are hardworking Texans proud of that smelter. Is there anything we can do to make your life easier while you are in El Paso?' I was very naive back then. I said, 'Thank you, our study is really going well and we are getting everything we need.' He repeated the offer: 'Is there anything at all we can do to make your time here more pleasant?' When I got back to the hotel and recounted the story to my colleague, he pointed out what was going on." What was going on was that the executive was trying to buy Landrigan off.

"That was the catalyst that made me stay in the game," Landrigan said. He stayed with the CDC for fifteen years and with two other doctors established an environmental epidemiology center there. It became the intellectual incubator for environmental pediatrics and is now a thousand-person unit.

He regards the state of children's health in America as "good news and bad news. The good news is that the classic diseases, cholera, dysentery, and so forth, that still stalk children in the third world have been eradicated here. For that reason, life expectancy has gone up and infant mortality down. The bad news: Today, children are subject to exposures of unknown toxicity to which children of one or two generations ago were not exposed." Landrigan initiated the concept of a large-scale, long-term National Children's Study, like the Framingham Heart Study, that will answer many of the questions about the social and environmental roots of childhood chronic disease.

The twenty-five-year-long $2.7 billion study, recently launched, will follow a cohort of one hundred thousand American children from their mothers' wombs through age twenty-five. It will periodically collect their blood, urine, and hair samples, monitor their physical and social environments, and collect data on other variables such as genetic susceptibilities and socioeconomic status. The study is expected to pin down the causes of disease such as childhood cancer, asthma, birth defects, neurological illness, diabetes, and obesity, and thus lead to ways to prevent them.

Lead poisoning has been recognized for many centuries, but it was long thought that the only problem with the metal was that at high doses it could kill. Herb Needleman, a professor of pediatrics and psychology at the University of Pittsburgh, is responsible for "one of the most successful campaigns in disease prevention of the past century— the removal of lead from the environment." Studying the residues of lead found in children's teeth, he found that even low levels correlated with lower IQs as well as cognitive and behavioral problems. More recently, he and others have been finding a strong association between childhood lead exposure and violent behavior that continues into adulthood. His research and his advocacy on behalf of children played a key role in the government decisions to ban lead in gasoline and paint.

But Dr. Needleman paid for his good deeds. In 1990, he was asked by the Justice Department to be a witness for plaintiffs against three lead polluters in Utah. One of the witnesses for industry was a psychologist who had written a paper associating lead with lower intelligence in children but later retracted her findings. Afterward, she became a grantee of the International Lead Zinc Research Organization and testified for Lead Industry Associates, which was trying to put lead back into gasoline. She and another scientist accused Needleman of manipulating data. Their accusation was forwarded to the National Institutes of Health's Office of Research Integrity by the law firm Hunton & Williams, which represented at least two companies that make or use lead (and have represented other polluting companies and industries).

"The next thing I knew," Needleman told us, "I was called in to the scientific integrity office at the university, who locked my files and wouldn't allow me access to them except in the presence of some other person." The office conducted an inquiry and came up with the odd finding that there was no evidence of scientific misconduct but that they could not rule out misconduct.

Needleman is a fighter. He demanded an open inquiry, which the university at first denied. But then he went to the faculty senate, which voted unanimously to recommend an open hearing. The university gave in. His accusers would agree to appear at the open session only if they were assured they would not have to answer questions they did not want to. When he asked the accuser who had attacked his findings who was paying for her lawyer, she would not answer. It turned out, Needleman found, that the lawyer was from Hunton & Williams.

The hearing board found unanimously that there was no evidence of any misconduct by Dr. Needleman. He is widely regarded as a hero in the public health community.

But Needleman is concerned that such industry attacks on the credibility of scientific research might frighten scientists away from controversial research. "If that happens, our craft will be in peril."

In fact, scientists serving as plaintiffs have not infrequently been entangled in acrimonious high-stakes battles. When epidemiologist Richard Clapp, a respected authority on environmentally related cancer, testified on behalf of IBM workers, his research was excoriated by defending lawyers as "giving junk science a bad name." Yet in the end, his study was published in a highly regarded journal, and workers facing the same illnesses at a New York State IBM plant have asked for his help. He has served as an adviser and expert witness for families of sick children, including families in Woburn, Massachusetts; Toms River, New Jersey; Anniston, Alabama; and as far afield as Ecuador, where families say they have been harmed by oil exploration practices. He also set up a cancer registry for the state of Massachusetts, a model for the nation. His hero is John Snow, a nineteenth-century English physician regarded as the father of modern epidemiology, who ended a cholera epidemic in a corner of London by removing the pump handle from the contaminated Broad Street well—or so the story goes. Clapp has been honored as "Public Scientist of the Year" by the Association for Science in the Public Interest.

Attacks on plaintiffs' scientists are more likely because of the rules set in motion in the Supreme Court's Daubert decision. When the state of Rhode Island brought a suit against companies that once made or sold lead-based paint, accusing them of knowing of its risks and asking that they pay for its cleanup across the state, lawyers for the defense demanded the raw data, even unpublished data and data from other communities, from scientists for the plaintiffs. Bruce Lanphear, a top lead researcher at the Cincinnati Children's Hospital Medical Center, was one of those scientists in this case, as was Herb Needleman. Under Daubert rules, they were forced to turn over data sets to the defense. Facing what he called the "extraordinary harassment," another researcher for the plaintiffs stepped down, only to be pursued under subpoena. When Dr. Lanphear refused to release some of his raw data for a lead case in Milwaukee, he was barred from testifying.

Another scientist seasoned in toxic lawsuits is Dr. David Ozonoff, founder of the Department of Environmental Health at Boston University. Trained as a physician, Ozonoff testified on behalf of the families in Woburn and Love Canal, New York, and is often a witness in suits brought against manufacturers of asbestos. An epidemiologist with a passion for higher mathematics, he is devising an approach to reach defensible conclusions even from studies with small data sets. He is also a pioneer in bringing public participation into the world of public health; he created a model "consensus conference," gathering together community members from the Boston area to learn about pressing environmental issues and procuring their judgments about what should be done to address them. Ozonoff says, "I am not a professional witness. I am mainly a scientist and a teacher of other scientists. I give my available time to people I think really need it, and that includes community groups, for whom I usually work for free."

Dr. Ted Schettler is a physician who works in community health clinics, a researcher, and an author and indefatigable compiler of scientific evidence of harm caused by toxics in the environment who devotes considerable effort to making sure that evidence is widely distributed. One of the ways he shares information is through the nonprofit organization Science, Ethics and Action in the Public Interest. He has been particularly dogged about showing how toxics damage the brain and is coauthor of two eye-opening books on children's health and the environment, *Generations at Risk* and *In Harm's Way*. Both Clapp and Schettler have been on the steering committee of Greater Boston Physicians for Social Responsibility, a group whose members live up to its name.

Other scientists among the posse work on behalf of public health through government. Dr. Richard Jackson was head of the National Center for Environmental Health at the Centers for Disease Control and Prevention until he left when the George W. Bush administration began using its wrecking ball on public health programs. He later served as director of California's public health system and is on the faculty of the School of Public Health at UC Berkeley. He has long been an outspoken advocate of a better, more active lifestyle for American children and, while at CDC, supervised the pioneering research that led to the launch of continuing studies of the body burden of toxic chemicals in Americans. He credits Dr. Lynn Goldman, now at Johns Hopkins, for helping to lower children's leukemia rates while she was an assistant administrator of the EPA during the Clinton administration and adopted policies that sharply reduced the number of toxics in the environment. Both he and Goldman, he said, "have been branded as troublemakers and are hated by the chemical manufacturers."

Dr. Philip R. Lee, professor emeritus of Social Medicine at the University of California at San Francisco, has spent a long career wedding medical science to public health and public policy, moving from the university to government and back. Among other posts, he was assistant secretary of the U.S. Department of Health and Human Services during the Clinton administration.

Research is the route through which Martha Herbert expresses her pursuit of scientific truth. The daughter of a machinist-turned–math teacher who emigrated from Ukraine to Brooklyn, she is an MD and PhD who works at Massachusetts General Hospital and Harvard Medical School with neurologically damaged children while doing intensive research into the causes of their illness. She has pioneered the understanding of neurological disorders, such as autism and ADHD, as environmentally triggered illnesses of the body, not just the brain. Many of the issues of neurotoxological disorders are not being studied by mainstream science, she said, and government regulatory agencies are not watching. "In the meantime, I'm dealing with this epidemic of neurobehavioral problems among my patients." But she believes diseases such as autism are preventable and curable. Dr. Herbert is a scientist and a healer and also an empathetic activist who offers guidance and support to families of sick children and to their organizations, speaking at their meetings and writing for their publications.

Many members of the posse are in academia, devoting their knowledge and professional training to the service of children's health. This book is full of stories about the discoveries coming from Frederica Perera's Columbia Center. There are untold others. John Wargo, professor of environmental risk analysis and policy at Yale University's School of Forestry and Scientific Studies, has investigated and written extensively on the vulnerability of children to toxics in their environment, particularly to pesticides and air pollution. He is a national expert on the effect of pesticides and heavy metals that harm the growth and development of children. "The EPA is asleep or looking the other way on these toxins," he said.

Michael Lerner, son of the late Max Lerner, journalist and political philosopher, founded an institute, Commonweal, which unites the physical and the spiritual to heal sick patients. An arm of Commonweal, the Collaborative on Health and the Environment, brings together scientists, activists, and laypeople from around the country to learn about, discuss, and find ways to address the environmental impact on health, particularly the health of children.

The *posse comitatus* constitutes a deep pool of talented, energetic scholars bringing intellectual firepower and scientific gravitas to bear on the problem of reducing the environmental dangers that threaten the health and well-being of our children as well as parents, physicians, and many others who care passionately about children and about justice. They persevere in their efforts, although they must contend with formidable, extravagantly well-funded foes who contest them every inch of the way, and are sometimes badly treated within their academic communities. It is hard to imagine what would happen to the nation's children without them.

Eleven

VALUES

CYNTHIA SAUER, WHO IS PLEASED TO BE CALLED CINDY, USED TO LIVE in Grundy County, Illinois, with her husband, Joseph, and her three daughters. They do not live there anymore.

Cindy contacted us when she heard about the book we were writing. She wanted us to know about one of her daughters, Sarah, who was diagnosed with a brain tumor when she was seven years old. She soon began hearing of other children in the area with brain tumors and babies born with rare birth defects. After Sarah underwent surgery to remove the tumor, Cindy related, several of the physicians caring for her daughter casually told her that her daughter's cancer was most likely environmentally induced.

The Sauers lived close to the Dresden 2 nuclear power plant and a fairly short distance from the Braidwood nuclear power plant, both owned by the Exelon Corporation, one of the nation's biggest electric utilities and the largest supplier of nuclear energy. Like most of their neighbors, they had assumed that the plants took adequate safeguards to protect their health and safety. What they did not know until well after Sarah was diagnosed with cancer was that, starting in the 1990s, the two nuclear plants had leaked millions of gallons of water contaminated with tritium, a radioactive isotope of hydrogen, into the surrounding environment. Some of it seeped into water supplies used by local residents for drinking, bathing, and cooking.

The reason the Sauers and other residents of the area did not know was that Exelon did not tell them.

When the Sauers brought Sarah home from the hospital, Cindy learned about an out-of-court settlement between Exelon and the Illinois attorney general of charges relating to violations of the Safe Drinking Water Act dating back to 1990. She was distraught. "I was thinking, How do we know we are not bringing her right back into environmental danger?" She suspected the danger might emanate from the nuclear reactors. "I called the EPA and they said, 'Don't worry.' Then a researcher from the EPA called back and said, 'I will deny I ever spoke to you, but I feel for you. You are right, there is something being injected into the earth in your area, and radiation is out of our jurisdiction. You need to contact the U.S. Nuclear Regulatory Commission.' "

Cindy began looking into what was happening to children in the area. She found that the rate of cancer in the county had increased dramatically in the late 1990s. She and her husband demanded an investigation of the reactors and called for public hearings. She started telephoning and writing to state and county health officials and to the Nuclear Regulatory Commission (NRC). She wrote to her congressman, Jerry Weller, a Republican, who wrote back with what appeared to be a boilerplate reply, saying he was a "strong supporter of the environment" and giving examples of legislation he had supported to protect open space and restore wetlands. In the most recent (as of this writing) scorecard by the League of Conservation Voters, which keeps track of voting records of members of Congress on environmental legislation, Mr. Weller scored a zero, meaning he had voted against pro-environmental bills every time in the first session of the 109th Congress. Mr. Weller has been the recipient of campaign contributions from Exelon's political action committee—but then so has just about every member of Illinois's congressional delegation as well as dozens of other senators and representatives.

Cindy wrote to President George W. Bush, asking for help in getting answers to her questions. Her letter to the president was answered by an official of the NRC, who assured her that the radiation limits set by the commission and by the EPA "are sufficiently protective of public health" and that measurements taken by NRC investigators found that "the effluents are within the limits set to protect the public." She later learned from Dr. Arjun Makhijani of the Institute for Energy and

Environmental Research that these supposed safe levels are set for the average adult male and do not protect the most vulnerable members of society: children, unborn babies, and pregnant women.

By the end of 2005, she said, facts about the reactors began to emerge. Over the next several months, she and other residents of the area would learn that numerous leaks from the plants in the 1990s had released over six million gallons of radioactive water. Moreover, it was happening not just at the Braidwood and Dresden plants, but at reactors across the country. As far as Cindy is concerned, "the nuclear industry chose to cover up the incidents, and government agencies simply looked the other way."

A fact sheet on tritium provided by the EPA notes that "as with all ionizing radiation, exposure to tritium increases the risk of cancer. However, tritium is one of the least dangerous radionuclides because it emits very weak radiation and leaves the body relatively quickly." A report commissioned by Exelon from a consultant who specializes in "regulatory and environmental solutions" concluded that drinking two liters of water from the most contaminated well near the Braidwood plant would be equivalent to "about one-eighth of the dose from eating a banana every day because of the naturally occurring radioactive potassium 40 in bananas." The Illinois Department of Health found that pediatric cancer was no higher in counties with nuclear facilities than in those without such facilities, although, as Cindy pointed out, the figures were countywide rather than for areas close to the plants.

A "definitive" report by the National Academy of Sciences in 1990, however, found that there are no safe thresholds for exposure to radiation. It stated that "the smallest dose of radiation is capable of causing mutations within DNA and therefore capable of causing cancer."

We spoke by telephone with Dr. Helen Caldicott, the physician and antinuclear activist who had spent time talking to Cindy Sauer and knew about the case. She told us that tritium is "extremely carcinogenic" and that children and fetuses are particularly vulnerable. Because it becomes radioactive water, she said, "the nuclear industry is scared to death of it." The U.S. government is failing to protect the public from radioactive contamination from nuclear plants, she said. In her opinion, "We are in the hands of people who are scientifically illiterate. The nuclear power industry runs the Nuclear Regulatory Commission—it's a conspiracy."

The reason local communities such as Grundy County do not protest about the dangers of radioactive contamination, she said, is that when a plant proposes to come into a community, "they pay for roads, kindergartens, schools, libraries—it's bribery."

In a press release announcing its lawsuit against Exelon for its leaks of radioactive tritium from the Braidwood plant, the Illinois Office of the Attorney General charged that "the method of operations put in place at the Braidwood Nuclear Plant since 1996 by Commonwealth Edison and their parent company as of 2000, Exelon, clearly placed their profit margin first with a callous disregard for the health, safety and welfare of local residents. Exelon was well aware that tritium increases the risk of cancer, miscarriages and birth defects and yet they made a conscious decision not to notify the public of their risk of exposure."

In July 2006, the NRC issued an "information notice" reporting radioactive contamination of groundwater not only from the Dresden and Braidwood nuclear plants, but also from other plants around the country, including Indian Point in New York State and the Connecticut Yankee plant in Haddam Neck, Connecticut. The commission report concluded by saying it was requiring "no specific action or written response." The nuclear industry issued a promise to monitor the leaks voluntarily.

Shortly thereafter, the Union of Concerned Scientists filed formal opposition to what it said was the commission's decision to do nothing about tritium leaks. According to David Lochbaum, director of the union's Nuclear Safety Project, "It is outrageous that the NRC would shirk its duties to protect public safety and rely—sight unseen—on an undocumented promise from industry with a long track record of broken promises."

The health effects of exposure to low levels of radiation to people living near nuclear plants remain controversial. Congressman Ed Markey, Democrat of Massachusetts, who has been in contact with the Sauers, introduced legislation that would authorize and modestly fund a study of such effects. But the proposal was killed by Representative Joe Barton, Republican of Texas and then chairman of the House Energy Committee. Barton received $1.84 million in contributions from the nuclear, oil, gas, and electric utility industries between 1997 and 2004, the most of any member of Congress.

Exelon now has a continuing program to monitor groundwater

near its plants. It has provided bottled water to some households with contaminated supplies and has bought a few of the contaminated houses. *Fortune* magazine named Exelon one of America's "Most Admired Companies" in 2005, and *Forbes* magazine named Exelon one of the best-managed utility companies in December 2004.

Cindy Sauer was not reassured by a belated display of responsibility from the big utility. She and her family moved out of Grundy County and out of Illinois. They now live in Indiana, where her husband began to build his medical practice anew and the girls entered new schools. But from her new home, Cindy has continued to tilt at the nuclear industry and the governments that are supposed to protect their citizens. "We are not going to give up," she said. "It is my daughters' future and the future of all the people we left behind. The most vulnerable members of our society are not being protected."

When we spoke to Cindy by telephone just after Sarah passed the five-year mark since her surgery, Sarah was "doing fine," she said, although with brain tumors you have to go ten years to be sure, she added. But "Sarah is not the same little girl. She still has her delightful personality, but she is by no means normal. She has disabilities. She will struggle for the rest of her life. She has poor balance, poor handwriting. Her speech is affected. She has to have special accommodations in class. It takes longer for her to complete a task. Her short-term memory is poor. She has difficulty with simple tasks. She cannot go out and run with other children—she falls down. She cannot pick up a basketball and throw it. She is twelve years old and only four feet six inches tall."

Cindy describes herself as a spiritual person and a devout Christian. What she finds profoundly troubling is what she sees as a dereliction by industry, by government, and by community and religious institutions of their moral duty to safeguard children from environmental dangers such as leaking nuclear plants. She was also disillusioned by neighbors whose only concerns were about how her complaints would affect their real estate values and jobs.

"I believe in God and keep putting our lives in His hands. The night before Sarah's surgery, I asked God that if she is going to die, to take her quickly while she is on the operating table. If not, let her live in glory."

The fact that children are getting sick and dying because of pollution in the environment, she believes, "reflects our values and priorities. Unfortunately, it comes down to the profit margin. Industry is turning its head away from this issue, and regulators are turning their heads away."

What "truly bothers" Cindy, however, is what she perceives to be the indifference of the right to life movement and officials who purport to support life and family values to the plight of children sickened by toxic substances and radioactive materials. "We talk about the right to life and family values. Come out of the box! We have the right to life across the board. We have the right to clean drinking water, the right to grow up knowing that natural resources around you and which you depend on are held in high esteem and protected. This is a life and family values issue that's much bigger than the abortion issue. I want them to stand up on all pro-life issues. The right to life movement absolutely ought to take up this issue and protect the health and lives of unborn babies not only before they are born, but throughout their entire lifetimes."

LIVING BY OUR VALUES

Questions of values are difficult, contentious, and profound. But this much seems clear: If we love our children, we should want to protect them, to give them a safe and healthy environment in which to be born and to grow up and in which to inhabit the future. Whether we are strict parents or nurturing parents, whether we are conservative or liberal, whether we are rich or poor, whether we are blue-collar workers or corporate executives, whether we are devoutly religious or not, our values should reflect a deep—indeed, elemental—desire to shield our kids from harm.

The United States is a fortunate nation. We possess a spacious and beautiful land. While there are still far too many American citizens living in poverty, it is not, for the most part, the starving, disease-ridden poverty afflicting millions of humans in much of Africa and parts of Latin America and Asia. Most of us enjoy relatively good health and increasing life spans. We have been spared the violent domestic sectarian and tribal strife that torments so much of the rest of the world. We are afloat in an overflowing sea of consumer goods and luxury items and awash in information. If America cannot take care of its children, who can? And many of the values we espouse—respect for the sanctity of life, our responsibility as stewards of God's creation, the centrality of the family, caring for the rights and needs of the vulnerable, the duty of service to the community—would seem to dictate that we give the highest priority to nurturing life and providing our children with an unpoisoned environment.

Yet we do not do so, at least not sufficiently. Instead, many of the institutions that intimately affect our lives, in their day-to-day operations, turn their backs on these values and place our children, including our unborn children, at unacceptable risk.

Why is this? Why do our society and its institutions and we as individual citizens not find this state of affairs intolerable and demand immediate change?

One answer would be that our professed values do not match the values we live by in the real world. Our economic system and our corporate culture rarely reflect values that would protect our children. As physician David S. Egilman and researcher on American civilization Susanna Rank Bohme commented, "It is not just a few bad apples in an otherwise healthy barrel" that are causing the problems. "It is the barrel itself, the current economic and political system that produces disease because political, economic, regulatory, and ideological norms prioritize values of wealth and profit over human health and environmental well-being."

Of all sectors of society, the scientific community has some of the most well-established and sharply defined set of values. Science requires an empirical search for knowledge and an objective process of finding truth through systematic experimentation and study and with verifiable results. Science should be free of bias, although like all human endeavors, it cannot escape vulnerability to personal and social values. But above all, science must be honest, as most scientists are; they are in general highly intelligent men and women dedicated to their disciplines and the pursuit of knowledge.

But as our stories have demonstrated, some scientists have been willing to trim their values to arrive at predetermined commercial or political goals. As Catherine D. DeAngelis, editor in chief of the *Journal of the American Medical Association,* noted, the marketing interests of companies can "dominate" their own scientific research, and individual scientists may exhibit "inappropriate or unethical behavior perhaps influenced by money or other factors." When science serves the master that pays for it without regard for the truth or the effect of its work on the broader society, it betrays its own values. When science serves up deception instead of truth, the consequences can be disastrous, particularly for the health and well-being of our children.

THE ROLE OF RELIGION

In recent years, religion, particularly Christianity, has been playing a prominent role in the politics and policies of the United States and in issues that traditionally have been left to science and education. Its influence has not always been benign. Despite Christian respect for the sanctity of life and its love of God's creation, Christian churches, along with Jewish and Islamic institutions, have been slow to address the degradation of the natural environment and its impact on public health and welfare.

Many reasons have been given for the lack of attentiveness on the part of churches. Christianity in particular has regarded nature as an instrument created for the service of human beings. In 1967, historian Lynn White Jr. published a widely discussed essay entitled "The Historical Roots of Our Ecological Crisis," in which he argued that the Judeo-Christian tradition "bears a huge burden of guilt for the destruction of nature by the science and technology of Western civilization." The Western version of Christianity, he wrote, "is the most anthropocentric the world has ever seen."

Christianity has long regarded this world as "a place you pass through" on the way to salvation outside the world, according to Mary Evelyn Tucker, co-director of the Forum on Religion and Ecology at Harvard University and a visiting professor at Yale. Those who regard the Bible as the source of all knowledge tend not to look at the book of nature for the truth, she noted. Moreover, the churches have been preoccupied with other issues, particularly issues of sexual morality, and have not lifted their gaze to what humanity is doing to the natural world and to itself.

Inattention to the effects of toxic substances on children, she said, "is a surprising lacuna in the thinking of churches. The lack of a sense of commitment to future generations is absolutely astonishing in American life." On the other hand, she added, the churches have the potential to raise a call to responsibility for future generations. "I think it is something they will do. The call to attend to the children of future generations is right there in religious circles."

The churches have traditionally been slow to insert themselves into social movements, Dr. Tucker said. They were late on civil rights, she noted, but added that once the religious community, particularly Dr. Martin Luther King, joined the struggle to secure those rights for all Americans, "it had a huge impact. Once civil rights were defined as a moral issue, it was transformative." The same could happen with the churches and a movement to protect children from environmental hazard, she said.

There is evidence of change within our religious institutions. Ecotheology has been spreading within the world's religions for several decades now. Christian churches in particular have been seeking a creation doctrine that embraces the love of nature. At least one commentator believes that "religion is now a leading voice telling us to respect the earth, love our nonhuman as well as our human neighbors, and think deeply about our social policies and economic priorities."

Though the Catholic Church was slow to embrace ecotheology, in 1989 Pope John Paul II issued a document entitled "Peace with God the Creator, Peace with All Creation." In it, he asserted, "The ecological crisis has assumed such proportions as to be everyone's responsibility," and added that "greed and selfishness, individual and collective, have gone against the order of creation." Earlier, the pope had stated, "The Church's commitment to life is seen in its willingness to collaborate with others to alleviate the causes of the high infant mortality rate and to provide adequate health care to mothers and their children before and after birth." The United States Conference of Catholic Bishops "has undertaken a broad program of children's health and the environment."

In its political and policy activities, however, the Catholic Church and much of its clergy have not appeared to give due attention to the environmental assault on our children as they have to other current issues, from abortion and contraception to stem cell research and gay marriage.

As with most such generalizations, there are many exceptions. There is, for example, Father Gerald Kleba, who, while serving a parish in Weldon Spring in St. Charles County, Missouri, found he was burying more infants and children than at any other time in his more than thirty-seven years as a priest. In the Catholic liturgy, he noted, when caskets are brought in for a funeral, they are covered with a pall made of white cloth. He realized that his and a neighboring church were the only churches in the St. Louis area that were having palls commissioned especially for the caskets of children.

In Weldon Spring, there is a very big Superfund site that processed the radium used for the Manhattan Project, Father Kleba told us. A number of creeks run out of the site area, and St. Charles County wells are a mile and a half down the road. The Department of Energy encapsulated the waste in what it said was an impermeable clay membrane. But a large number of workers at the site have died of cancer, and the parents of the sick and dead children suspected that the cause was leaking radioactivity.

The parents of some of the dead children and Father Kleba began to speak out about what was happening and asking for an investigation. Several officials of the Missouri Department of Health did come out and talk to the parents, but there were no autopsies and no medical studies, he said.

The parents and Kleba found that speaking out about links between the nuclear waste site and the dead babies created "a very difficult situation. First it's very difficult to lose a baby. If the government, from the county to the state to the federal government, says there is absolutely no problem, the implication is that the real fault lies with you who are inept parents. You are paying the cost because you smoked pot in college or drank wine during pregnancy. It is the implication that says that the environment is perfectly good—the problem is close to home, and you are the responsible ones."

Not all of the parents spoke up. People buy homes in St. Charles County, Kleba said, "with the notion that this is the garden of paradise. So if you are a struggling family wanting to move here, you expect your housing to appreciate rather than depreciate. So along with losing a child, you worry, 'I'll lose my investment, too.' It's a huge financial pressure."

Father Kleba, who now has a parish in St. Louis, did not win popularity contests for trying to do something about the waste site. "Housing developers and politicians wanted me to leave." Diocesan officials apparently heard the complaints. "As you see," he said, "I am not there anymore."

Before he left, he was invited to bless a monument that had been erected to honor workers who had died at the waste site. "I prayed for our government, which can get so attached to untruths and misstatements while others lay down their lives for our country."

Mainstream and liberal Protestantism have generally been engaging with issues of public health and the environment in recent years. The United Church of Christ has been a vocal leader in this and other social causes. Bob Edgar, its leader as this is being written, was an activist on environmental legislation when he was a member of Congress from Pennsylvania. Though the Protestant evangelicals long remained aloof from efforts to protect the environment and public health, in 1994 the Evangelical Environmental Network issued a document entitled "On the Care of Creation," which laid out a biblically based call for Christians to commit themselves to reverse the degradation of the earth. Quoting from Psalm 24—"The Earth is the Lord's and the fullness thereof"—the declaration spelled out the various ways God's creation is being degraded by humans, including global toxification. It warned, "We and our children face a growing crisis in the health of the creation in which we are embedded and through which, by God's Grace, we are sustained. Yet we continue to degrade that creation." The network has also prepared a detailed agenda for protecting the environment, along with instructions for keeping a healthy home. The instructions include avoiding the use of toxic products in the home and offers a list of "common toxins," among them chlorpyrifos, dioxin, lead, formaldehyde, mercury, methylene chloride, PCBs, perchloroethylene, phthalates, polyvinyl chloride, and a range of pesticides.

More recently, the leaders of eighty-five evangelical organizations issued a statement calling on government to act to curb the emission of gases that cause global warming. The signers included the heads of the Salvation Army and World Vision, two of the largest charitable organizations in the country. The statement asserted, "This is God's world and any damage that we do to God's world is an offense against God himself."

THE VIEW FROM THE RIGHT

The religious Right, a large and influential force within the evangelical churches, has begun to embrace ecotheology, though at times some of its leadership have been overtly hostile to initiatives to protect the environment and public health. In 2005, for example, the National Association of Evangelicals, an organization that represents forty-five thousand churches serving thirty million people in the United States, circulated a draft policy statement encouraging the government to create mandatory controls for carbon dioxide emissions, the chief greenhouse gas. But the statement was dropped after a number of powerful evangelical leaders, including James C. Dobson, chairman of Focus on the Family, Richard Land of the Southern Baptist Convention, and Donald E. Wildmon, chairman of the American Family Association, wrote a letter to the association urging it not to adopt any official position on global warming, saying that not all "Bible-believing evangelicals" agree about the causes and severity of climate change.

Richard Cizik, vice president for governmental affairs of the National Association of Evangelicals and one of the prime movers of the draft statement on climate, added that Dobson's organization also was opposed to his call for reducing mercury pollution from power plants.

A number of evangelical organizations remain focused on other objectives to the exclusion of children's and family environmental health. As an example, Concerned Women for America, founded and chaired by Beverly LaHaye, wife of Tim LaHaye, who coauthored the best-selling *Left Behind* series of novels about the end of days, has as its mission promoting "biblical values . . . thereby reversing the decline in moral values in our nation." It focuses on six "core values," one of which is the "Sanctity of Human Life," supporting "the protection of all innocent human life from conception until natural death. This includes consequences resulting from abortion."

None of these values, however, includes protecting children, born or unborn, from the consequences of exposure of mothers and children to toxic substances. "We don't have any policies on the environment," said Janice Shaw Crouse, a senior fellow of the organization. "We don't have anyone on our staff to handle those kinds of issues. We look at things like social issues and hazardous things in our culture like single parenting and teenage pregnancy." As evangelicals, she said, they are aware that "the Bible has a lot to say about taking care of children. That is certainly what we as individuals care about. But none of us is qualified to go into those issues in depth. Our limited resources are focused on our six core issues. We have to target because our resources are spread thin."

James Dobson and conservative activist Gary Bauer wrote a book in 1990 entitled *Children at Risk*. However, the book had nothing to do with toxic threats to the physical health and well-being of kids or unborn children. It was intended, said Dobson in his newsletter, to warn parents, teachers, and church leaders "that homosexual activists on the far left had crafted an alarming new strategy to gain control of children."

The most formidable obstacle standing in the way of the religious Right rallying to the cause of children's health and well-being is its entanglement in a mean and dangerous intermixture with partisan politics and corporate oligarchy.

As journalist and social commentator Bill Moyers stated: "If the Green Revolution is a bloody pulp today, it is not just because the environmental movement mugged itself. It is because the corporate, political, and religious Right ganged up on it in the back alleys of power. Big companies fund a relentless assault on green values and policies. Political ideologues launch countless campaigns to strip from government all its functions except those that reward their rich benefactors. And homegrown ayatollahs are more set on savaging gay people than saving the green earth."

The ties between the right-wing churches, the corporate sector, especially the energy sector, and the right-wing politics that dominated national policy during the early years of the twenty-first century were clear to see. In his book *American Theocracy,* Kevin Phillips noted that as the evangelical movement grew in importance, corporations hired more Washington lobbyists "with biblical worldviews or Christian right connections. In Texas and Oklahoma and across the South and some Rocky Mountain states, the connections among the boardrooms, petroleum clubs and conservative preachers are well established." Phillips noted that the economic conservatives in the oil and gas, coal, and automobile industries that play a dominant role in the Republican Party "may not believe in end times, but their opposition to regulatory environmental prescriptions and tougher fuel-efficiency standards makes them ally with the economically undemanding religious right."

He also points out that corporations, particularly in the energy industry, provide direct financial support for their supporters in the Christian Right. Such groups as the Interfaith Council for Environmental Stewardship, which has pronounced as unfounded current fears about such environmental problems as global warming and species extinction, and the Acton Institute for the Study of Religion and Liberty, which emphasizes market mechanisms and private property rights, receive "corporate support for what amounts to a pro-business, pro-development explanation of Christian stewardship." He reports that ExxonMobil provided financial aid for years to the Acton Institute, whose leader contends that "left-leaning environmentalism is idolatrous in its substitution of nature for God."

We hope that the information in this book will help contribute to an understanding that, rather than pitting nature against God, God is reflected in nature and in the well-being of God's creatures, our children. We hope members and churches of the religious Right will join in the effort to save our children from poisons placed in the environment by human activity.

LOOKING AT OURSELVES

And what about the rest of us? Do the values by which Americans live their daily lives demonstrate a real concern for the health and future of the nation's children?

Certainly most families that find their own children and grandchildren are victims of toxic pollution will do all in their power to save them, to attack the source of danger, to challenge the system that permits such danger. When children are stricken, the families teach themselves the science and medical details of the illness, they cry the alarm and demand a response. They want polluters to be held accountable, and they force their way into the political system, trying to make government at all levels do the right thing for their kids. And there also are the thousands of other Americans, the *posse comitatus,* who are rallying to the aid of the children—the scientists, physicians, environmentalists, teachers, and ordinary citizens we wrote about earlier, whose values tell them that the assault on our children is wrong and must be stopped and they must play a role in stopping it.

Many other Americans, however, whatever the values they profess or even think they hold, appear to be indifferent to the plight of children and, in their indifference abet the assault on our kids. Some, when their own economic self-interest is involved, react with overt hostility to efforts by families to protect their kids and stop the poisoning of their community.

Recall Doc Shelton, Walter West, and Diane Avrilett, who have spent years trying to get industry and local government to clean up industrial effluents polluting streams and a lake and causing cancer and learning disabilities in Lufkin, Texas. The town government, according to their account, went out of its way to keep the offending plant in operation, and their neighbors turned against them, once going so far as to shoot bullets at Avrilett's restaurant. Recall the mothers of disfigured children in Dickson, Tennessee, who, when they blamed pollution from a nearby toxic waste dump, were told by fellow townspeople they had harmed their own children by drinking or drugs or by inbred marriages, none of which was true. Think about Ann Tillery, whose neighbors laughed at her when she became an environmental activist collecting obituaries of children after her own grandson died of leukemia. And think about Hilton Kelley's story of how the NAACP and clergy of Port Arthur, Texas, declined to take on polluting refineries because the companies were making small financial contributions to their institutions. Or Juan and Brian Parras, who were ignored when they tried to block a high school for poor Latino kids in Houston from being built in the middle of a giant complex of refineries and petrochemical plants.

Most Americans would not dream of exposing their children to poisonous chemicals or criticizing neighbors who were trying to protect their children. Their values would not permit it. But it is probably fair to say that, by and large, people running the daily marathon of life and work give little thought to the dangers that surround the children. How many of us check the ingredients of our household cleaners or heed the warnings on the labels of garden pesticides? Or go ahead and use those products even if we do because of their producers' assurances that they are safe or out of the naive belief that the government would not allow dangerous products on the market? How many of us spend the extra few dollars to give our children chemical-free food and drink even if we can afford to? Do our farmers worry about what the chemicals they put into their fields to maximize crop yields are doing to their children, now and later in their lives? As parents, do we make sure our children's schools provide a safe, healthy environment for small bodies? As consumers, do we stop to think about the impact of the endless ephemera we purchase, what kind of pollution is produced by their production, use, and disposal? Why do so few of us take the trouble to demand that polluting factories and commercial facilities clean up their acts or leave our communities? Why do so many of us vote for politicians who do the bidding of the corporations that are poisoning our communities and sickening our kids?

Do our family values truly encompass the health and welfare of our children? Do we care enough to challenge the system that puts them at risk? Do we care about justice for our children?

Twelve

JUSTICE

It need not be this way.

We can curb the flow of poisons into the environment and into their bodies. We can change the way we live. We can be far more cautious about what is put into the air, the water, the land, our houses, our schools, our food. We can change the rules and make them work. We can make products that do not harm the young or the rest of us. We can elect honest, ethical, intelligent people to govern and deploy the nation's resources to preserve the environment and public health. Our science can again be focused on the public good and less on private gain. The corporations and their managers can be held accountable for the harm they do.

We can obtain justice for the children.

But will we? Will we as parents, as grandparents, as a society, finally acknowledge the peril in the world around us and how it is affecting children, born and unborn? Will we at last overcome our indifference and inertia and demand the profound changes that will be required? Will we act decisively to see that those changes are made?

Or will we, in the words of Commonweal's Michael Lerner, a leader of the emerging environmental health movement, continue to avert our eyes and "stumble toward a future we are unwilling to face honestly"?

If we are to act, it must be soon. The poisoning of the environment that threatens our children is only one of a number of human assaults on the natural world—others include accelerating climate change, the massive extinction of plant and animal species, the destruction of soil and water resources—that now put into question the continued viability of life on this planet. As the Millennium Ecosystem Assessment by 1,300 scientists enlisted by the United Nations stated in 2005, "Human activity is putting such strain on the natural functions of Earth, that the ability of the planet's ecosystems to sustain future generations can no longer be taken for granted."

This view is no doubt derided as alarmist by critics, especially those whose activities are causing the crisis. But we need to be alarmed; we must overcome disbelief, if not for ourselves, then for our children and grandchildren who will inhabit the future. We cannot, given what we know already, simply assume the worst will not happen. It may not, but the stakes are much too high to take such a gamble. For the sake of our children, we had better take steps to ensure it does not.

We believe that the changes made to save our children will go far toward saving the planet.

PARENTAL DUTY

The first line of defense for the children is, of course, their families. Parents can do much to shield their children, including providing them as toxic-free an environment as possible. As Elizabeth Sword, former director of Healthy Child Healthy World, remarked, "Parents are the CEO in their own home." They can remove poison-containing products from their homes and not use them on lawns or gardens. They can insist that their houses be free of lead and that their children not be exposed to asbestos in their schools. They can give their kids a nourishing, balanced diet of unadulterated food, starting before they are conceived. Children can be given nurturing home and school social settings and encouraged to keep their minds and bodies fit with physical exercise and by avoiding harmful substances such as tobacco, alcohol, and recreational drugs.

The precautions that parents can take to minimize the toxic threat to their children could fill a book. In fact, a number of books contain-

ing just such advice are already on the market, and we list some of
them in an appendix. Another appendix includes suggested immediate
actions that parents should consider for their homes as well as websites
of organizations that offer advice on children's environmental health.

But given the overwhelming momentum of our technology-based,
consumption-driven economy, our industrial agriculture, the sea of
chemicals already out there, the new ones pouring into the environment every day, and the dearth of information about the nature of the
threats to the children and how to deal with them, there is only so
much parents and local communities can do on their own—and that is
not remotely enough. We will have to change the way the world now
works.

This will be an immensely challenging task. But if we lift our eyes
a bit, we can see that our man-made environment was once a much
safer place for our children to inhabit, and not so long ago. As a democratic society, we have it within ourselves to make the necessary re-
forms in our science, our medicine, our industry, our economics, and
our politics and re-create a safer, healthier, fresher environment. The
effort will be met with determined, well-financed resistance. But we
believe it can be done.

LEARNING AND KNOWING

First, we need to become more educated about the hazards that surround us and may harm our children. And it is not only parents who
need to learn. While our chemists are often superbly trained scientists,
for the most part they receive little instruction about the toxic effects of
the substances they work with. Schools at all levels pay little or no attention to the interactions of chemistry with ecology and evolutionary
biology. Instruction in environmental health ought to begin in elementary school and continue through high school and college. Most medical schools still give short shrift to environmental pediatrics, and there
are still too few practicing physicians who devote much thought to
what in the environment may be causing illness in their patients. Worst
of all, medical science continues to devote most of its resources to curing illness and very little to preventing it. The medical profession
ought to devote more education to preventing illness by removing the
environmental hazards that trigger so much of it. And it needs to pay
special heed to the environmental health of children.

We need and should demand more knowledge about the dangers to our children than is now available. We should have a national system to track the links between environmental hazards and chronic disease, identifying sick children by illness, location, and the hazards of the environment in which they live, the way we now identify and track infectious disease.

Many of the scientists and physicians we spoke with told us that the precipitous cutbacks in federal spending on scientific research have reduced the capacity of science to respond to public health needs. Federal spending on science research should be restored and expanded, and a sizable portion of that budget should be earmarked for children's environmental health. The National Children's Study, the massive epidemiological project that will follow one hundred thousand American children from conception into their twenties, will provide many of the missing answers. This study is the linchpin to understanding the social and environmental causes of illness, the linchpin to prevention. Its funding must be protected over the length of its execution.

We should not allow manufacturers to conceal the dangers their products and manufacturing processes create behind the convenient facade of "proprietary information." The government should place more data, not less (as it has now succeeded in doing, at the behest of industry), on its Toxics Release Inventory. California's Proposition 65 puts information in the hands of parents and consumers by requiring all products containing more than a minimal amount of a carefully considered list of toxic substances to display what they are clearly on the label. That it is a start and should be a national requirement. But much more is needed.

Companies that produce hazardous substances should be required to produce regular reports on the production and releases of such products, and those reports should be subject to outside audits, much as financial statements are today, and made public, especially to the communities near their facilities. The location and contents of warehouses for products such as pesticides should be made known to residents.

As we reported earlier in this book, only a fraction of the chemicals to which our kids are exposed have been tested for their toxicity and health effects. As the past decades have demonstrated, this situation has led to dereliction and tremendous harm. The testing of chemicals cannot be left to the companies that produce, market, and use them, whether they do it voluntarily or in obedience to government fiat. The temptation to find results that will enable products to remain on the market or new products to be sold with inadequate information about health effects makes commercial interests an unreliable source of test results. Even ethical corporate managers are constrained by their duty and accountability to shareholders to act in ways that maximize profits, not protect public health. Companies that make hazardous products should bear the costs of testing those products, perhaps by paying a fee into a general fund. But the chemicals should be selected and the tests must be conducted by those with no financial interest in the results— either by government scientists or by an independent body of experts with no ties to the company or industry. Once those tests are completed, they could be subject to review by company scientists and also by reviewers with wider social interests, including community groups, environmental health groups, physicians, ecologists, ethicists, and others with a stake in the ultimate fate of the chemical.

The testing and review of individual chemicals will be a long, drawn-out, painstaking process. Even more mind-boggling is the prospect of testing the effect of chemicals in combination for their synergistic effects on the body. As Dr. Martha Herbert of Massachusetts General Hospital and Harvard Medical School noted, if the three thousand so-called high-production chemicals now in commerce were tested in combinations and permutations of only three of them, it would require eighty-five *billion* tests.

Our research institutions, including the federal agencies, should switch their testing methods to innovations based on new science and technologies, such as the use of lab test tube studies that make use of new understanding of cellular pathways, as suggested by the National Research Council.

Even so, it clearly will take many years before we acquire the information and knowledge, not to mention action by companies and government, to remove one by one the hazards from the environment. Meanwhile, the children will continue to be exposed and many of them will suffer. So what can we do in the meantime?

PRECAUTION

One thing we can do is be careful—a lot more careful. We can demand that even in the absence of conclusive scientific proof, hazardous substances be withheld or withdrawn from the market whenever there is substantial evidence of serious or irreversible threats to health.

This commonsense solution has a formal name: It is called "the Precautionary Principle." The term is fairly recent and the concept still somewhat amorphous. But its practice has long antecedents. The phrase *Premum non nocere* ("First do no harm") has been embedded in medical education and, one would hope, in practice for centuries. Its premise is captured in folk aphorisms such as "Better safe than sorry" and "Look before you leap," although as employed, the principle turns out to be more complex and sophisticated.

A widely used definition of the principle was formulated by a group of scientists and physicians convened at the Wingspread Conference Center in Wisconsin in 1998: "When an activity raises threats of harm to human health or the environment, precautionary measures should be taken even if some cause and effect relationships are not fully established scientifically."

The principle acknowledges, in effect, the limitations of current science in identifying links between hazards in the environment and illness and environmental degradation. It establishes a means of devising policies to protect our children in the absence of conclusive scientific evidence and with broader democratic participation in the decision-making process.

Another, usually unspoken, premise of the principle is that many (if not most) polluters will not concede that their activities cause harm, even in the face of reasonable evidence. And, as several scientists noted, "a requirement of absolute 'proof' of harm before action can be taken is either ideologically motivated or deprived of a fundamental understanding of the limits of science." The precautionary principle enables—indeed, would require—action to protect children from harm in the face of uncertainty, whether the uncertainty is real or manufactured by polluters. Advocates of the principle do not call for the complete elimination of risk, a goal they know to be unobtainable, but they do insist that risk be lowered as much as possible.

Joel Tickner, a scientist with the Lowell Center for Sustainable Development at the Lowell campus of the University of Massachusetts, noted that the Precautionary Principle asks a very different set of questions about regulating toxics in the environment than the current regulatory regime. Instead of asking "What level of risk is acceptable?" it asks questions such as "How much risk can be avoided while still maintaining necessary values?" "Are there alternatives that achieve a desired goal (a service, product, etc.)?" or "Do we need this product in the first place?"

The answers to these questions would require a fundamental shift in the way the nation—and the rest of the world, for that matter— deals with toxics in the environment. The burden of proof that toxics cause harm would be reversed. Instead of those exposed to the haz- ard—potential victims—having to prove their danger, the maker of the substance would have to prove its safety. Chemicals would cease to be regarded as innocent until proven guilty. As Tickner points out, "Humans and the environment receive the benefit of the doubt under terms of scientific uncertainty and ignorance, rather than a particular substance or action."

Dr. Kenneth Olden, former director of the National Institute of Environmental Health Sciences, is convinced there is sufficient evi- dence of the harm caused by many substances in the environment to apply the Precautionary Principle now where "it makes sense." As an example, he said, "You know it can't be good to have thimerosal, a de- rivative of mercury, in vaccinations. Some people say it is a cause of autism. Whether I believe that or not, I do know that mercury is poiso- nous. Why would we put a poison in little kids? Nowadays kids have many immunizations, and it adds up to a lot. The Precautionary Prin- ciple says to me: Whether it causes anything or not, it just seems like a dumb process, so let's stop it."

The European Union has been moving somewhat gingerly toward this approach with a program called REACH—Registration, Evalua- tion, Authorisation, and Restriction of Chemicals. The program would do what the name says for chemicals suspected of causing health and environmental harm that are produced in volumes of one thousand metric tons or more a year. In support of the program, the European Environmental Agency commissioned a report entitled *Late Lessons from Early Warnings: The Precautionary Principle 1896–2000,* which presents a series of examples of how environmental and public health fiascoes might have been avoided if the Precautionary Principle had been adopted in the twentieth century.

The report noted, for example, that there were warnings of the toxicity of polychlorinated biphenyls as early as the end of the nine- teenth century, when workers exposed to the substance developed chloracne, a severe skin affliction. By the 1930s, there was a "low level of proof" that PCBs posed dangers and by the 1960s a "high level of proof." One dramatic example of such proof was an incident in Japan in 1968 when rice oil contaminated with PCB was consumed by some 1,800 people. Those who ingested the oil developed severe cases of chloracne, and many of the children of women who were pregnant when they ate it were growth retarded, had discolored skin, had lower than average IQs, and exhibited an apathetic manner. But it took more than another decade before countries, including the United States, began to regulate PCBs. In the United States, there has been no general requirement to clean up the widespread residue of its use. Meanwhile, many American workers, pregnant women, and their children were exposed to the chemicals while they were in use and continue to be exposed to residues still in the environment around the country, as told in the story earlier in this book about Pittsfield, Massachusetts.

The Food Quality Protection Act, which requires a tenfold margin of safety on behalf of children when regulating the amount of pesticides on food, is an example of the Precautionary Principle in action—or would be if it were adhered to.

There are many opportunities in sight for deploying the Precautionary Principle. Trichloroethylene is already regarded as the cause of a multitude of dangers to children. But TCE is not yet regulated as a toxic substance. Surely, while the evidence may not be 100 percent conclusive, there is more than enough to warrant keeping the stuff out of the environment and away from pregnant women and from children.

"Suppose the EPA made a decision it was going to reduce the use of organophosphate pesticides by 80 percent over ten years?" Dr. Philip Landrigan offered as a bold example. "That would cut the Gordian knot of risk assessment. Now industry has us where they want us: tied down in this endless process of risk assessment, which is like a big tangled mess of spaghetti."

Precaution can be pursued in ways other than simply barring the manufacture of chemicals. It could, for example, include integrated pest management techniques that slash the amount of pesticides used on crops and gardens; these techniques are already employed successfully in many areas across the United States. It could involve changing children's diets, which at the same time would combat the obesity that afflicts so many American children and renders them more susceptible to environmentally induced illness. It could involve the treatment of children already known to have been exposed to a cancer-causing agent even before symptoms appear, which might not happen until many years later.

Some scientists, even some who care about children and public health, regard the Precautionary Principle as merely risk assessment in a different guise. Perhaps that is so, but it would still throw a mantle of protection over our children that is less permeable than today's threadbare net.

As might be expected, many in industry react to the Precautionary Principle as if they saw the Apocalypse approaching. An article assailing the principle in *Chief Executive,* a magazine for corporate managers, carried the title "Invasion of the Fearmongers" and then, presumably without irony, quoted a chemical industry executive as saying, "The precautionary principle is probably the worst thing that could happen to this country." The critics complain that it would cause severe economic hardship, block the innovation of new products, and cause many essential commodities and products to be removed from the market.

The critics also warn that removing chemicals from the market on suspicion that they cause harm could have drastic consequences for public health. There is one particular story they use frequently to frighten off support for the principle. Wirthlin Worldwide, a public relations concern, presents the story as follows: "When studies in the early 90's [*sic*] suggested that the chlorination of public water supplies might expose consumers to a negligible risk of cancer, environmental activists called for a complete ban on chlorine. Peru responded by dechlorinating its drinking water. As a result, 300,000 people contracted cholera and over 3,500 died." "Kip" Howlett, director of the U.S. Chlorine Chemistry Council, had an even more "horrific" account, saying that there were 1.3 million cases of illness and 13,000 deaths.

A cautionary tale indeed. Except for one detail: It is not true. U.S. and Peruvian scientists who investigated the cholera outbreak reported in the British medical journal *Lancet* that the problem had been caused by a complex set of circumstances, including the absence of chlorination systems in most of Peru, illegal tapping of water lines, absence of sewage systems, and the storage of water in rain barrels. An editorial accompanying the article in *Lancet* said that the report should put an end to the "rumour" that the lack of chlorination was a deliberate decision by the Peruvian government. An article in the U.S. journal *Risk Analysis* concluded that "the Peruvian cholera epidemic was not caused by a failure of precaution but rather by an inadequate public health infrastructure unable to control a known risk: that of microbial contamination of water supplies."

Implementation of the Precautionary Principle is a work in progress. A number of problems need to be resolved, such as the danger of employing precaution where there are "false positives"—incorrect estimates of danger that might remove valuable products from commerce or divert resources from real risk. Dr. Bernard Goldstein, a physician and dean of the University of Pittsburgh's School of Public Health, has suggested that the cost of taking precautionary action be justified after adoption "with appropriate research to determine if the action does in fact deal effectively with the 'threat of serious or irreversible damage' that led to its adoption." In other words, once a product is removed from the market as a precaution, accelerated testing would be done to determine if it was indeed dangerous.

But as Ted Schettler and his colleagues affirmed in their book, *Generations at Risk,* decisions about protecting the health of our children cannot be left to scientists alone; they must be made by the broader society. "There always will be scientists who call for more research to 'prove' that some exposure results in an effect. Certainly research should continue, but if our interest is in protecting health, we cannot wait for science to provide proof."

Application of the Precautionary Principle inevitably would waste money and resources from time to time. Better that than wasting a child's life.

CHEMISTRY WITHOUT HARM

We cannot do without chemicals. They are the basic building blocks of our industrial civilization. The modern world could not exist without them. If we suddenly removed large numbers of chemicals from commerce without replacing them, the economy would be devastated, and so probably would public health. So if we implemented the Precautionary Principle and started to pull chemicals off the shelf, how would we keep things going, how would we hold the world intact? The obvious answer, it seems to us as well as to a growing number of scientists and even some in industry, would be to replace them with new and better chemicals that do not harm health and the environment.

That is already starting to happen. There is a whole new field of research and application called "green chemistry" or sometimes "sustainable chemistry" that is beginning to gain traction in university laboratories and corporate planning. It is happening slowly. It is not yet a revolution. But it could become one.

Green chemistry, as defined by the Center for Green Chemistry at the University of Massachusetts in Lowell, "reduces or eliminates the use or generation of hazardous substances in the design, manufacture and application of chemical products." In other words, green chemicals are designed from the start to serve a function without presenting any threat (or presenting less of a threat) of harm to the environment or to human health.

"The hazardous chemical in a red dye has nothing to do with the function of the dye," explained Professor John Warner, former director of the center. "Plasticizers don't have to be toxic—they are that way because the people who made them had no training about whether chemicals are toxic. If we understand what shapes of molecules and what groups of molecules contribute to a particular disease, then when I am designing a product I can recognize as a carcinogen, I better not use it. We can formulate better base materials at the molecular level."

Warner, who speaks with the fervor of an evangelist saving souls, is one of the acknowledged fathers of green chemistry. Another is Paul Anastas, who now heads the Green Chemistry Institute of the American Chemical Society (no relation to the American Chemistry Council, the industry trade association). The new chemistry differs from the old "where we've created a mess and then come up with bandages to make it less bad," Anastas explained.

Warner was working for the Polaroid company when he developed ways of making nontoxic materials for Polaroid film. He took the discovery to the EPA, where Anastas, an old friend of his, was in charge of regulating such materials. They discussed how there was no language to talk about toxicology or environmental health in the normal practice of chemistry. Chemists, said Warner, have "no clue about what makes a chemical toxic. We are dealing with monkeys typing Shakespeare: Every so often something comes out safe, but we have no idea why."

Warner told us that his two-year-old son died when a liver transplant failed. "Why did he die? I had probably synthesized more new molecules than anyone else on the planet, yet I didn't know why he died. What if something I touched in the lab somehow caused my son's disease? I realized something had to be done. So I started a green chemistry doctoral program [at the University of Massachusetts] to train scientists to develop new materials in a safe way."

Warner's center is practicing what it preaches. Its experimental lab, the usual organized chaos of benches, beakers, retorts, plastic tubes, freezers, and gas-inhaling hoods, is a beehive of innovation. One student is cloning an enzyme that will be able to degrade a plastic polymer so that it can be reclaimed and reused; another is working on an alternative to what has now been recognized as a hazardous chemical in Teflon; still another is looking at a biodegradable alternative to glues now based on petrochemicals.

Warner said that companies are contacting him all the time to ask him to come up with safe chemicals for their products. The wake-up call, he said, was the Bhopal tragedy in India, when a chemical release caused thousands of deaths. But for the scientific community, "it is a huge mind-set change to accept that chemicals can be made safe."

The emergence of green chemistry could create a win-win-win situation for the United States and the world. Most important, it could greatly reduce the toll that man-made toxics are taking on children and the rest of us, including our environment. Great care would have to be exercised to assure the new chemical is not toxic or at least substantially less toxic than the chemical it replaces. We don't want to make an expensive leap from the frying pan into the fire. But Warner said that toxic molecules can be identified and eliminated as building blocks of the new chemical. And rather than imposing a financial burden, sharply reducing the threat of toxic hazards to the health of our children could save billions of dollars for the nation by reducing medical costs, eliminating lost work time for parents, and removing the rest of the economic toll now imposed by environmentally induced illness.

Green chemistry could reduce the need for complex environmental rules and eliminate many of the adversarial confrontations between industry and environmentalists, industry and public health advocates, industry and government. It would diminish the dependence on the world's dwindling supplies of oil required to fabricate the petrochemicals that form the bulk of today's chemical production. It holds out the prospect of enormous benefits for the chemical industry and industries that use its products. First of all, it would sharply cut the large sums that industry now spends on complying with environmental regulations and penalties for violating those regulations. It could also cut down appreciably on industry spending on raw materials and hazardous waste disposal by creating products that are not hazardous and can be continuously recycled. A study by the Toxics Use Reduction Institute at the University of Massachusetts at Lowell found that nontoxic products could be made less expensively than materials now used in many products, such as formaldehyde, perchloroethylene used in dry-cleaning, and lead in a wide variety of uses.

There is no reason, for example, that cancer-causing polyvinyl chlorides cannot be phased out as an ingredient in plastic in the near future. Researchers at Tufts University have found affordable alternatives for just about every commercial application. Alternatives made from compositions of cardboard or of cornstarch, sugar, and vegetable could replace PVC and many other products.

The demand for safe processes, materials, and products could open broad new markets for corporations that could lead to a surge of economic expansion similar to the burst of affluence that arose out of Silicon Valley in the 1990s. The quest for new materials, processes, and products could stimulate a wave of invention and innovation. Certainly no jobs would be lost as new chemicals are produced to replace the hazardous substances now on the market, and probably many new jobs would be created to meet the demand for safe chemicals. (Well, some jobs might be lost in companies created to defend dangerous products or to manufacture doubt about scientific evidence showing that they cause harm. There would be fewer chemicals in need of defending and dwindling evidence of harm.)

These moves may help push the corporate and investor world away from current short-term thinking, obsessed with quarterly earnings expectations, that hobbles long-term decisions. Evolution into nontoxic manufacture requires a longer horizon for planning and production, and this, in turn, will prove better for the American economy.

Some have suggested that green chemistry could be the basis of a new industrial revolution. It could make obsolete the methods of the first industrial revolution, which, according to architect and green design guru William McDonough, was based on the premise "If brute force doesn't work, we are not using enough of it." The toxification of the environment, McDonough has asserted, is "a tragedy of our own making because we have no other plan. We are not so smart. It took us five thousand years to put wheels on luggage. . . . The toxic age will not end until we plan the end of the toxic age." He envisions "a world of abundance and good design—a delightful, safe world that our children can play in."

So far, industry has not made any loud objection to green chemistry. In fact, there are glimmerings that suggest some business leaders understand the benefits of safer technologies, and some companies have dipped their toes into the waters of green chemistry. Dow Chemical and Cargill, the agribusiness giant, have joined forces to produce nontoxic products made from grain. Samsung Electronics has committed to phasing out packaging using some applications of polyvinyl chloride, which, it said, "could have very adverse environmental effects." The rapidly rising demand for organic foods, cosmetics, and household products suggests that the public is ready for green chemistry. Corporate titans such as General Electric and Wal-Mart are making gestures toward using cleaner technologies and products, although some environmentalists voiced suspicion that what is happening amounts to what they call "greenwashing" of environmental and health records.

MOTIVATING INDUSTRY

History suggests, however, that industry will not embrace the new chemistry quickly or eagerly. Some of our major industrial sectors have resisted new technologies, particularly those designed for health, safety, and the environment. The sad state of the U.S. automotive industry, for example, is in part, at least, a result of its stubborn resistance to efforts to make safer, more fuel-efficient, and environmentally benign vehicles. When seat belts were first proposed, auto executives said Americans would not use them. When the government moved toward requiring catalytic converters to clean tailpipe exhaust, they complained it would hamper the efficient functioning of engines and make cars unaffordable. They said the same when the government mandated vapor-capturing equipment to reduce the pollution caused by fueling of cars and trucks. Airbags? Impractical. Fuel efficiency standards? Americans want big cars, so we are going to give them Hummers. As a result, the European and Asian auto manufacturers, who were quicker to adopt new technologies and innovative designs, are taking over a lion's share of our own American market.

Safe chemicals would help shift the chief mission of the medical community from curing patients to preventing illness. This, however, might not sit well, as Kenneth Olden observed, with those who now profit from commodifying cures and commercializing them into profitable products.

Public pressure, if it can be mobilized, might move American industry toward making safer chemicals and products. More likely it will take a combination of such pressure plus a government mandate to prod the business community into overcoming the kind of no-can-do spirit exhibited by the auto companies and some other industries. But if government requires and supports green chemistry, it would not only provide a safer environment for our children, it could also help save U.S. industry from itself and increase its competitiveness in world markets.

Decisive government action could create market conditions that, in the words of Michael Wilson, a scientist with the Center for Occupational and Environmental Health at UC Berkeley, "will begin to motivate industry to focus its enormous talent and technical capacity on innovating green chemistry at a level commensurate with the scale and pace of chemical production." The current law, Wilson said, has created an uneven playing field for sustainable chemistry because it encourages the continuation of toxic chemicals in commerce. Strong rules governing toxic chemicals, vigorously enforced, would offer a persuasive incentive for manufacturers to come up with greener chemicals.

Even if industry leaped to adopt the new chemistry, however, it would be many years before enough toxics were drained from the environment to substantially ease the assault on children's health. There are just too many bad actors out there now and too few new safe chemicals ready to go. Government subsidies and a true commitment by industry backed by aggressive advertising and marketing could help speed up the pace. But our kids need more protection now.

What is needed to move quickly is a fundamental restructuring and strengthening of U.S. laws governing the manufacture, use, and sale of chemicals and strong government enforcement of those laws. As we have seen, the Toxic Substances Control Act, as currently constituted, is inadequate to the task of protecting public health and the environment. Some have suggested a system similar to that used to screen new drugs, where the Food and Drug Administration tests all new products before allowing them on the market. Even that system, however, fails to catch all dangerous drugs.

CORPORATE REFORM

No matter what principles or laws are adopted, there are those who, driven by greed or ambition, will seek to evade them, to cheat and deceive. Sometimes corporations are caught in the act and punished, but usually the punishment consists of monetary fines that are mere pinpricks to multibillion-dollar companies. Perhaps what is needed to encourage honesty and caution by corporate managers is to give some prison time to those who know about the harm their products are causing, conceal that harm, and continue to market those products to an unsuspecting public. Our nation must eliminate the common corporate practice of settling lawsuits with a denial of guilt and, in essence, blackmailing families into silence.

Taming corporate power and its abuses will take far more than ineffectual regulatory Band-Aids, fines, and a few prison sentences. The time has come to rethink the position of corporations in American society. Their wealth and privileged legal status, their enormous resources for influencing public opinion and government policy, and their global reach have made them the dominant force not only in our economy, but in our scientific, political, and social affairs as well. Some of them have abused that power, often at the expense of the environment and our children. Corporations are now largely beyond the reach of other sectors of society to control, including governments and, in many if not most cases, their own shareholders. They have become, in the words of former Supreme Court justice Louis Brandeis, a "Frankenstein's monster."

It was not always so. The Founding Fathers, given the experience of oppression by great British corporations such as the East India Company, "hated corporations as much as they hated the King," according to corporate anthropologist Jane Anne Morris. In the first decades of the new country, she noted, few corporations were chartered, and those had strictly constrained powers. Less than a century later, President

Grover Cleveland would complain that "corporations, which should be the carefully restrained creatures of the law and the servants of the people, are fast becoming the people's masters."

That was in 1888, and Cleveland now appears to have been all too prescient. Recall President Eisenhower's accurate warning about the military-industrial complex. Perhaps Americans now ought to be thinking about returning corporations to their role as creatures of the law and servants of the people.

We ought to reconsider our current practice of granting corporate charters in perpetuity. States used to charter them for specific times and purposes and renewed the charters only if those purposes were fulfilled. Now corporate charters can be virtually eternal, no matter how the company performs or what harm it may cause. We should discard nineteenth-century law that grants corporations the same rights as individual citizens under the Constitution. Corporations are not people; that is bad fiction. They are far richer and more powerful than all but a handful of individual citizens. Given the same rights as individuals, they are able to exert decisive influence over the electoral process and place politicians in office who represent them, not the public.

As we have pointed out, market capitalism places no responsibility on corporations to act for the public good. Perhaps what would work would be some form of social capitalism, where the absolutes of the marketplace are tempered by a built-in ethic of responsibility for the public good. Corporations could of their own accord keep their operations open, honest, and accountable, voluntarily making sure that they do not endanger children by any of their operations or products. They could be required to absorb the costs their pollution imposes on public health and welfare and not simply pass those costs on to the taxpayers and to families who must pay the bills. Shareholders could demand that the management of their companies act in the public interest and the interest of children. That is starting to happen, although infrequently. But the government of a democratic nation can require corporations to act in socially responsible ways.

POLITICAL ACTION

It begins at the local level. Parents and grandparents whose children are threatened with environmental hazards can join with other citizens in their community to demand political action to remove the hazards and if that doesn't happen remove the politicians who refuse to act. Lois Gibbs, the housewife who organized her community in Love Canal, New York, to obtain justice for families whose houses were built on top of a toxic waste dump, formed a national organization now called the Center for Health, Environment & Justice. It helps communities around the country organize politically to address toxic threats to their homes and children. She told us a story about Franklin, Ohio, where a new school was built improperly so that the walls remained wet, and the children grew sick from the mold, the gases, and the chemicals used in construction. When the school board refused to do anything about it, the parents went to court and got a judgment removing the board from office and replaced with a board that would act.

In New York, Peggy Shepard's community organization, West Harlem Environmental Action, has successfully pressed City Hall to grant relief from environmental hazards, including putting additional safeguards on a wastewater treatment plant. Ms. Shepard says, "With a little access and smarts you can do something. You have to confront and challenge."

Confrontation need not be carried out with antagonism. Charlotte Brody, a registered nurse who was a founder of Health Care Without Harm, a group seeking to take toxics out of medical practice, told us about the successful campaign that (as related earlier) led to shutting down thousands of medical water incinerators. Instead of blaming the facilities that burned the waste, instead of making them feel guilty, the campaign told the incinerator operators that they may not know they were polluting people and explained that they could be part of doing something about it. "To understand strategically how you are going to win, you have to stand down from your own fury," Brody said. To get a polluter or politician to change, she explained, "you have to give him a story in which he is the hero."

She added, however, that people cannot be passive in the face of threats to their children or community. "Being a victim is not acceptable. You cannot find solace in losing."

JOINING FORCES

Achieving change will require a strong social movement. Such a movement has yet to take shape.

Parents and the organizations they have formed locally and nationally to help their sick children would be the natural core for a new
movement to coalesce around. There are hundreds of these groups,
each focusing on a specific illness, such as autism or asthma, and seeking relief in specific communities from polluting facilities or waste
sites. Families across the nation in communities threatened by TCE
contamination have begun to network and organize as a putative force
for change. There are also many organizations devoted to children's
health generally and to environmental health in particular. Some of
these organizations consist of a few families, but many of them have
hundreds or even thousands of members. They have publications,
websites, and chat rooms, and some of them have lobbyists. They have
a voice. And they have voters. To exercise their potential strength,
however, they would have to join forces, to form a national organization or network that would speak and act for all of them.

If they could accomplish that, the children's health groups could
form alliances with other movements whose goals have been blocked
and accomplishments eroded over recent years. Natural alliances for
those seeking to protect children from environmental ills would be
with the national environmental, environmental health, and environmental justice movements and with children's welfare organizations.
The public health community, with decades of know-how in prevention of illness, would be a strong force in a new movement. For that
matter, the scientific community, which has seen federal funding dry
up and has witnessed the distortion of science, cannot be satisfied with
the current status quo. Those whose religious beliefs focus on the sanctity of life should rally to the cause of children's health.

Natural does not mean easy. These groups sometimes have trouble
closing ranks even within themselves. But together, they could form
the base for a broad coalition of those sectors of American society that
are dismayed and, in effect, disenfranchised by the direction the country has taken in recent years. This would include the civil and human
rights organizations, which have watched those rights being trashed in
the name of fighting terrorism. The American labor movement has
been beaten bloody by an unrelenting antiunion offensive by corporations, an offensive tacitly supported by the right-wing politicians in
power. Labor has also suffered from the hemorrhaging of manufacturing jobs to distant continents. Some unions, such as the United Steel-

workers, are already wise about the nature of environmental hazards, not only in the workplace, but in communities, and are addressing the problem effectively. Across the nation, in rural, suburban, and urban communities, thousands of grassroots community development organizations have sprung up over recent decades to repair physical, social, and economic problems at the local level; they are all family-based, a natural force to join a coalition and carry out its goals.

The antiwar movement has had trouble being heard and attracting support because the end of the draft has meant that it is chiefly the sons and daughters of the poor and minorities who fight and die in places like Iraq and Afghanistan. But its members are passionate and active. Women's organizations, with agendas including reproductive health that have been ignored or opposed by Congress and the White House, should be ready to sign on.

There are also millions of Americans who are watching with alarm as the wall between church and state is being dismantled and would be glad to change the politics that are tearing it down. Millions more seem dissatisfied and angered at the direction the country is taking in the early years of the twenty-first century but have been given no cause to rally around.

Forging a new popular coalition and welding it into a potent social force would be a tall order.

But a new movement built around the central principle of environmental health could, in the words of Michael Lerner, "be the greatest unifying force since the great progressive and social movements that swept the world over the past two centuries. . . ."

RETHINKING OUR PRIORITIES

None of this will happen until all of us—parents, grandparents, and the rest of us—make up our minds that it is finally time to take care of the children. We will have to overcome our disbelief that they are in danger, our credulity in the assurances that polluters and their apologists give us that they are causing no harm. We can no longer be passive citizens, accepting what is thrust on us and our children.

For starters, we will have to become more intelligent consumers, cutting down on the piles of stuff we are lured into buying and especially informing ourselves about what is in that stuff to make sure it is safe and does not pose a hazard to our children. The physician and neuroscientist Martha Herbert complained that "we've given in to a consumer mentality where we take whatever we're given. The stores contain unspeakable quantities of junk—aisles and aisles of things like fake plants. Our natural resources are disappearing to make this stuff, and they are full of chemicals. Why do we need it?"

We need to rethink our economic priorities. Do we live in "a world of scarce health resources," as the dean of the Harvard School of Public Health wrote a few years ago? Or are we just misdirecting our resources? We spend billions of dollars a year on alcohol and tobacco and electronic toys and other ephemera and hundreds of billions on questionable wars. We spend $5 billion a year on Halloween costumes and candy, and $80 billion is thrown away on gambling. Surely some of these resources could be used to take better care of our kids. And, of course, we could save the nation many billions of dollars each year that it now spends on medical care for sick children. Where there is a conflict, maximizing the health and safety of our children should take clear precedence over maximizing production and profits.

If it takes a village to raise a child, it also takes a village to protect a child. We must regain our sense of community, our understanding that what affects one of us can affect us all. If a neighbor's child has cancer, or asthma, or autism, or a cleft palate, our own children and children to come may also be at risk. When the mother next door whose child is sick with leukemia starts asking loud questions about pollution from the biggest employer in town, we should not think first about our own property values. Ask not for whom the bell tolls.

"Everyone is feeling anxious about the loss of a sense of the commonwealth, the common good," says Dr. Richard Jackson. "Before World War II, we built libraries and public spaces that were for the public good. Suddenly, it's all about the individual self." We have been led to believe over the past few decades that individual opportunity and fulfillment is everything and takes priority over our duty to the community. As a result, the social bonds that keep us together, that lead us to help and support one another, to care for our neighbors' children as well as our own, have become badly frayed. Those bonds need to be repaired.

The very design of our communities often erodes those bonds. We need to reinvent the American community. Homes near one another and within walking distance of jobs and stores. Sidewalks. Community spaces. Community activities. Family farms nearby and farmers markets in town. And living, green nature. In a study of schoolchildren with ADHD, those who had higher contact with nature showed better concentration, task completion, and ability to follow directions; the more natural the setting, the greater the reduction in ADHD symptoms. "The way we build our communities, the extent to which nature survives and we have contact with nature, the psychosocial environments that we offer through our communities and families, the food environment . . . there are many factors that comprise the environment. We need to be thinking about all of those, if we want to pursue a global vision of safe, healthy environments for the American people," affirms Dr. Howard Frumkin, director of the National Center for Environmental Health of the Centers for Disease Control and Prevention.

TAKING CONTROL

A social movement whose mission is community and the health of American children can pursue changes in consumer action, present a unified body to confront corporations singly or by industry, locally or nationally, use the unbounded capacities of the Internet to get out news and information overlooked by the corporate media, provide practical advice to parents, learn from one another's lawsuits.

This social movement can turn into a new, irresistible political force. The assault on the children is across so broad a front, the problem so complex, the potential solutions so far-reaching, the opposition so intransigent or indifferent, that solutions must also be found at the national level and probably at the international level as well. And that means returning to a responsive, nonideological politics that puts partisanship in a backseat in order to deal with the real, pressing needs of the American people, including and especially taking care of its children. We urgently need decent, intelligent people to run this country.

In a democracy, political fashion and governments can change with frequency. In our lifetime, the political zeitgeist has zigzagged from the Left to the center to the Right and again to the center and to the Right. Of course, the electoral process has to be honest. If elections are rigged or stolen, there is little prospect for political reform.

The American people will have to find a way to place into office politicians who will act for the public good, will act to protect our children. As Mark Miller, a pediatric environmental health physician in San Francisco, observed, "All health problems are political system problems." And Philip Landrigan believes that "with the proper people in the White House, we could set some very bold goals—a bold vision like Franklin Roosevelt had in 1932 that could turn the country around. The nation has a real responsibility to protect the environment for our kids and the next generation. They are the future of our nation."

We must, must become active, informed participants in our democracy. We have to vote—too many of us don't. We must vote wisely to elect governments at all levels that will work hard for our welfare and happiness and the future of our children, not for special interests whose only goals are wealth and power. We need to end the ability of those special interests to control the political process with their money. There is no reason we cannot have public funding of elections; other countries do. We should listen carefully to what political candidates are saying and doing and not respond to slogans built around vicious attacks, ads, or hot-button issues such as abortion and gun control, which are cynically used to inflame voters. Those are important, but there are other important issues that have a greater impact on our children that those same politicians ignore or actively suppress.

When candidates appeal to our narrow self-interest by saying they will not raise taxes, we should understand what they mean. What they mean is that they will deny us the vital services a government owes its citizens in order not to raise the taxes of those who are already wealthy. It means shutting down the libraries of the Environmental Protection Agency in the same year the Exxon Corporation announces record profits of over $30 billion. It means eliminating or starving federal research programs that could provide the means to protect our children. We cannot put and keep in power those who care nothing about the children, who, after all, are the nation's future.

Nearly half a century ago, Rachel Carson published her great work, *Silent Spring,* which warned of the dangers chemicals were causing in the natural world. The book was ferociously attacked by the chemical industry, but it became a best seller and one of the most influential works of the twentieth century. Toward the end of the book, she noted that our industrial civilization is on a "smooth superhighway on which we progress with great speed, but at its end lies disaster." But there is another road we can take, she said, that would take us to a safe destination.

"The choice, after all, is ours to make. If, having endured much, we have at last asserted our 'right to know,' and if knowing, we have concluded that we are being asked to take senseless and frightening risks, then we should no longer accept the counsel of those who tell us that we must fill our world with poisonous chemicals; we should look about and see what other course is open to us."

Should we not send our children down that safer road?

ACKNOWLEDGMENTS
<hr>

We are particularly grateful to Greater Boston Physicians for Social Responsibility and to Maria Valenti, Fay Reich, Richard Clapp, and Ted Schettler, who not only served as our fiscal sponsors, but also were unfailingly generous with their advice and assistance. Dr. Clapp, professor of public health, Boston University School of Public Health, and Dr. Schettler, science director of the Environmental Health Network, were also among our team of advisers who provided technical assistance and helped guide us through the complexities of our subject. Others who advised us included Dr. Philip Landrigan, chair, Department of Community Medicine, Mt. Sinai Hospital, and of the Center for Children's Health and the Environment, Mt. Sinai School of Medicine, whose encouragement persuaded us to undertake this book; Dr. Paul Epstein, associate director of the Center for Health and the Global Environment, Harvard Medical School; Dr. Martha R. Herbert, pediatric neurologist, Massachusetts General Hospital; Dr. Robert Jackson, former director of the National Center for Environmental Health of the Centers for Disease Control and Prevention; Dr. Michael Jacobson, executive director, Center for Science in the Public Interest; Dr. Philip Lee, Stanford University, and former assistant secretary of the Department of Health, Education and Welfare; Dr. John Peterson Myers, CEO of Environmental Health Sciences and coauthor of *Our Stolen Future;* Dr. Herbert Needleman, professor of pediatrics and child psychiatry, University of Pittsburgh; Dr. David Ozonoff, professor of environmental epidemiology, Boston University School of

Public Health; Dr. Frederica Perera, director, Center for Children's Health, Columbia University Mailman School of Public Health; Dr. Ellen Silbergeld, professor of toxicology, Johns Hopkins School of Public Health; Peggy Shepard, executive director of West Harlem Environmental Action; and Dr. John Wargo, professor, School of Forestry and Environmental Studies, Yale University. We are deeply indebted to them for sharing their wisdom and knowledge.

Others who gave us valuable suggestions and advice included Ken Cook, head of the Environmental Working Group; Jennifer Sass of the Natural Resources Defense Council; Michael Lerner, head of the Collaborative on Health and the Environment; Dr. Devra A. Davis, head of the environmental oncology department of the University of Pittsburgh; Dr. David Michaels of George Washington University; Dr. Karen Zelan; and Kathy Burns. Susan Edelmann, former director of the liberal arts program at New York University's School of Continuing Education, gave our manuscript her usual tough and thorough reading. Sophia Bertling, Becca Katz, and Gerardo Sanchez provided valuable research assistance.

Whatever errors of fact or analysis may remain in the book are ours alone.

Our sincere thanks to the Ford Foundation and especially to Mil Duncan, to Martin Kaplan of the V. Kann Rasmussen Foundation, and to Jeff Lewis and Teresa Heinz and the Heinz Family Foundation for providing us with the funding essential to gather material for and write this book. We are grateful to the Rockefeller Foundation for awarding us a productive residency at its beautiful study center in Bellagio, Italy, and to our gracious hostess, Pilar Palacia.

Wendy Strothman is our talented and energetic agent. Caroline Sutton is our patient and thoughtful editor at Random House. Our sincere thanks to Jonathan Karp, former editor in chief of Random House, for his decision to publish this book.

Finally, we thank Alan Jones, our computer genius, without whose cheerful emergency services this book would never have been completed.

How to Reduce Your Child's Risk and Change the Future

You don't have to be rigid about the following suggestions—just do what you can.

FOR PREGNANT WOMEN AND PREGNANCY-PLANNING COUPLES

It should go without saying that men and women looking forward to conceiving a child will avoid tobacco, alcohol, and recreational drugs.

Eat low on the food chain—grains, beans, vegetables, and fruits—and avoid foods high in saturated fat—butter, cheese, meat, and processed foods such as lunch meats made from ground meat and animal parts, because many persistent chemicals concentrate in animal fat. Foods high in protein also appear to lower the risk of having a child who develops leukemia.

See the suggestions about organic food in the section "Food" (page 267).

Women need to get enough folic acid (0.4 mg a day) for at least three months before conception and three months into pregnancy, to lower the risk of birth defects. Both prospective parents should take omega fatty acid supplements and eat foods with these healthful fatty acids (such as flaxseed or flaxseed oil, soybeans, organic tofu in moderation, and walnuts); these acids may reduce your child's risk of learning or behavioral problems. For the same reason, a multivitamin is probably a good idea. Pregnant and nursing women should also have a diet high in calcium and iron, which reduces lead uptake. Use iodized salt to combat chemical interference with the thyroid.

Eat fish lowest in mercury (see p. 268).

Don't work at a job or in a workplace that uses hazardous substances, such as lead, pesticides, asbestos, and solvents. Pregnant women and babies should not inhale fumes from art materials. Avoid exposure to dry-cleaning.

Water is a major pathway for toxics to enter your body and harm your fetus.

See the section on "Water" (page 266).

Since your child is most vulnerable during the prenatal stage, follow the suggestions under "Cosmetics and Other Personal Care Products" (page 271); if you have to set priorities, don't use nail polish or wear drugstore sunscreen, because of their hormone-disrupting ingredients. It's wise to wait until after your baby is born and finished nursing to dye your hair or resume using brands of a questionable nature. Avoid perfume and other products with added fragrance.

See the suggestions under "Household Cleaning and Laundry" (page 272).

Go to www.birthdefects.org, a website of the nonprofit Birth Defects Research for Children, and click on their "Healthy Baby Resource," a comprehensive guide to food, garden, drugs, and other sources of possible toxic exposure.

If you deliver in a hospital or if you or your baby is hospitalized for any reason, demand the use of equipment such as intravenous tubes without hormone-disrupting phthalates. For more information, contact Health Care Without Harm, www.noharm.org. It may be wise before delivery to read and print out their material and to consult your obstetrician, pediatrician, and the hospital about this issue.

FOR PARENTS WITH SMALL CHILDREN

Your Neighborhood

If you have a choice, find a relatively unpolluted community to live in and a residence as far from major highways as possible. A number of sources can give you information on the level of pollution in a neighborhood.

The Environmental Protection Agency maintains a database using information from the federal Toxics Release Inventory of chemical releases, locality by locality, at www.epa.gov/triexplorer. A similar database has been created by the nonprofit organization OMB Watch, at www.rtknet.org ("rtk" stands for "right to know"). With this and other databases offered free by OMB Watch, you can identify specific factories and their environmental effects and assess the people and communities affected.

Install Google Earth and search www.turboperl.com to check data from the EPA on air pollution down to the neighborhood level.

To identify all sources of pollutants in your neighborhood, use www.scorecard.org and www.epa.gov/enviro. The site http://toxmap .nlm.nih.gov can help identify the numbers and types of chemical-manufacturing facilities in cities and states across the United States. It is a geographic information system maintained by the National Library of Medicine, containing visual representations of data from the EPA's Toxics Release Inventory and Superfund programs.

Your Home

Take off outdoor shoes at the front door; at least your family should do so. For visitors, put down a mat at the front door with "leaves" of thin, sticky pages that are discarded when dirt accumulates on each. Websites for these products include sticky-mats.com and TackMat.com.

Wash your hands after coming home, after cleaning or gardening, and before meal preparation. Teach your children to wash their hands before they eat or snack. (It's understood that washing hands after bodily functions is already a strong American tradition.) Use soap rather than detergents and antibacterial products (see "Household Cleaning and Laundry," page 272).

If you work at a job or in a workplace that uses hazardous chemicals, such as lead, pesticides, asbestos, or solvents, leave your clothes at work or remove them immediately upon coming home and store them in an area away from children. And shower.

Open a window in every room for at least five minutes a day (or night) to let toxic particles escape.

Finish renovations well before you bring your baby home. If you need to repaint, use paints low in volatile organic compounds. If your house is pre-1950s, be cautious about possibly uncovering lead in the old paint. If you have to engage in a substantial renovation, move to a rental apartment if you can afford it, or seal off the space under renovation.

Have your clothes "wet-cleaned," a relatively new process that works just as well as dry-cleaning. If clothes are dry-cleaned with perchloroethylene, let them air outside for a few days or in a ventilated space your children do not use.

Don't use pesticides. If you are renting, ask the property manager to stop using pesticides, at least in your apartment. If unwanted animals or insects invade your home, use the least toxic methods such as traps and gels. Northwest Coalition for Alternatives to Pesticides, www .pesticide.org, has online fact sheets about alternatives. Beyond Pesticides, www.beyondpesticides.org, offers online fact sheets about the risks of various pesticides and a national directory of less toxic pest control companies. The Pesticide Action Network, www.panna.org, maintains a searchable database of pesticide chemicals, at www.pesticideinfo .org/Search_Chemicals.jsp, and information on nontoxic alternatives.

Head lice? Consult www.headlice.org. Insects? Apply the lowest DEET concentration possible.

Chemical tick-and-flea collars and baths for your pets are dangerous to your children and pets. There are new oral and topical flea treatments. And the Web has further advice about alternatives.

Water
Drink and shower in the best water possible. (Boiling water does not remove synthetic toxicants.) If you can afford it, install a whole-house filter that filters out contaminants such as trichloroethylene. If that's not feasible, use a carbon-based filter that removes lead, chlorine, and other contaminants. Removing smaller-molecule contaminants such as TCE or perchlorate will require a reverse osmosis or granular acti-

vated carbon filter, installed under your kitchen sink; the company comes to your home to change the filter once a year. These alternatives are better for the environment (and health) than water delivered in glass or plastic bottles. Carbon filtration is ten to twenty times less expensive than bottled water.

Tap water is often better regulated than bottled water and must meet more stringent standards at both federal and local levels, a study by the Natural Resources Defense Council found; see www.nrdc.org/water/ drinking/bw/bwinx.asp. The Environmental Working Group is currently studying different bottled water for its purity; see www.ewg .org/issues/bottledwater/index.php to participate or for the results.

Your local water utility is required annually to mail you a copy of their annual report on contaminants in your water. If this report omits pollutants of concern to you, such as solvents, look at their website or phone them. You can pay to have your tap water tested privately. For a list of labs certified by your state to analyze drinking water, see www.epa.gov/safewater/labs or contact the EPA's Safe Drinking Water Hotline at 1–800–426–4791.

You may be able to find out about the contaminants in your tap water through the Environmental Working Group's website, www.ewg.org/ tapwater. This searchable database covers water-testing data from forty-two states.

If you draw water from a private well, have the water tested, including for solvents, pesticides, perchlorate, and volatile organic compounds.

The American Dental Association cautions parents to avoid fluoridated water when reconstituting infant formula.

If you suspect or know of trichloroethylene in your water, you can find out what other families and communities are doing about it through a network of fellow communities; to join, go to TCE-request@list.cpeo .org.

Food

Breast milk is the best food for an infant. Avoid soy-based formula, unless your baby is allergic to milk. This period of a child's life is an especially important time for organic products. For bottle-fed babies, use silicone nipples and glass bottles. Choose a breast pump manufactured without harmful chemicals (phthalates and bisphenol-A).

Buy baby foods packed in glass containers.

Fish is a great food, a source of omega fatty acids and protein . . . except that mercury finds its way into the fat of many varieties of fish. For information about levels of mercury in various types of seafood, consult the Environmental Defense's Seafood Selector, www.environmentaldefense .org/go/seafood.

Fruits and vegetables are great foods, too . . . except for the residues of pesticides that may linger on them. For information about residue levels on various fruits and vegetables, consult the Environmental Working Group's website, www.foodnews.org. You can also subscribe to their monthly e-newsletter, www.ewg.org/newsletter/foodnews. Fruits and vegetables out of season are likely imported from southern countries where use of pesticides is widespread.

Buy organic foods as much as possible, especially milk, eggs, meat, and your children's most heavily consumed items. (Organic produce also needs washing, to clean off possible bacteria.)

"Natural" is a word without any real meaning. No standards or oversight back the use of the term.

Scrub nonorganic produce with a vegetable brush to remove at least some of the pesticides from the outside skin of the fruit or vegetable. Consult *Fresh Choices: More Than 100 Easy Recipes for Pure Food When You Can't Buy 100% Organic*, published by Rodale Press.

Since even chemists cannot decipher labels on foods or other products such as cosmetics (and "trade secrets" keep other ingredients from being listed at all), the only rule of thumb is that the fewer the ingredients with names you cannot recognize, the better.

To learn about the ingredients in manufactured foods, you can join the Feingold Association, a nonprofit group focusing on the harmful effects of certain additives and preservatives; you will receive a book deciphering hundreds of brand-name foods, plus a guide, *Healthier Food for Busy People,* explaining how to spot clues to what makes a food unhealthful: www.feingold.org.

Organic meat, poultry, eggs, and dairy products come from animals raised without antibiotics and growth hormones; no genetically modified products are involved in any part of production, from seed to feed. Animals must be given time to graze on organic pastures, though so far guidelines do not specify the amount of time. Organic produce is raised without using most conventional pesticides, without fertilizers made with synthetic ingredients or sewage sludge, and without bioengineering or ionizing radiation. Before produce can be labeled "organic," the farms where the food is grown are inspected to ensure they meet U.S. Department of Agriculture organic standards, then certified. Companies that handle or process organic food before it gets to your local supermarket or restaurant must be certified, too. These national standards were put in place in October 2002; see www.ams.usda.gov/nop/Consumers/brochure.html. To find or check out products from baby food and bedding to personal care, consult the Organic Trade Association, www.ota.com; search "Organic Facts," then "Directories," then "Organic Pages Online."

Avoid highly processed junk foods, which crowd out healthful foods and actively contribute to childhood illnesses.

Avoid nonstick cookware coated with perfluorochemicals, such as Teflon. Stainless steel is a fine alternative, especially for hot liquids. Cast-iron frying pans are inexpensive, are great for cooking, and add iron, an important nutrient, to cooked food. Ceramic titanium and porcelain-enameled cast iron work well as nonstick cookware. *New York Times* journalist Marion Burros recommends Le Creuset enameled cast-iron pans with a matte black interior. Ceramic and glass/ Pyrex are good products for baking. Note that microwave popcorn bags are also coated with the same grease-repellent chemicals as Teflon.

Never microwave in a plastic container or with plastic wrap unless the product says it's microwave safe (and maybe not even then). Ask the deli counter to wrap your purchases in paper. Choose plastic wrap made of polyethylene, such as Glad. For more information, download the short, clear *Smart Plastics Guide to Healthier Food Uses of Plastics* from www.iatp.org/foodandhealth and look at the eye-popping information at www.ecologycenter.org/ptf/toxins.html. Why not use waxed paper, ceramic metal canisters, and glass containers? Or Cool Totes, a lunch box made of lead-free nylon and cloth?

For portable liquid containers, stainless steel is a good alternative, especially for hot liquids. SIGG is one reliable company that manufactures food-grade steel bottles and produces plastic products without bisphenol-A.

For the infrequent times when nothing but a lightweight unbreakable material will do, set aside one plastic container, made of either #1, #2, #4, or #5 plastic. Keep an eye out for biodegradable plastics, which should be available soon.

For food storage, try glass Pyrex containers; ceramic, porcelain, and glazed stoneware; and stainless steel.

Canned foods: the fewer the better, because the chemical bisphenol-A, which lines cans, may leach into the liquids. Some organic food companies, such as Eden Foods, use cans without bisphenol-A linings.

Furnishings

The more natural the components, the healthier. If possible, buy furniture, kitchen cabinets, and flooring made from solid wood, not pressed wood or particleboard, to avoid formaldehyde and other chemicals. Why not substitute linoleum or chlorine-free plastic flooring for vinyl or the numerous siding and plumbing alternatives to vinyl PVC?

You can see which plastics offer the most harm through Greenpeace's "Pyramid of Plastics," http://archive.greenpeace.org/toxics/pvddatabase/bad.html, and find alternatives for almost any use at the PVC Alternatives Database, http://archive.greenpeace.org/toxics/pvcdatabase, and the Healthy Building Network, www.healthybuilding.net/pvc/alternatives.html. Another good guide to healthier building materials is www.pollutioninpeople.org/safer/products/building materials.

Read www.besafenet.com/pvc for information on the hazards of PVC, examples of common PVC products in the home, and safer alterna-

tives. To find out which companies have phased out PVC, see www
.besafenet.com/pvc/companypolicies.htm.

Use natural fibers such as wool and cotton, which are made without
styrene-butadiene or polyurethane and are naturally fire resistant. The
permanent press and stain resistance on bedding usually comes from a
dose of formaldehyde; in other cases, the sheet may be made with
nanoparticles, whose safety is far from tested at this time. Consult
www.ewg.org/pbdefree for a searchable database of household furnishings free of brominated flame retardants. Mattresses and pads
made of natural latex can be found at local stores and on the Web (see,
for example, www.nontoxic.com).

Decline optional treatments for stain and dirt resistance on furniture or
carpet.

To find companies that manufacture products without hazardous materials (such as flame retardants, phthalates, and PVC), search the interactive "Chemical House" maintained and updated by Clean Production
Action, www.cleanproduction.org. Another great source of information
is Healthy Child Healthy World's HeathEHouse, www.healthychild
.org, an interactive website displaying the various rooms of a typical
home, the possible contaminants lurking in each, and advice about
simple ways to avoid or minimize risk.

Baby Green Marketplace, developed by the Birth Defects Research
nonprofit, offers 1,060 listings of companies selling natural, organic, or
green products. Click on www.birthdefects.org to access this resource.

A database disclosing the ingredients in some brands of baby mattresses, bath books, baby bottles, and teethers is available from the
Environment California Research and Policy Center website, www
.environmentcalifornia.org, at "Product Lists."

The Web carries lots of sites with ratings of products for their health-
fulness or lack thereof. One of the most reliable is www.thegreenguide
.com.

Cosmetics and Other Personal Care Products
Even behind the simple word *soap* lurks the probability of toxic chemicals. The words *fragrance, preservatives,* and *colors* in the ingredients

list are tip-offs of potentially harmful chemicals. Personal care products, the purview of the FDA, do not have to meet standards for use of the term *organic*. But the Organic Trade Association screens many items; see www.organicconsumers.org/bodycare. Another source of information, listing both the most harmful ingredients to avoid in soap and shampoo and names of product lines without them, is www.thegreenguide.com/doc/99/soap.

To understand the toxins that lurk within cosmetics, look at the Alliance for a Healthy Tomorrow's fact sheet, at www.healthytomorrow.org/PDF/CosmeticsFactSheet.pdf.

To avoid harmful ingredients, consult the searchable database of over twenty-five thousand brand-name personal care products and the seven thousand ingredients they contain, www.ewg.org/skindeep, developed by the Environmental Working Group and the Campaign for Safe Cosmetics. For a list of companies that have pledged to eliminate hazardous ingredients from their personal care products, see www.safecosmetics.org/companies.

Don't give your child fluoridated toothpaste, which little children often swallow (notice the poison control warning on the tube), until he or she is six years old.

Some sunscreens include estrogen-disrupting ingredients. For a database of nonestrogenic products with UVA and UVB protection, consult www.ewg.org/sunscreen.

Household Cleaning and Laundry
You don't need industrial-strength, synthetics-based cleansers to have white laundry and clean countertops. Nor do you need a different type of cleaner for each chore. You can make your own cleaning solutions with ingredients off your kitchen shelves. An example: Pour baking soda mixed with vinegar down the drain to clear pipes.

The site www.care2.com carries recipes for easy-to-assemble, nontoxic homemade cleaning products that are far cheaper and just as effective as commercial products. Consult www.lesstoxicguide.ca for substances to avoid, self-help recipes, and nontoxic products by brand name. Search the Web for "homemade household cleaning products" for other informative sites.

For information on how five types of chemicals commonly found in household cleaners present health risks, read "Household Hazards: Potential Hazards of Home Cleaning Products," www.womenand environment.org. This site also offers advice about actions you can take.

Microwave scrubbing pads to kill bacteria. Microfiber cleaning cloths are nontoxic, long-lasting, and efficient.

Avoid antibacterial ingredients, especially triclosan, currently added to handwashing and dishwashing fluids, toothpaste, and even socks!

Use nontoxic laundry products without a chlorine base. Fragrance-laden laundry softener fluff-dry sheets are among the most toxic products you can bring into your home. An alternative is a small rubber ball with little spikes you toss into the dryer along with the laundry (from www.gaiamliving.com).

Use a vacuum with a HEPA (high-efficiency particulate air) filter that traps dust and pollutants. Wall-to-wall carpets are a welcome mat for toxic-laden dust; area rugs of natural materials are a better choice.

Air fresheners are made from a number of chemicals, including formaldehyde (a carcinogen), naphthalene (a suspected carcinogen), xylene (a neurotoxin and possible reproductive toxin), butane gas (a neurotoxin), cresol, ethanol, phenol, and strong phthalate-laden fragrances. Some solid deodorizers include the pesticide paradichlorobenzene (a carcinogen that can also cause liver and kidney damage). Aerosol air fresheners release chemicals as tiny particles that can be inhaled deeply into lungs and transferred into the bloodstream. Plug-in air fresheners break chemicals into even smaller particles. Make your own using the simple recipe at www.eartheasy.com.

Toys

Children's toys, especially imported toys but also some made in the United States, may contain toxic ingredients. These substances can enter their bodies if the child mouths the object, as well as through the skin or inhalation. To check out toys that may pose a risk, consult www.kindersafe.org, which links to numerous information-packed sources, and www.HealthyToys.org, where the Michigan-based Ecology Center offers their toxic-toy-testing results and advice. Avoid soft vinyl teethers and toys, which may leach phthalates into your baby's mouth.

Outdoors

Use green lawn techniques and a green lawn service. In do-it-yourself lawn care, avoid pesticides, herbicides, and insecticides. For more information, see www.ehhi.org, a report on pesticides in lawn care products.

Give your children as much time as possible in the great outdoors—proved to lower levels of hyperactivity—slathered with UVA-UVB nonestrogenic sunscreen (see earlier) and topped with a hat, of course.

Your Car

If you want to avoid the toxics most common inside cars, consult the report *Toxic at Any Speed,* www.ecocenter.org/toxicatanyspeed.shtml.

Your Child's School

Find out if your child's school uses pesticides. If they do, ask them to notify you before a pesticide application so that you can keep your child home that day (some states mandate parental prenotification). Ask them to use alternative products. Introduce them to Beyond Pesticides, Inc.

Similarly, find out if the school uses industrial-strength toxic cleaners. Introduce the school to the materials produced by Inform, Inc., www.informinc.org, such as their free booklet, *Cleaning for Health.*

The Healthy Schools Campaign, www.healthyschoolscampaign.org, offers a free booklet, *The Quick and Easy Guide to Green Cleaning in Schools.*

Consult the nonprofit organizations formed to give parents and citizens proven tools to reduce toxic exposure in their children's schools. The Healthy Schools Network offers lots of ideas and fellowship, as well as a clearinghouse of information, www.healthyschools.org/clearinghouse.html. The Center for Health, Environment & Justice also runs an initiative to help parents and communities organize and take action for a healthier school environment: www.greenflagschools .org. Kid-appealing information, also informative for parents, is available from the Natural Resources Defense Council, www.nrdc.org/ greensquad.

If you're concerned about school lunch offerings or junk food vending machines, find answers from www.parentsagainstjunkfood.org and www.healthyschoolscampaign.org/issues/school.nutrition.

The Center for Science in the Public Interest has great suggestions for improving school food choices. Check them out at www.cspinet.org/ nutritionpolicy/policy_options.html#ImproveSchoolFoods.

If your school or town plans to renovate a playing field, make them aware of the hazards of artificial turf; consult playgrounds@ sciencecorps.org.

Your Pediatrician
To find a pediatrician trained in children's environmental health, use the directory developed by the Association of Occupational and Environmental Clinics, www.aoec.org/directory.htm.

Proactively ask your pediatrician if the vaccinations he or she proposes for your child contain thimerosal, a mercury-based preservative. If they do, decline those vaccinations. For more information, see www.nomercury.org. To minimize the impact of vaccinations (which may also contain aluminum), organize a schedule that avoids multiple doses on any given day. For guidance on essential vaccinations and how to minimize ill effects, consult *Healing the New Childhood Epidemics* by Kenneth Bock, published by Random House.

If your pediatrician does not query you, yet you think it's relevant, tell him or her about possible toxics your child may have been exposed to.

Suggest your pediatrician become knowledgeable about environmental health issues by participating in a training seminar from the Boston chapter of Physicians for Social Responsibility or by reading the *Pediatric Environmental Health Toolkit* of downloadable materials at http://psr.igc.org.

Take Action

Our nation needs an active citizenry. It is the American way. Repairing our children's health requires all parents to take on some active role.

Write to your local and federal legislators.

Join a local environmental group and/or organize your community or other parents. For the personal assistance and tools you'll need, look to the Center for Health, Environment & Justice, founded decades ago by Lois Gibbs, the mother who sparked community action in Love Canal, New York. Look up the center's national grassroots campaign, Childproofing Our Communities, www.childproofing.org, or phone 1–703–237–2249. They will help you form a local campaign to fight and win.

An excellent source of advice and ideas for collaborative action is *What We Can Do: Community Efforts to Protect Our Health,* www .womenshealthandenvironment.org/toolkit. Consult the resources described in appendix B, including the CHE Scientist Registry, which will connect your group with a knowledgeable scientist willing to assist your work.

Run for local office to move forward the right of children to live in a toxic-free world. Read *How to Run for Political Office* from the Center for Health, Environment & Justice. It's listed on their website, www.chej.org.

Join and donate to one of the national children's health groups to add your power to theirs. To locate one, look at the Center for Health and Environment website, www.healthandenvironment.org.

FOR MORE INFORMATION

See appendix B for publications and websites with general information.

Sources of information for this section include the Environmental Working Group; Greater Boston Physicians for Social Responsibility's "Fact Sheet for Parents and Future Parents"; Environment and Human Health, Inc.'s *12 Steps to Reducing Carcinogenic Exposures; Child Health and the Environment—A Primer,* August 2005, www.healthyenvironmentforkids.ca; and *What You Can Do,* www .womenshealthandenvironment.org.

Resources

These are books, organizations, and websites that offer information about children and the environment and practical advice for protecting children from environmental hazards in their daily lives. They are only a fraction of the materials in print or online that can help educate and assist parents, grandparents, pediatricians, and others who care about children's health.

BOOKS

Raising Healthy Children in a Toxic World—101 Smart Solutions for Every Family, by Philip J. Landrigan, Herbert Needleman, and Mary Landrigan, Rodale Press, 2001, is one of the best, most complete, and most useful guides for reducing children's exposures to toxic substances. Written by experienced and respected environmental pediatricians and a public health professional, it contains detailed solutions for identifying and protecting kids from hazards in their surroundings and is presented in clear and appealing fashion.

Raising Children Toxic Free, by pediatricians Needleman and Landrigan, is also an excellent guide for parents and physicians about realistic ways to keep children safe from environmental pollutants.

In Harm's Way is the companion and equally outstanding report to *Generations at Risk* by Greater Boston Physicians for Social Responsibility, written by Ted Schettler, Jill Stein, Fay Reich, and Maria Valenti. It deals with toxic threats to neurodevelopment. A bit technical, it is nonetheless a valuable guide to the environmental causes of neurological diseases such as ADHD and autism and the pathways of toxic exposure that contribute to them.

Child Health and the Environment—A Primer, published by the Canadian Partnership for Children's Health and Environment, is an introduction to the physiology of fetal and childhood development and how those can be affected by the environment. Nicely illustrated and with helpful charts and tables.

Children's Environmental Health: Reducing Risk in a Dangerous World, by Dona Schneider and Natalie Freeman for the American Public Health Association, is organized by environmental health risks facing children and includes dangers such as infectious disease and accidents as well as risks from toxic exposures. It offers detailed strategies for safeguarding children.

Generations at Risk, by Ted Schettler et al., is a report from Greater Boston Physicians for Social Responsibility that covers environmental threats to reproductive health.

Healthy Child Healthy World, written and produced by the organization of the same name (www.healthychild.org), offers easy steps for making the home safe.

Our Stolen Future, by Theo Colborn, Dianne Dumanoski, and John Peterson Myers, is a broad exposé and highly readable explanation of how chemicals disrupt the endocrine systems of living organisms, including humans. Its chapter entitled "Defending Ourselves" offers pointed advice, useful for couples who plan on having children, on how to avoid the disruptions caused by those chemicals.

Pediatric Environmental Health, edited by Ruth A. Etzel for the American Academy of Pediatrics, is directed primarily at physicians but is an accessible if encyclopedic storehouse of information about environmentally caused diseases in children, what causes them, and what to do about them.

Autism Is Treatable! comprises proceedings of the Defeat Autism Now! conference, spring 2005, sponsored by the Autism Research Institute. The materials from this conference offer information, advice, and hope to parents of children with autism. It covers a wide range of topics from emerging science to therapies and foods.

ORGANIZATIONS AND WEBSITES

Governmental Agencies

The Agency for Toxic Substances and Disease Registry of the U.S. Department of Health and Human Services offers detailed information about the health effects of toxics. Its website, aimed at children and parents, is user-friendly: www.atsdr.cdc.gov/child/atsdrpage2. friendly .html.

The Children's Environmental Health Initiative of the National Institute of Environmental Health Sciences conducts extensive research on children and the environment and makes much of its work available to the public in formats usable by parents and even children: www .niehs.nih.gov/external/resinits/ri-28.htm.

The U.S. Environmental Protection Agency's Office of Children's Health Protection has collected a storehouse of useful information about environmental dangers to children's health, where they can occur, and how to avoid them: http://yosemite.epa.gov/ochp/ochpweb .nsf/content/homepage.htm.

The site http://householdproducts.nlm.nih.gov (or Google "Household Products Database") tracks seven thousand products by brand name (but information is missing on ingredients withheld as trade secrets by the manufacturers).

A federal database of toxic substances, where they can be encountered, and what various federal agency research says about them is available at http://toxtown.nlm.nih.gov (or Google "Tox Town").

Nongovernmental Organizations

Beyond Pesticides provides a wealth of information in a variety of formats about the risks of insecticides, herbicides, fungicides, and other toxic pesticides to human health and suggests alternatives to their use. It also provides insights into pesticide policies of government and industry: www.beyondpesticides.org.

The Center for Children's Health and the Environment at the Mt. Sinai School of Medicine in New York City does research into environmental sources of childhood illness and ways to combat such illness through science and public policy. It offers a variety of practical information for parents and physicians: www.childenvironment.org.

The Center for Health, Environment & Justice offers information on toxic substances and technical and organizational support to help communities deal with environmental threats to health: www.chej.org.

CHE Scientist Registry is a resource organized by the Collaborative on Health and the Environment. CHE has identified scientists and clinicians who are knowledgeable about environmental health science and are willing to volunteer their services to communities and patient groups. The purpose is to provide groups with selected scientific resources with accurate and understandable information. The registry specializes in information about reproductive health, the fetal origins of disease, and childhood illnesses. The registry works like a personal referral service. To make a request for referral, contact Julia Varshavsky at Julia@healthandenvironment.org, with as specific a request as possible.

The Children's Environmental Health Network is a national organization of physicians, scientists, environmentalists, community organizers, and other citizens. It monitors the science and public policy aspects of children's environmental health and proposes courses of action for individual citizens and communities as well as national policy initiatives: www.cehn.org.

Healthy Child Healthy World (formerly called Children's Health Environmental Coalition), a national group formed by parents of sick kids, provides information about preventable childhood illness caused by exposure to toxics in the home, school, and community and what parents can do: http://healthychild.org. Its HealthEHouse is an interactive website displaying the various rooms of a typical home, the possible contaminants lurking in each, and advice about simple ways to avoid or minimize risk.

The Collaborative on Health and the Environment, a leader in the emerging environmental health movement, is an international network of individuals and organizations working together to collect and advance knowledge about links between health and the environment. It pays special attention to children: www.healthandenvironment.org.

The Columbia Center for Children's Environmental Health does research to study the effects of environmental pollutants and their impact on infant and child health. The results help communities improve environmental health through science and public policy: www.ccceh.org.

Environmental Health Sciences is a nonprofit group that publishes free of charge *Environmental Health News* and *Above the Fold* online, both valuable summaries of new research articles and media stories with hyperlinks to the original sources. Users can go to stories about children's health with a click: www.environmentalhealthnews.org.

The Environmental Working Group is a Washington, D.C.–based nonprofit that investigates and publicizes environmental contamination, who is doing it, and what the consequences are. It collects and publishes online and in print a trove of information about how children are exposed to toxic substances and offers practical advice about what parents might do about such things as chemicals in cosmetics, flame retardants, playground equipment, food, and frying pans: www.ewg.org.

The Institute for Children's Environmental Health leads a national collaborative working to eliminate environmental contributors to learning and developmental disabilities. It has also developed an extensive database of informational resources, searchable by medium (i.e., book, database, etc.), health problem (i.e., cancer, autism, etc.), and toxic (i.e., arsenic, formaldehyde, etc.) at: www.iceh.org, click on "Resources."

The Natural Resources Defense Council, a national nonprofit research, lobbying, and litigation group, maintains a very usable website devoted to the health of children. It monitors toxic dangers to kids and also keeps track of what the government is doing or not doing to protect children: www.nrdc.org/health/kids. NRDC maintains a question-and-answer website, www.simplesteps.org, through which the public can ask Dr. Gina Solomon questions.

Physicians for Social Responsibility, founded in the 1960s to address the medical consequences of nuclear warfare, now produces research, policy reports, and training materials to address other serious environmental threats. It holds seminars across the nation to train pediatricians in children's environmental health: www.psr.org.

The Science & Environmental Health Network is a consortium of North American environmental organizations and a leading proponent of the Precautionary Principle as a new basis for environmental and public health policy, to protect and prevent illness: www.sehn.org.

Nonprofit Organizations Representing Specific Childhood Illnesses

The Learning Disabilities Association of America provides support to children, teachers, and physicians who help children with learning problems. Its Healthy Children Project (www.healthychildrenproject .org/exposures/index.html) offers information about the link between toxics in the environment and those problems: www.ldanatl.org. Consult also the Institute for Children's Environmental Health.

The Autism Research Institute (ARI), founded to conduct and sponsor scientific research, collaborates with the Autism Society of America, a parent advocacy organization, following the theory that autism is a whole-body condition that is treatable. They offer the Defeat Autism Now! project for parents and physicians. The ARI maintains a website that offers advice and support to parents and information for physicians and has published a manual for use by doctors for helping kids with autism: www.autismresearchinstitute.com.

Birth Defect Research for Children, a national nonprofit organization founded by a parent of a child with birth defects, offers free information, parent networking, and birth defect research. It operates the only National Birth Defect Registry, which collects information on all categories of structural and functional birth defects as well as the health, genetic, and environmental exposure histories of the mothers and fathers of the affected child: www.birthdefects.org.

Families Against Cancer and Toxics, founded by parents of children with cancer, is both a network and a source of information and research news: www.familiesagainstcancer.org.

Web Resources for Scientific Research

A gold mine of information for the layman (or -woman) on various types of toxics, their effects on the body, the settings where you encounter toxics, regulatory agencies, and references to further sources, all in an interactive format, created by a toxicologist, can be found at www.asmalldoseof.org.

Based on years of research by leading scientists, a Web interactive database of toxicants and diseases is available at http://database .healthandenvironment.org. Maintained and updated by the Collaborative on Health and the Environment, it summarizes the evidence of exposure to chemical contaminants and over 180 associated human diseases or conditions. It is a useful tool for researchers, health professionals, health-affected groups, and others interested in reviewing the weight of evidence between toxicants and diseases. The database also features links to other useful databases and resources.

For science-related questions about the toxicity of a chemical or situation, e-mail ask@busbrp.org. This will bring you to "Ask the Researcher," an initiative of Boston University's School of Public Health; scientists of various disciplines will take on a question every few months and post their answer at http://busbrp.org/ask.html.

INTERVIEWS

The following women and men were interviewed in person or by telephone in the period 2004–2007. All direct or indirect quotations from them are from those interviews, unless otherwise noted. The affiliations of those quoted are identified in the text.

Andrews, William
Agee, Lynn
Apple, Mike
Ashwood, Paul
Avrilett, Diane
Baltz, Davis
Behm, Jo
Berger, Martha
Bernard, Sally
Bethell, Cindy
Blackburn, Elizabeth
Blaxill, Elyse and Mark
Boulanger, Aimee
Bove, Frank
Brody, Charlotte
Browner, Carol
Buffler, Patricia
Burns, Kathy
Cagle, Jimmy
Caldicott, Helen

Capiello, Dina
Carpenter, David
Casteel, Jenny
Clapp, Richard
Cochran, Debbie
Colangelo, Aaron
Collman, Gwen
Condon, Suzanne
Cook, Ken
Corn, Don
Cude, Judy
Dansereau, Carol
Davis, Devra
DeLoach, Kim
Denison, Richard
Deth, Richard
Epstein, Paul
Eskenazi, Brenda
Fagliano, Jerald
Feldman, Jay

Finney, Karen
Firestone, Michael
Flake, Cathy
Florini, Karen
Gallagher, Bobbie
Geiser, Kenneth
Gibbs, Lois
Gillick, Linda
Goldman, Lynn
Gomez, Basilia
Gonzalez, Stacey
Goozner, Merrill
Grandjean, Philippe
Gray, Tim
Groff, Jeff
Hagerman, Randi
Halpern, Michael
Hanks, Dale
Harris, John
Hawes, Amanda

Hayes, Tyrone
Heilig, Steve
Heminway, Diane
Hendren, Robert
Herbert, Martha
Hersey, Jane
Hertz-Picciotto, Irva
Hill, Ed
Holt-Orsted, Sheila
Honein, Peggy
Hunter, Allen
Jackson, Douglas
Jackson, Richard
Jacobson, Michael
Jameson, William
Karstadt, Myra
Kelley, Hilton
Kleba, Gerald
Klein, Rachel
Kochran, Joe
Koplan, Jeff
Kowalczyk, William
Kreuger, Elaine
Landrigan, Philip
Lanphear, Bruce
Lee, Philip
Legator, Marvin
Lerner, Michael
Lewis, Charles
Londer, Sonya
Marmagas, Susan
Masters, Roger
Mateo, Sandra
McNally, Siobhan
Mekdeci, Betty
Melnick, Ron

Michaels, David
Midkiff, Ken
Miller, Claudia
Miller, Elise
Miller, Mark
Montgomery, Latisha
Moran, Michael
Murray, Jeff
Myers, John Peterson
Natan, Tom
Needleman, Herbert
Nordbrock, Terry
Olden, Kenneth
Olney, Richard
Olson, Eric
Orsi, Bobbie
Ozonoff, David
Ozonoff, Sally
Pardieck, Roger
Parras, Brian
Parras, Juan
Paules, Richard
Perera, Frederica
Pessah, Ira
Piland, Christie
Praglin, Gary
Rasmussen, Sonya
Risotto, Steven
Robinson, Judith
Rodman, Joanne
Romitti, Paul
Rosario, Filidia
Rosenblatt, Martin
Ruch, Jeff
Sass, Jennifer
Sauer, Cindy

Schade, Mike
Schaeffer, Eric
Schettler, Ted
Sharp, Renee
Shelby, Michael
Shelton, Bill
Shepard, Peggy
Siegel, Lenny
Silbergeld, Ellen
Silverman, Chloe
Slotkin, Theodore
Smyth, Martin
Solomon, Gina
Swan, Shanna
Sword, Elizabeth
Tennant, Raymond
Tickner, Joel
Tillery, Ann
Tozzi, Jim
Tucker, Mary Evelyn
Veloz, Claria
Walker, Kim and Brad
Wallinga, David
Ward, Jonathan
Wargo, John
Warner, John C.
Warren, Jacqueline
West, Walter
Willett, Walter
Wilson, Michael
Wood, Amy and Travis
Woodruff, Tracey
Zeldin, Daryl
Zoeller, Tom

One INQUEST

6 **the available data is limited**   Sarah Janssen, Gina Solomon, and Ted Schettler, *Chemical Contaminants and Human Disease: A Summary of Evidence,* report prepared for the Collaborative on Health and the Environment, 2004.

6 **according to local residents**   Tennessee Department of Environment and Conservation, Community Meeting Questions and Answers, September 23, 2003.

7 **contamination came from other sources**   Letter from C. Jason Resher of the Division of Solid Waste Management, Tennessee Department of Environment and Conservation, to Mr. Jim Lunn, manager of the Dickson County Landfill, September 9, 1994.

8 **zero money, one county attorney said**   Patricia Lynch Kimbro, "County, City Settle Landfill Lawsuits with Families, Fight Against Companies Continues," *Dickson Herald,* November 7, 2006.

9 **"distracting from the environmental concerns"**   Letter from David M. Borowski, environmental specialist, Tennessee Department of Health, Communicable and Environmental Disease Services, to Mike Apple, September 17, 2003.

Two INDICTMENT

16 **of *A Civil Action* notoriety**   Jonathan Harr, *A Civil Action* (New York: Random House, 1995).

17 **fetal death, and birth defects**   Frank Bove et al., "Drinking Water Contaminants and Adverse Pregnancy Outcomes: A Review," *Environmental Health Perspectives* 110, Suppl. 1 (February 2002): 61–70.

17 **protecting citizens from TCE's dangers**   ATSDR, National Exposure Registry, *Trichloroethylene Technical Report,* undated, www.atsdr.cdc.gov/NER/TCE/tc9402.html.

17 **nearly three hundred synthetic chemicals**   Tim Kropp et al., *Body Burden:*

The Pollution in Newborns (Washington, DC: Environmental Working Group, July 14, 2005).

17 **carried through succeeding generations**   M. D. Amway et al., "Epigenetic Transgenerational Actions of Endocrine Disruptors and Male Fertility," *Science* 308, no. 1 (2005): 1466–1469.

17 **violence within their communities**   Philip J. Landrigan et al., "Children's Health and the Environment: A New Agenda for Prevention," *Environmental Health Perspectives Supplements* 105, no. S3 (June 1998), http://ehp.niehs .nih.gov/members/1998/Suppl-3/787–794landrigan/la.

18 **persistent toxic chemicals**   *Arctic Pollution 2002* (Oslo: Arctic Monitoring and Assessment Programme, 2002).

18 **"We have erred on the side"**   Bruce P. Lanphear et al., "Protecting Children from Environmental Toxins," *PLoS Science* 2, no. 3 (March 2005): 7.

19 **impact on adults, not children**   Ibid.

19 **our children's bodies and brains**   *Polluting the Future: Chemical Pollution in the U.S. That Affects Child Development and Learning* (Washington, DC: National Environmental Trust, Physicians for Social Responsibility, Learning Disabilities Association of America, September 2000), 1.

19 **overwhelm its critics**   "REACHing for Chemical Safety," *Environmental Health Perspectives* 111, no. 14 (November 2003), www.ehponline.org.

20 **how those hazards can affect children**   Ruth A. Etzel, ed., *Pediatric Environmental Health,* 2nd ed. (Elk Grove, IL: American Academy of Pediatrics, 2003), 1.

20 **"Insurance companies often refuse"**   Paul Krugman, "First, Do More Harm," *New York Times,* January 16, 2006, op-ed page.

20 **"We are by default conducting"**   Herbert L. Needleman, MD, and Philip J. Landrigan, MD, *Raising Children Toxic Free* (New York: Farrar, Straus & Giroux, 1994), 3.

21 **"Patterns of illness have changed"**   Philip J. Landrigan, "Disease of Environmental Origin in American Children: Prospects for Research and Prevention," testimony before the Committee on Appropriations, U.S. House of Representatives, May 2, 2000.

22 **was a turning point**   Bruce P. Lanphear, "Origins and Evolution of Children's Environmental Health," *Essays on the Future of Environmental Health Research* (National Institute of Environmental Health Sciences, 2005), 24ff.

23 **"instrumental in highlighting"**   Etzel, *Pediatric Environmental Health,* 4.

23 **exposed in the womb**   Ted Schettler et al., *In Harm's Way: Toxic Threats to Child Development* (Cambridge, MA: Greater Boston Physicians for Social Responsibility, May 2000).

24 **doubled every seven to eight years**   Needleman and Landrigan, *Raising Children Toxic Free,* 55.

Three VICTIMS

26 **A teenager named Jane**   Not her actual name. Because of her family's request for privacy, we have used a pseudonym.

27 **particularly among children**   Erin Koenig, "Cancer & Industry Incomplete Stats Hinder Investigations," *The Examiner,* August 20, 2003, 8.

27 **years or even decades**   Needleman and Landrigan, *Raising Children Toxic Free,* 43.

27 **who had attended the high school**   J. E. Loughlin et al., "Lymphatic and Haematopoietic Cancer Mortality in a Population Attending School Adjacent to Styrene-Butadiene Facilities, 1963–1993," *Journal of Epidemiology and Community Health* 53 (1999): 283–287.

28 **high exposures to butadiene**   E-mail exchange with Jim McGraw.

28 **increased risk of leukemia**   Ann L. Coker, "University of Texas School of Public Health Reports Possible Link Between Ship Channel Air Pollutants and Cancer Risks," January 19, 2007, www.houstontx.gov/health/UT.html.

29 **international report, by the World Health Organization**   "Principles for Evaluating Health Risks in Children" (Geneva: World Health Organization), September 22, 2007, whqlibdoc.who.int/publication/2006/924157237X_eng.pdf.

29 **toxic action of many chemicals**   Etzel, *Pediatric Environmental Health,* 9.

30 **resist and recover than adults**   Robert L. Brent, "A Pediatric Perspective on the Unique Vulnerability and Resilience of the Embryo and the Child to Environmental Toxicants," *Pediatrics* 113, no. 4 (April 2004): 937.

30 **"irreversible structural and/or functional"**   Donald T. Wigle, *Child Health and the Environment* (New York: Oxford University Press, 2003), 9.

30 **leading to maleness**   Needleman and Landrigan, *Raising Children Toxic Free,* 37.

30 **every organ in the body**   Ibid., 60.

31 **fine-tuning neural circuits**   Steven G. Gilbert, *A Small Dose of Toxicology* (Boca Raton, FL: CRC Press, 2004), 219.

31 **forever diminish the child**   Theodore Slotkin, presentation at an American Association on Mental Retardation conference, October 20, 2005.

31 **lead, pesticides, and PCBs**   Ted Schettler et al., *Generations at Risk* (Cambridge, MA: MIT Press, 1999).

31 **permanent changes in a child's**   Needleman and Landrigan, *Raising Children Toxic Free,* 35.

31 **an increase in autism**   M.I.N.D. is the name of the institute. It is shorthand for Medical Investigation of Neurodevelopmental Disorders.

31 **"If something happens to the brain"**   Elizabeth Weise, "Mercury Damage 'Irreversible,' " *USA Today,* http://USATODAY.COM+-+Mer.

31 **which would cause excitability**   Patricia Rodier, "Environmental Causes of CNS Maldevelopment, *Pediatrics* 113, no. 4 (April 2004): 1076–1083; Schettler et al., *Generations at Risk,* 60.

31 **"any of the organs or processes"**   Schetttler et al., *Generations at Risk.*

32 **health and quality of life**   John Peterson Myers, "Good Genes Gone Bad." www.ourstolenfuture.org/Commentary/JPM/2006–0401goodgenesgonebad .html. Originally published in *American Prospect,* April 2006.

32 **to childhood cancer**   "Critical Periods in Development," Environmental Protection Agency, February 2003, www.epa.gov/cgi-bin/epaprintonly.cgi.

33 **"emerging as the main"**   See www.sanger.ac.uk.

33 **nervous, immune, and endocrine systems**   Theo Colborn, Dianne Dumanoski, and John Peterson Myers, *Our Stolen Future* (New York: Penguin Books USA, 1997), 33.

33 **greatly reduced sperm counts**   Ibid., 48–67. It's estimated that five million women globally were medicated with DES, which was first synthesized in 1938; it was also added to animal feed to speed up fattening.

33 **cause other disorders**   T. W. Sadler, *Langman's Medical Embryology,* 9th ed. (Philadelphia: Lippincott Williams & Wilkins, 2004), 157.

33 **reach the developing child**   EPA, "Critical Periods in Development."

34 **is six months old**   Gertrude S. Rabinowitz et al., "The Rationale for a National Prospective Cohort Study of Environmental Exposure and Childhood Development," *Environmental Research,* Section A85 (2001): 61.

34 **at adults' breathing level**   Shelia Hoar Zahm and Mary H. Ward, "Pesticides and Childhood Cancer," *Environmental Health Perspectives* 106, no. S3 (June 1998), www.ehponline.org, accessed March 2, 2005.

34 **arsenic or other poisons**   Rabinowitz et al.,"The Rationale," 9.

34 **"There's gooey stuff "**   Fred Setterberg and Lony Shavelson, *Toxic Nation* (New York: John Wiley & Sons, 1993), 51.

35 **far more sensitive to air pollution**   Ibid., 179–180.

35 **one-tenth of that lead**   Needleman and Landrigan, *Raising Children Toxic Free,* 5.

35 **more sensitive to toxic assaults**   R. R. Dietert and M. P. Holsapple, "Methodologies for Developmental Immunotoxicity Testing," *Methods* 1 (January 2007): 123–131.

35 **can suffer oxidative stress**   Veronika Pašková et al., "Toxic Effects and Oxidative Stress in Higher Plants Exposed to Polycyclic Aromatic Hydrocarbons," *Environmental Toxicology and Chemistry* 25, no. 12 (December 2006): 3238–3245.

35 **according to EPA estimates**   EPA, *Guidelines for Carcinogenic Risk Assessment (Final)* (Washington, DC: U.S. Environmental Protection Agency, EPA/630/P-03/001F, 2005).

36 **toddlers than for adults**   Ibid.

36 **well-being of the child**   Arnold J. Sameroff, "Environmental Risk Factors in Infancy," *Pediatrics* 103 (November 1998): e1287.

36 **during her pregnancy, it's been discovered**   Ted Schettler, "Human Health and the Environment: An Ecological View of Health," www.sehn.org.

36 **further stresses on their bodies**   Ronald W. Patra et al., "The Effects of Three Organic Chemicals on the Upper Thermal Tolerances of Four Freshwater Fishes," *Environmental Toxicology and Chemistry* 26, no. 7 (2007): 1454–1459.

36 **blood type A rather than O**   The DNA strands of the double helix are made up of the nucleotides A, T, C, and G. When the sequence of the nucleotides is altered, it may change the way the gene expresses itself. See "Polymorphisms and Genes" fact sheet (Washington, DC: National Institute of Environmental Health Sciences, National Institutes of Health, undated), www.niehs.nih.gov/envirogenom.polymorph.htm.

37 **inept at detoxifying pollutants**   S. Jill James et al., "Metabolic Biomarkers of Increased Oxidative Stress and Impaired Methylation Capacity in Children with Autism," *American Journal of Clinic Nutrition* 80 (2000): 1611–1617.

37 **blood reaching their brain**   "Researchers Find Link Between Autism and Abnormal Blood-Vessel Function and Oxidative Stress," press release, University of Pennsylvania School of Medicine, August 15, 2006, www.uphs.upenn.edu/news/news_releases/aug06/autbldvsl.htm.

37 **"scientists and lay individuals believe"**   Robert J. Brent and Michael Weitzman, Preface to "The Vulnerability, Sensitivity and Resiliency of the Developing Embryo, Infant, Child, and Adolescent to the Effects of Environmental Chemicals, Drugs, and Physical Agents as Compared to the Adult," *Pediatrics*

213, no. 4 (April 2004): 993. Dr. Brent also opines that TCE "does not appear to represent a teratogenic risk for the developing embryo."

Four EVIDENCE

38 **58,000 children**   As of 2003, the most recent statistic, fifty-eight thousand children age nineteen or younger have had a diagnosis of invasive cancer at some point in the last five years. See http://seer.cancer.gov/csr/1975–2003/ results_merged/topic_prevalence.pdf, 32.

38 **disfiguring, debilitating birth defects**   Each year, 1 out of 33 children is born with one or more birth defects. See www.nlm.nih.gov/medlineplus/ birthdefects.html.

38 **poisoned with lead number 310,000**   See www.cdc.gov/nceh/lead/faq/about .htm. The current CDC standard of 10 micrograms of lead per deciliter of blood as "lead poisoning" is arbitrary; there is no level of lead that does not offer injury to a developing child.

38 **die from asthma**   Tracey Woodruff et al., *America's Children and the Environment: Measures of Contaminants, Body Burdens, and Illnesses* (Washington, DC: U.S. Environmental Protection Agency, Office of Children's Health Protection, EPA/240/R-03/001, 2003), 11.

38 **torment their behavior**   Schettler et al., *In Harm's Way;* John Wargo and Linda Evenson Wargo, *The State of Children's Health and Environment 2002* (Los Angeles: Children's Health Environmental Coalition, February 2002), 6. The United Kingdom and other developed countries report parallel figures. See *Living in Britain* (UK: Office for National Statistics, 2005).

39 **one generation to the next**   Theo Colborn, "Neurodevelopment and Endocrine Disruption," *Environmental Health Perspectives* 112, no. 9 (June 2004): 944–949.

39 **once a medical rarity**   Rachel Carson, *Silent Spring,* 25th anniversary ed. (Boston: Houghton-Mifflin, 1987), 221.

39 **leapt 67.1 percent**   *Summary of Changes in Cancer Incidence and Mortality, 1950–2001, SEER Cancer Statistics Review, NCI, 1975–2001, Table 1–3, Summary of Changes in Cancer Incidence and Mortality, 1950–2001* (Bethesda, MD: National Cancer Institute). This data is "temporarily unavailable" on the Web; its former URL was http://seer.cancer.gov/csr/1975–2001/index.html. The statistics trace incidence, not mortality, since new treatments have reduced the death rate for newly diagnosed cancer cases.

39 **exceeding rates among adults**   Ted Schettler and Richard Clapp, personal communication, January 9, 2004. Like adult cancers, childhood cancers are permutations of molecularly different diseases with likely different causes. Children, however, are prone to different cancers from those detected in adults. They suffer mainly from various types of leukemia, brain and central nervous system tumors, sarcomas, lymphomas, tumors of the abdomen and kidney (known as Wilms' tumors), and cancer of the eye. Leukemias are a collection of cancers of the blood cells. Lymphomas are cancers that develop in the lymphatic system. Neuroblastomas are cancers of the nervous system. Carcinomas are cancers that develop in the epithelial tissue that lines the surfaces of certain organs, such as the lung, liver, skin, or breast. Sarcomas are cancers that arise from cells in bone, cartilage, fat, connective tissue, and muscle. See *Cancer and the Environment* (Raleigh, NC: National Cancer Institute

and National Institute of Environmental Health Sciences, publ. 03–2039, September 2003), 6.

39 **childhood cancer in the world** *Childhood Cancer: Rising to the Challenge* (Switzerland: International Union Against Childhood Cancer, 2006), 10, www.uicc.org. The countries with higher rates are Brazil, Australia, and New Zealand; actually, Uganda has the top rate, but it's not comparable as it reflects the endemic of HIV-AIDS, a condition that raises a child's risk of developing lymphomas.

39 **increased about 35 percent** L. A. G. Ries et al., *CNS and Miscellaneous Intracranial and Intraspinal Neoplasm, Cancer Incidence and Survival Among Children and Adolescents: United States SEER Program 1975–1995* (Bethesda, MD: National Cancer Institute, 1999), 51–63; Kropp et al., *Body Burden.*

39 **leukemia over 47 percent** L. A. G. Reis, *SEER Cancer Statistics Review, 1975–2003* (Bethesda, MD: National Cancer Institute), http://seer.cancer.gov/csr/1975–2003.

39 **the trend will continue** *Cancer Query Systems: Delay-Adjusted SEER Incidence Rates,* http://srab.cancer.gov/delay/cancques.html.

39 **face a second bout** Elizabeth Sword, personal communication, December 11, 2003.

39 **immediately apparent defects** March of Dimes, www.marchofdimes.com; *Children's Environmental Health* (Raleigh, NC: National Institute of Environmental Health Sciences), www.niehs.nih.gov/oc/factsheets/ceh/bckgrnd.htm.

40 **probably the true rate** Betty Mekdeci and Ted Schettler, *Birth Defects and the Environment* (Bolinas, CA: Collaborative on Health and the Environment, May 1, 2004), www.protectingourhealth.org.

40 **twelve months of age** Tom Natan, *Polluting Our Future: Chemical Pollution in the U.S. That Affects Child Development and Learning* (Washington, DC: National Environmental Trust, September 2000), 18; Lynn Goldman et al., "Environmental Pediatrics and Its Impact on Government Health Policy," *Pediatrics* 113, no. 4 (April 2004): 1146–1157; "Birth Defects: The Leading Cause of Infant Mortality," fact sheet (Atlanta: Centers for Disease Control and Prevention, undated).

40 **the tenements of Harlem** Daryl Zeldin, personal communication, May 2, 2005.

40 **8.7 percent in 2001** L. Akinbami, "The State of Childhood Asthma, US 1980–2005," www.cdc.gov/nchs/dataad/ad381; Woodruff et al., *America's Children and the Environment.*

40 **beyond the newborn period** Needleman and Landrigan, *Raising Children Toxic Free.*

40 **exploded by 160 percent** *Toxic Chemicals and Public Health* (Washington, DC: National Environmental Trust, July 2004).

40 **except among black children** Centers for Disease Control and Prevention, www.cdc.gov/nceh/airpollution/asthma/children.htm.

40 **been diagnosed with ADHD** "Mental Health in the United States: Prevalence of Diagnosis and Medication Treatment for Attention-Deficit/Hyperactivity Disorder," *Morbidity and Mortality Weekly Report* 54 (September 2, 2005): 842–847.

40 **delays in growth and development** Fact sheets (New York: Mt. Sinai School of Medicine Center for Children's Health and the Environment, June 2002), www.childenvironment.org.

40 **are mentally retarded**   D. Croser, executive director, American Association on Intellectual and Developmental Disabilities, personal communication, March 29, 2007.

40 **diagnosed as having autism**   "CDC Releases New Data on Autism Spectrum Disorders from Multiple Communities in the United States," press release, Centers for Disease Control and Prevention, February 8, 2007, www.cdc.gov. The CDC states that depending upon different methods used, studies indicate that autism spectrum disorder afflicts between 1 in 500 and 1 in 166 children.

40 **no sign of abating**   Dona Schneider and Natalie Freeman, *Children's Environmental Health* (Washington, DC: American Public Health Association, 2000), 12.

40 **over the past two decades**   Amy M. Branum and Kenneth C. Schoendorf, "Changing Patterns of Low Birthweight and Preterm Birth in the United States, 1981–98," *Paediatric & Perinatal Epidemiology* 16, no. 1 (January 2002): 8–15. According to the March of Dimes, prematurity has increased 30 percent since 1981, with five hundred thousand babies born preterm in 2004 (9.4 percent of babies were premature in 1981, whereas by 2004 the figure was 12 percent). Jo Merrill, personal communication, September 16, 2005.

40 **and continue to rise**   Brady E. Hamilton et al., *Births: Preliminary Data for 2005* (Atlanta: Centers for Disease Control and Prevention, Division of Vital Statistics), November 17, 2006.

40 **prematurity from multiple births**   Eighty-three percent of premature births are singletons, indicating that artificial processes, which often result in multiple births, are not a major factor in the rise of prematurity rates. Coca Masters, "Ahead of Their Time," *Time,* November 13, 2006, www.time.com.

40 **weighing less than five pounds**   Institute of Medicine, *The Role of Environmental Hazards in Premature Birth* (Washington, DC: National Academies Press, 2003).

40 **by 8.2 percent in 2005**   Hamilton et al., *Births: Preliminary Data*.

40 **autism, asthma, and diabetes**   *Healthy from the Start: Why America Needs a Better System to Track and Understand Birth Defects and the Environment* (Washington, DC: Pew Environmental Health Commission, undated); Wargo and Wargo, *State of Children's Health and Environment,* 17; David O. Carpenter, "Health Effects of Polychlorinated Biphenyls," *Reviews on Environmental Health* 21, no. 1 (2006).

41 **among all developed nations**   EPA, *Overview of the Special Vulnerabilities and Health Problems of Children* (Washington, DC: Office of Children's Health Protection, EPA, February 2003), chapter 3.1, 1.; Nicholas D. Kristof, "The Larger Shame," *New York Times,* September 6, 2005, A31.

41 **less than healthy child**   National Research Center, *Scientific Frontiers in Developmental Toxicology and Risk Assessment* (Washington, DC: National Academies of Science Press, 2000).

41 **34 percent of all pregnancies**   A. J. Wilcox, "Incidence of Early Loss of Pregnancy," *New England Journal of Medicine* 319, no. 4 (July 28, 1988): 189–194.

41 **a wake-up call for action**   Devra Lee Davis et al., "Declines in Sex Ratios at Birth and Fetal Deaths in Japan, and in U.S. Whites but Not African Americans," *Environmental Health Perspectives* 115, no. 6 (June 2007): 941–946; Devra Lee Davis et al., "Reduced Ratio of Male to Female Births in Several

Industrial Countries: A Sentinel Health Indicator?" *Journal of the American Medical Association* 279, no. 13 (April 1998): 1018–1023. The U.S. data covered the years 1970–1990; other countries reported decreases in the time frame from 1950 to 1990.

41 **"cryptorchidism," and testicular cancer**   C. Mauduit et al., "Long-Term Effects of Environmental Endocrine Disruptors on Male Fertility," *Journal of Gynecology, Obstetrics and Fertility* 34, no. 10 (September 25, 2006): 978–984; C. P. Nelson et al., "The Increasing Incidence of Congenital Penile Anomalies in the United States," *Journal of Urology* 174, no. 4, pt. 2 (October 2005): 1673–1676.

41 **higher incidence in some than in others**   Niels Erik Skakkebaek et al., "Testicular Dysgenesis Syndrome: An Increasingly Common Development Disorder with Environmental Aspects," *Human Reproduction* 16 (2001): 972–978; Mauduit et al., "Long-Term Effects of Environmental Endocrine Disruptors on Male Fertility."

41 **sharper in some countries than in others**   Shanna H. Swan et al., "The Question of Declining Sperm Density Revisited," *Environmental Health Perspectives* 108, no. 10 (October 2000): 961–966. The largest study of its kind reported a decline of 1.2 percent per year, or about 17 percent from 1987 to 2004, in Massachusetts's men's testosterone levels. T. G. Travison et al., "A Population-Level Decline in Serum Testosterone Levels in American Men," *Journal of Clinical Endocrinology and Metabolism* 92 (2007): 196–202; Schettler et al., *Generations at Risk,* 164.

41 **at a younger age**   Ivelisse Colon et al., "Identification of Phthalate Esters in the Serum of Young Puerto Rican Girls with Premature Breast Development," *Environmental Health Perspectives* 108, no. 9 (September 2000), www.ehponline.org.

41 **breast cancer risks later in life**   EPA, *Overview of the Special Vulnerabilities,* chap. 5, 4.

41 **as a generation ago**   Judy Foreman, "Endometriosis Can Afflict Young Women, Too," *Boston Globe,* October 30, 2006, C1.

41 **reports problems conceiving**   Schettler et al., *Generations at Risk,* xv.

42 **carrying a baby to term**   This study covered the years 1982 to 1995. *Survey of Family Growth* (Hyattsville, MD: National Center for Health Statistics, 2002).

42 **as well as for mild cases**   "Male Reproductive Health and the Environment," fact sheet (New York: Mt. Sinai Center for Children's Health and the Environment, June 2002).

42 **are developing the disease**   *The Cases of Asthma, Learning Disabilities and Breast Cancer* (Bolinas, CA: Collaborative on Health and the Environment, undated), www.cheforhealth.org/articles/doc/110.

42 **the tests are so accurate**   Joceyln Kaiser, "No Meeting of Minds on Childhood Cancer," *Science* 286, no. 5446 (December 1999): 1832–1834.

42 **trends for adult cancers**   Tracey Woodruff et al., "Trends in Environmentally Related Childhood Illnesses," *Pediatrics* 113, no. 4 (April 2004): 1133–1140.

42 **as skeptics claim**   Dr. Rachel Gittleman Klein, personal communication, August 30, 2005.

43 **"a near pathological denial"**   Martha Herbert and Chloe Silverman, "Autism and Genetics," *GeneWatch* 16, no. 1 (January 2003), www.gene-watch.org.

43 **avail themselves of special services** *In Harm's Way Training Manual* (Cambridge, MA: Greater Boston Physicians for Social Responsibility, September 2002), 2.

43 **"even though we may have"** As quoted in Michael Spzir, "New Thinking on Neurodevelopment," *Environmental Health Perspectives* 114, no. 2 (February 2006), www.ehponline.org.

43 **pounds of chemicals per day** Michael P. Wilson et al., *Green Chemistry in California: A Framework for Leadership in Chemicals Policy and Innovation* (Berkeley: California Policy Research Center, University of California, 2006), xii, www.ucop.edu/cprc/documents/greenchemistryrpt.pdf. The Toxic Substances Control Act required manufacturers to report to the EPA all chemicals produced or imported into the nation; from this data, the EPA compiles the Chemical Inventory Database. The law also requires manufacturers to report their yearly releases of some of those chemicals into the air, land, and water, from which the EPA compiles the Toxics Release Inventory.

43 **two hundred billion pounds of chemicals** Sandra Steingraber, *Living Downstream* (New York: Vintage Books, 1997), 90.

43 **grew to revenues of $635 billion in 2006** "Essential2economy" (Washington, DC: American Chemistry Council), www.americanchemistry.com.

43 **$484 billion a year in 2004** "REACHing for Chemical Safety."

44 **Rachel Carson wrote *Silent Spring*** That would be $300 million in 1962 dollars.

44 **$180 million a day in 2007** Clifford Krauss, "Record Profit for Exxon and Shell," *New York Times,* February 2, 2007, http://select.nytimes.com.

44 **eighteen thousand different pesticide products** EPA, *Promoting Safety for America's Future,* annual report (Washington, DC: Office of Pesticide Programs, EPA, 2002).

44 **households and agriculture** Bernard Weiss et al., "Pesticides," *Pediatrics* 113, no. 4 (April 2004): 1030–1036.

44 **used in the 1960s** Lynn R. Goldman and Sudha Koduru, "Chemicals in the Environment and Development Toxicity to Children: A Public Health and Policy Perspective," *Environmental Health Perspectives* 108, no. S3 (June 2000), http://ehp.niehs.nih.gov.

44 **more than ten million products** Priscilla Flattery, chief of staff, EPA's Office of Pollution Prevention & Toxics, personal communication, December 15, 2006.

44 **piling up in waste dumps** *Pollution Is Personal* (Lowell: Massachusetts Precautionary Principle Project, fall 1999), www.sehn.org/ppfactsh.html.

44 **at least 4.4 *billion* pounds** *U.S. EPA Toxics Release Inventory Reporting Year 2003 Public Release Data, Summary of Key Findings*, www.epa.gov/tri/tridata/tri03/KeyFind.pdp.

44 **probably a vast undercount** After passage of the Toxics Release Inventory, strongly opposed by U.S. industry, various industries lobbied successfully to exempt the chemicals they produce from inclusion, among them asbestos, vinyl, and pesticides. Only 7 percent of the chemicals produced in greatest volume are included in the inventory. So instead of 4.4 billion pounds, a realistic count of releases from manufacturing plants would be more like 200 billion pounds, counting only pollutants released into the air annually, not even pollutants seeping into soil and water, according to Congress's watchdog agency, the General Accounting Office. See *Air Pollution: EPA Should Improve*

Emissions Reporting by Large Facilities (Washington, DC: GAO-01–46, April 2001), 4.

45 **before the babies' parents were born** Kropp et al., *Body Burden*. For this study, the Environmental Working Group and its collaborator, Commonweal, commissioned two major laboratories to undertake these tests in August and September 2004. The labs used umbilical cord blood collected just after the cord was cut from newborn babies selected randomly from across the country by the Red Cross under its voluntary national cord blood collection program. Among the 287 chemicals looked for in these infants were 217 toxic to the brain and nervous system, 208 that can cause birth defects or abnormal development in animals tests, 180 that cause cancer in humans or animals, and 195 that affect the reproductive system. Had these newborns been tested for a broader array of chemicals than the 287, more would have been detected, concludes the Environmental Working Group.

45 **most popular type of pesticide, chlorpyrifos** Robin Whyatt and D. B. Barr, "Measurement of Organophosphate Metabolites in Postpartum Meconium as a Potential Biomarker of Prenatal Exposure," *Environmental Health Perspectives* 109, no. 4 (April 2001): 417–420.

45 **also invaded their unborn bodies** Robin Whyatt et al., "Contemporary-Use Pesticides in Personal Air Samples During Pregnancy and Blood Samples at Delivery Among Urban Minority Mothers and Newborns," *Environmental Health Perspectives* 111, no. 5 (May 2003), www.ehponline.org.

45 **in the bodies of the children** *3rd National Report on Human Exposure to Environmental Chemicals, 2005* (Atlanta: National Center for Environmental Health, Centers for Disease Control and Prevention, NCEH Pub. No. 05–0570, July 2005).

45 **traces of it in every child** Jane Houlihan, "Pollution Gets Personal," presentation during the Human Body Burden of Synthetic Toxic Chemicals conference, Seattle, February 17, 2004, www.iceh.org/pdfs/SBLF/ SBLFHoulihanTranscript.pdf.

46 **"acceptable" for a long-term exposure** See www.panna.org/campaigns/ docsTrespass.

46 **630,000 newborns, 1 in 6** Dr. Mahaffey explained that because mercury concentrates in the umbilical cord, a newborn could exceed the safety level of 5.8 parts per billion in a mother whose mercury level was just 3.5 parts per billion. Nearly 16 percent of women have levels that high. There is, however, a cushion between the safety level and the level known to show harm. Jennifer 8. Lee, "EPA Raises Estimate of Babies Affected by Mercury Exposure," *New York Times,* February 10, 2004; Joan Lowy, "Mercury Risk to Newborns Alarming," *Seattle Post-Intelligencer,* February 4, 2004, http://seattlepi .newsource.com.

46 **used in plastics and cosmetics** *3rd National Report on Human Exposure to Environmental Chemicals.*

46 **sometimes twice as high** Ibid.

46 **handling these chemicals for a living** Douglas Fischer, "A Body's Burden: Our Chemical Legacy," *Oakland Tribune* (Oakland, CA), March 2005: 1.

46 **at or above the "acceptable" standard** *3rd National Report on Human Exposure to Environmental Chemicals;* Woodruff et al., *America's Children and the Environment,* 59.

46 **disrupting the fetal hormone system** Benjamin C. Blount et al., "Levels of

Seven Urinary Phthalate Metabolites in a Human Reference Population," *Environmental Health Perspectives* 108, no. 10 (October 2000): 979–982.

47 **"The public should take assurance"**   "CropLife America Statement: CDC Biomonitoring Report on Environmental Chemicals," press release, CropLife America, July 21, 2005.

47 **"Saying 'just because you can' "**   Bruce Lanphear, personal communication, March 20, 2006.

47 **9.4 million to support**   Schettler et al., *In Harm's Way,* 16.

48 **diminish intelligence and disturb behavior**   Natan, *Polluting Our Future,* 4, 7; Phillippe Grandjean and Philip J. Landrigan, "Chemical Exposure Creating a Silent Pandemic of Neurodevelopmental Disorders," *Lancet* 368 (November 8, 2006), www.thelancet.com and www.hsph.harvard.edu/neurotixc/appendix.doc.

48 **$43.8 billion annually**   Leonardo Trasande et al., "Public Health and Economic Consequences of Methyl Mercury Toxicity to the Developing Brain," *Environmental Health Perspectives* 113, no. 5 (May 2005): 590–596.

48 **societal costs to trillions**   Bernard Weiss, "Vulnerability of Children and the Developing Brain to Neurotoxic Hazards," *Environmental Health Perspectives* 108, Suppl. 3 (June 2000): 375–381.

48 **children and their families**   Rachel Massey and Frank Ackerman, *Costs of Preventable Childhood Illness: The Price We Pay for Pollution* (Medford, MA: Global Development and Environment Institute, Working Paper No. 03–09, September 2003); Natan, *Polluting Our Future,* 6; Philip J. Landrigan et al., "Environmental Pollutants and Disease in American Children: Estimates of Morbidity, Mortality, and Costs for Lead Poisoning, Asthma, Cancer, and Developmental Disabilities," *Environmental Health Perspectives* 110, no. 7 (July 2002): 721–728. A similar analysis just for the state of Massachusetts came up with a cost figure of between $1.1 billion and $1.6 billion annually; see Massey and Ackerman, *Costs of Preventable Childhood Illness.* Montana's yearly cost is $400 million, from asthma, neurodevelopment disorders, birth defects, cancer, and lead poisoning; in Montana, lead comes not from urban peeling paint, but from tailings from mining.

48 **another $26.2 billion a year**   Masters, "Ahead of Their Time."

48 **"A corporation tends to be"**   Robert Monks and Newton Minow, *Corporate Governance,* 3rd ed. (Malden, MA: Blackwell Publishers, 2003); J. Bakan, *The Corporation* (New York: Free Press, 2004).

49 **before they are sold and used**   GAO, *Chemical Regulation: Options Exist to Improve EPA's Ability to Assess Health Risks and Manage Its Chemical Research Program* (Washington, DC: Government Accountability Office, GAO-05–458, June 2005).

49 **basic information was missing for 92 percent**   See www.epa.gov/oppt/ntppac/pubs/document.htm.

49 **chemical plants for antiterrorist planning**   GAO, *Chemical Regulation. Actions Are Needed to Improve the Effectiveness of EPA's Chemical Review Program* (Washington, DC: Government Accountability Office, GAO-06–1032T, August 2006).

50 **"Medicines are the only chemicals"**   "Medicines Are the Only Chemicals That Have to Be Proven Safe; Why?" *New York Times,* June 26, 2002, A5, ad from New York City's Mt. Sinai School of Medicine Center for Children's Health and the Environment, www.childenvironment.org.

50 **two hundred of these chemicals**   GAO, *Chemical Regulation: Actions Are Needed,* 1.

50 **of the chemicals we use**   Richard Denison, vice president, Environmental Defense, personal communication.

50 **have hidden the facts**   In 1991 and 1992, when the EPA offered amnesty from fines to any manufacturer that turned in reports about harm that should have been submitted earlier, chemical companies suddenly produced more than ten thousand studies on products already on the market. See *Unreasonable Risk: The Politics of Pesticides* (Washington, DC: Center for Public Integrity, 1998), 2.

51 **new chemical will be harmful**   GAO, *Chemical Regulation: Actions Are Needed.*

51 **a handful of exceptions**   If a chemical formerly manufactured in smaller quantities reaches the one-billion-pound-a-year production mark, the manufacturer must return for further EPA approval. Richard Denison, personal communication, October 3, 2006; and GAO, *Chemical Regulation: Actions Are Needed,* 1.

51 **PCBs and dioxin**   The five grandfathered chemicals banned or restricted are PCBs, dioxin, hexavalent chromium, asbestos, and chlorofluorocarbons; the Fifth Circuit Court of Appeals vacated the ban against asbestos. GAO, *Chemical Regulation: Options Exist,* 18, 27. The EPA has also limited the uses of four new chemicals introduced after the TSCA was passed: mixed mono- and diamides of an organic acid, triethanolamine salts of a substituted organic acid, triethanolamine salts of tricarboxylic acid, and tricarboxylic acid. GAO, *Chemical Regulation: Options Exist,* 18.

51 **chemical was in 1990**   GAO, *Chemical Regulation: Options Exist,* 18, 27.

51 **remove a hazard voluntarily**   Ibid., 29.

52 **scientific evidence demonstrating toxicity**   David Wallinga, "Failure to Protect: Why Current Laws Don't Protect Against Chemical Exposures," presentation at the American Association on Mental Retardation Pollution, Toxic Chemicals and Mental Retardation, report prepared for Pollution, Toxic Chemicals and Mental Retardation conference, July 22–24, 2003, www .aamr.org/Reading_Room/pdf/Wingspread.pdf.

52 **innovate to find safer ones**   Michael P. Wilson, *Green Chemistry in California* (Berkeley: California Policy Research Center, University of California, 2006).

52 **before the FQPA was enacted**   By 2006, the EPA had completed its review of all pesticides in use before the FQPA was passed in 1996, as the FQPA required.

52 **information about basic toxicity**   Called the Screening Information Data Set, these tests cover acute toxicity, chronic toxicity, developmental and reproductive toxicity, mutagenicity, ecotoxicity, and environmental fate.

53 **but were rejected**   Lonnie Arnold, "States Take Legal Action to Protect Children from Pesticides," January 9, 2005, www.canyon-news.com.

53 **active and dangerous**   Caroline Cox and M. Surgan, "Unidentified Inert Ingredients in Pesticides: Implications for Human and Environmental Health," *Environmental Health Perspectives* 114, no. 2 (December 2006): 1803–1806.

53 **require disclosure on labels**   "Controversy Surrounds EPA Review of Pesticides," *HealthDay,* August 2, 2006, www.nlm.nih/gov/medlineplus.

53 **seldom been imposed**   Consumers Union, "A Report Card for the EPA: Successes and Failures in Implementing the FQPA," 2001. For eighty-two

reviews of organophosphate pesticides, the tenfold safety factor was imposed on only 16 percent.

53 **nearly one hundred junkets** *Unreasonable Risk,* 34–35.

53 **inefficient rule-making tools** GAO, *Chemical Regulation: Options Exist,* and *Chemical Regulation: Actions Are Needed;* Marla Cone, "Senate Panel Weighs Toxic Chemicals Law," *Los Angeles Times,* August 3, 2006, www.latimes.com.

53 **EPA desks a year** Enesta Jones, EPA press officer, personal communication, April 11, 2007; Priscilla Flattery, chief of staff, Office of Pollution Prevention & Toxics, EPA, personal communication, April 13, 2007.

53 **"Any new chemical"** Steve Mitchell, "Chemicals Alleged to Cause Health Problems," UPI, June 11, 2002, www.upi.com/NewsTrack/Science.

54 **it failed to understand** For ten years, Dow hid from the EPA the 302 lawsuits and other claims alleging Dursban poisoning. The EPA's response was to give Dow a slap on the wrist in the form of an $876,000 fine. *Unreasonable Risk,* 27.

54 **single active ingredients** Schettler et al., *Generations at Risk,* 246.

54 **than its active ingredient** Sophie Richard et al., "Differential Effects of Glyphosate and Roundup on Human Placental Cells and Aromatase," *Environmental Health Perspectives* 113, no. 6 (June 2005): 716–720.

54 **fetuses, infants, and children** "Registering Skepticism: Does EPA's Pesticide Review Protect Children?" *Environmental Health Perspectives* 114, no. 10 (October 2006), www.ehponline.org.

54 **heart disease, Alzheimer's, or Parkinson's** Bruce P. Lanphear et al., "Protecting Children from Environmental Toxins," *PLoS Medicine* 2, no. 3 (March 2005), http://medicine.plosjournals.org; Philip J. Landrigan et al., "Children's Health and the Environment: Public Health Issues and Challenges for Risk Assessment," *Environmental Health Perspectives* 112, no. 2 (February 2004): 257–265.

54 **sensitive to harmful agents** Lanphear, "Protecting Children from Environmental Toxins."

54 **"If we know there are"** P. Grandjean and Philip Landrigan, "Developmental Neurotoxicity of Industrial Chemicals," *Lancet* 368 (December 2006): 2167–2178.

55 **"voluntary development of health"** *Bill Moyers's Trade Secrets,* Public Broadcasting System, March 26, 2001, www.pbs.org/tradesecrets.

55 **deeper analysis in the future** Richard Denison, personal communication, August 5, 2006.

55 **a "D" for its foot dragging** Richard Denison, *High Hopes, Low Marks* (Washington, DC: Environmental Defense), 2007, www.environmentaldefense.org/hpvreportcard.

55 **"Cosmetic products and ingredients"** See www.cfan.fda.gov.

55 **food dye it certifies** *Behavior, Learning & Health: The Dietary Connection* (Riverhead, NY: Feingold Association of the United States, 2003), www.feingold.com.

56 **animals, or the environment** Steven Gilbert, *A Small Dose of Toxicology* (Boca Raton, FL: CRC Press, 2004), 17.

56 **change or withdraw the chemical** This process started in the Nixon administration, was mandated by President Reagan, and reached new heights of application in the George W. Bush administration.

56 **"a world of scarce"** Barry R. Bloom, "The Scientific Basis of Health Deci-

sion Making," *Harvard Public Health Review* (Fall 2004), www.hsph.harvard
.edu/review/review_fall_04/risk_deantxt.html.

56 **childhood lead poisoning?** "Pricing the Priceless," fact sheet (Washington, DC: OMB Watch, March 20, 2002), www.ombwatch.org.

56 **occurring in the future** Ibid.

56 **future life is worth** Charles W. Schmidt, "Subjective Science: Environmental Cost-Benefit Analysis," *Environmental Health Perspectives* 111, no. 10 (August 2003).

57 **assumes that's acceptable** Alliance for a Healthy Tomorrow, www
.healthytomorrow.org/flaw3.html.

57 **A panel making recommendations** Jennifer Beth Sass, "Industry Efforts to Weaken the EPA's Classification of the Carcinogenicity of 1,3-butadiene," *International Journal of Occupational and Environmental Health* 11 (2005): 378–383.

58 **though it is "limited"** Gina Solomon, MD, MPH: Natural Resources Defense Council; University of California at San Francisco; Ted Schettler, MD: Science and Environmental Health Network; Boston Medical Center; Sarah Janssen, MD, PhD. The database that captures this information is at www.healthandenvironment.org, click on "CHE Toxicant and Disease Database."

58 **that beset children** See the last chapter of this book for information about these databases.

58 **"known human carcinogens"** To receive a copy of the report on carcinogens or to suggest agents suspected of causing cancer, go to http://ntp-server
.niehs.nih.gov.

58 **475 chemicals as carcinogenic** Environmental Working Group, *Children's Health Policy Review, 2003* (Washington, DC: Environmental Working Group, 2003).

59 **pollutants in the womb** Frederica P. Perera et al., "Effects of Transplacental Exposure to Environment Pollutants on Birth Outcomes in a Multiethnic Population," *Environmental Health Perspectives* 111, no. 2 (February 2003): 201–205.

59 **economic interests in seeing this happen** Devra Lee Davis, *The Secret History of the War on Cancer* (New York: Basic Books, 2007).

59 **Lowell Center for Sustainable Production** Tami Gouveia-Vigeant and Joel Tickner, *Toxic Chemicals and Childhood Cancer* (Lowell: Lowell Center for Sustainable Production, University of Massachusetts, May 2003); Joel Tickner and Richard W. Clapp, *Environmental and Occupational Causes of Cancer: A Review of Recent Scientific Evidence* (Lowell: Lowell Center for Sustainable Production, University of Massachusetts, September 2005).

59 **including children's cancers** Tickner and Clapp, *Environmental and Occupational Causes of Cancer,* 1.

60 **use of pesticides increases** M. S. Sanborn, et al., *Systematic Review of Pesticide Human Health Effects* (Toronto: Ontario College of Family Physicians, April 2004), http://www.ocfp.on.ca; K. B. Flower, "Cancer Risk and Parental Pesticide Application," *Environmental Health Perspectives* 112: 631–635.

60 **with the least protection** Carol Dansereau, Farm Worker Pesticide Project, Seattle, WA, personal communication, November 15, 2005.

60 **a child's direct exposure** J. M. Pogoda and S. Preston-Martin, "Household Pesticides and Risk of Pediatric Brain Tumors," *Environmental Health Perspectives* 105, no. 11 (November 1997): 1214–1220.

60 **associated with non-Hodgkin's lymphoma?** These five are 2,4-D, glyphosate, MCPP, dicamba, and diazinon; diazinon, no longer in use inside our homes, may still be used on our lawns. John P. Wargo et al., *Risks from Lawn-Care Products* (North Haven, CT: Environment & Human Health, June 2003).

60 **at risk of developing asthma** M. T. Salam et al., "Early Life Environmental Risk Factors for Asthma," *Environmental Health Perspectives* 112, no. 6 (May 2004): 760–765.

60 **seasonal patterns in prematurity** "U.S. Premature Births Linked to Increase in Pesticides and Nitrates in Water," *Medical News Today,* May 9, 2007, www.medicalnewstoday.com.

61 **more than home use** John Harris, director, California Birth Defects Monitoring Program, personal communication, November 15, 2004.

61 **brain, skull, and scalp** M. Cacasana et al., "Maternal and Paternal Occupational Exposure to Agricultural Work and the Risk of Anencephaly," *Journal of Occupational and Environmental Medicine* 63, no. 10 (October 2006): 649–656.

61 **penal urinary tract deformity** M. R. Araneta et al., "Prevalence of Birth Defects Among Infants of Gulf War Veterans, 1989–1993," *Birth Defects Research, Clinical Teratology* 67, no. 4 (June 2003): 246–260; Suzanne Gamboa, "Gulf War Vets' Children Have More Birth Defects," Associated Press, June 4, 2003, www.notinourname.net.

61 **College of Family Physicians finds** Sanborn, *Systematic Review of Pesticide Human Health Effects.*

61 **Baltimore-Washington Infant Study** Ellen K. Silbergeld and Thelma E. Patrick, "Environmental Exposures, Toxicologic Mechanisms, and Adverse Pregnancy Outcomes," *American Journal of Obstetrics and Gynecology* 192 (May 2006): S11–S21.

61 **wives of men exposed to solvents** Schettler et al., *Generations at Risk,* 83.

61 **from dioxin to PCBs** UCSF-CHE Fertility Summit, www .healthandenvironment.org/fertilitysummitblog/907. UCSF stands for University of California at San Francisco; CHE stands for Collaboration on Health and the Environment.

61 **in the United States and Europe** Shanna H. Swan et al., "Have Sperm Densities Declined?: A Reanalysis of Global Trend Data," *Environmental Health Perspectives* 105, no. 11 (1997): 1228–1232; "The Question of Declining Sperm Density Revisited," *Environmental Health Perspectives* 108, no. 10 (October 2000): 961–966.

61–62 **one area and environment to another** Shanna H. Swan et al., "Geographic Differences in Semen Quality of Fertile US Males," *Environmental Health Perspectives* 111, no. 4 (2003): 414–420.

62 **due to exposure to pesticides** Shanna H. Swan, "Biomarkers of Pesticide Exposure in Relation to Semen Quality," *Environmental Health Perspectives* 111, no. 12 (2003): 1478–1484.

62 **incompletely descended testicles** Shanna H. Swan et al., "Decrease in Anogenital Distance Among Male Infants with Prenatal Phthalate Exposure," *Environmental Health Perspectives* 113, no 8 (August 2005), http://ehp .niehs.nih.gov/docs. Studies are beginning to correlate phthalate exposure with obesity and diabetes in men. See R. W. Stahlhut et al., "Phthalates and Metabolism," *Environmental Health Perspectives* 115, no. 6 (June 2007): 876–882.

62 **well below the EPA safety margin**   "What Are Phthalates?" Phthalates Information Center, www.phthalates.org.

62 **caused by natural variability**   Marla Cone, "Study Finds Genital Abnormalities in Boys," *Los Angeles Times,* May 27, 2005, www.latimes.com.

62 **"Exposure of children one to six"**   "NTP-CERHR Expert Panel Update on Reproductive and Developmental Toxicity of Di(2-ethylhexyl) Phthalate," http://cerhr.niehs.nih.gov/index.html, November 2005.

62 **higher in children than in adults**   See www.cdc.gov/exposurereport.

63 **boys and girls more feminine**   Hestien J. Vreugdenhil et al., "Effects of Perinatal Exposure to PCBs and Dioxins on Play Behavior in Dutch Children at School Age," *Environmental Health Perspectives*, 110, no. 10 (October 2002): 593–598.

63 **has demasculinized them**   Colborn et al., *Our Stolen Future.*

63 **in a string of 7,400 bathtubs**   Jane Houlihan, *The Pollution in People* (Washington, DC: Environmental Working Group, 2002), 44, archive.ewg .org/reports/bodyburden1.

63 **overall fertility returned to normal**   P. Mocarelli et al., "Paternal Concentrations of Dioxin and Sex Ratio of Offspring," *Lancet* 355 (2000): 1858–1863; Richard Clapp and David Ozonoff, "Where the Boys Aren't: Dioxin and the Sex Ratio: Commentary," *Lancet* 355 (2000): 1838–1839.

63 **the community is collaborating**   C. A. Mackenzie et al., "Declining Sex Ratio in a First Nation Community," *Environmental Health Perspectives* 113, no. 10 (October 2005): 1295–1298.

63 **villages from Russia to Greenland**   Paul Brown, "Man-Made Chemicals Blamed as Many More Girls than Boys Are Born in Arctic," *Guardian,* September 12, 2007, www.guardian.co.uk/print/0,,330722948–110592,00.html.

63 **researchers find time after time**   Fifty-three percent of the chemicals released into the air and water and 20 percent of the most used chemicals are neurotoxins.

64 **micrograms per deciliter today**   Woodruff et al., *America's Children and the Environment,* 55.

64 **"You might not notice"**   David Bellinger, as quoted in Charles W. Schmidt, "Poisoning Young Minds," *Environmental Health Perspectives* 107, no. 6 (June 1999), www.mindfully.org.

64 **impaired language ability**   W. Yuan et al., "The Impact of Early Childhood Lead Exposure on Brain Organization," *Pediatrics* 118 (2006): 971–977, www.ehponline.org.

64 **months, even years, to appear**   "Mercury and Learning Disabilities," fact sheet, Learning Disabilities Association of America, undated, www.ldaa.org, accessed May 5, 2004.

64 **five-to-seven-IQ-point loss**   David Carpenter, School of Public Health, SUNY Albany, personal communication, February 10, 2005.

64 **$1.3 billion each year**   Trasande et al., "Public Health and Economic Consequences."

64 **"In contrast to the costs"**   Leonardo Trasande et al., "Mental Retardation and Prenatal Methylmercury Toxicity," *American Journal of Industrial Medicine* 49 (2006): 153–158.

64 **"Mercury does not reduce"**   Joel Schwartz, "Mercury from Fish Does Not Reduce Children's IQs," *Environmental Health Perspectives* 114, no. 7 (July 2006), correspondence, A399–A400.

65 **verbal intelligence test scores** Wally Kennedy, "Study Raises Health Questions: Miami Children's Hair Tested for Heavy Metals," *Joplin Globe* (Joplin, MO), November 18, 2005, www.joplinglobe.com/story.php?story_id= 212161&c=87, accessed November 29, 2005.

65 **grew by 43 percent** Raymond F. Palmer and Claudia Miller et al., "Environmental Mercury Release, Special Education Rates, and Autism Disorder: An Ecological Study of Texas," *Health & Place* (2005), www.elsevier.com/ healthandplace.

65 **than any other state** Llan Levine, *Dirty Kilowatts* (Washington, DC: Environmental Integrity Group), July 2006, 32 ff; and Dina Capiello, "New Mercury Limits Raise Toxic Debate, Especially Here," *Houston Chronicle,* March 16, 2005, www.chron.com/disp/story.mpl/special/04/toxic/3087382.html.

65 **vinyl chloride, and the solvent TCE** Marla Cone, "Study Links Air Pollutants with Autism," *Los Angeles Times,* June 23, 2006, www.latimes.com, accessed June 23, 2006.

65 **federally mandated childhood vaccines** The drug manufacturers added thimerosal as a preservative so that vaccines could be packaged in 10-cc multi-use vials rather than 1-cc single-use containers that don't need a preservative because they are used as soon as they are opened but cost the manufacturers a bit more. See L. K. Ball et al., "An Assessment of the Use of Thimerosal in Childhood Vaccines," *Pediatrics* 107, no. 5 (May 2001): 1147–1154, www.fda .gov/cber/vaccine/thimerosal.htm.

65 **mandated for children under age two** Arthur Allen, "The Not-So-Crackpot Autism Theory," *New York Times Magazine,* November 10, 2002, 66–69.

65 **far exceeds federal health guidelines** John Wargo, *Our Children's Toxic Legacy,* 2nd ed. (New Haven, CT: Yale University Press), 36. Merck & Co. had been aware for nearly a decade earlier that infants were getting an elevated dose with their vaccinations (up to eighty-seven times higher than guidelines for the maximum daily consumption of mercury from fish) but did not disclose this information. See Myron Levin, "'91 Memo Warned of Mercury in Shots," *Los Angeles Times,* February 8, 2005, www.latimes.com/business/ la-fi-vaccine. Merck's handling of the problems with its painkiller Vioxx seems to repeat this behavior.

65 **autism in 2004** *Immunization Safety Review: Vaccines and Autism* (Washington, DC: Institute of Medicine, May 17, 2004).

65 **showed no such effects** Mady Hornig et al., "Neurotoxic Effects of Postnatal Thimerosal Are Mouse Strain Dependent," *Molecular Psychiatry* 9 (September 2004): 833–845; Michael Szpir, "Tracing the Origins of Autism: A Spectrum of New Studies," *Environmental Health Perspectives* 114, no. 7 (July 2006), www.ehponline.org. In a second study, a team of cell biologists, toxicologists, and molecular bioscientists at the University of California at Davis discovered that, in mice, thimerosal even at very low concentrations weakens the white blood cells, the T-cells, that help find and kill external agents that attack the immune system and can turn them into "rogue" cells.

66 **agricultural fields . . . were autistic** Eric M. Roberts, "Maternal Residence Near Agricultural Pesticide Applications and Autism Spectrum Disorders Among Children in the California Central Valley," *Environmental Health Perspectives,* 115, no. 10 (October 2007), www.ehponline.org/docs2007/10168/ abstract.html.

66 **behind in reading comprehension**   Joseph Jacobson and Sandra Jacobson, "Intellectual Impairment in Children Exposed to Polychlorinated Biphenyls," *New England Journal of Medicine* 335 (September 12, 1996): 783–789.

66 **acting on the thyroid**   "Endocrine Disruption and Flame Retardants," *Environmental Health Perspectives* 114, no. 2 (February 2006), www.ehponline .org.

66 **another chemical cousin**   Steve Curwood, "The Dangers of Triclosan," *Living on Earth,* November 3, 2006, www.loe.org.

66 **"a silent pandemic"**   Grandjean and Landrigan, "Chemical Exposure Creating a Silent Pandemic of Neurodevelopmental Disorders."

Five SCENE OF THE CRIME

68 **it's free of phthalates**   See www.environmentcalifornia.org/uploads/B0/av/ B0avehMELtJWs0ZmzXiK4w/Product_List.pdf. This is a database, compiled in 2005 by Environment California, of children's products with and without toxic ingredients.

68 **glued together with formaldehyde**   www.healthehouse.org.

69 **$1.72 billion-a-year industry**   Gina Solomon, "Common Air Fresheners Contain Chemicals That May Affect Human Reproductive Development" (New York: Natural Resources Defense Council), September 19, 2007.

69 **show up in your urine**   Shanna Swan, personal communication, October 13, 2006; see also www.nottoopretty.org.

69 **some toxic, others benign**   Environmental Working Group, *Skin Deep* (Washington, DC: Environmental Working Group, October 2005); see also www.skindeep.org.

69 **Moondrops lipsticks contain phthalates**   Tom Natan, *Cabinet Confidential* (Washington, DC: National Environmental Trust, undated).

69 **hair mousse usually contain two**   Julie Sevrens Lyons, "Phthalates Are Used in Personal Products," *Mercury News,* May 18, 2005, www.safecosmetics .org/newsroom; Natan, *Cabinet Confidential*.

69 **Oil of Olay, and CoverGirl**   Environmental Working Group, *Beauty Secrets* (Washington, DC: Environmental Working Group, June 2004), www.ewg .org/reports/beautysecrets.

69 **that chemical than the FDA standard**   Mark Rossi, *Neonatal Exposure to DEHP and Opportunities for Prevention* (Falls Church, VA: Health Care Without Harm, June 2001).

70 **and a probable human carcinogen**   See www.cnt.org/wetcleaning.

70 **made of PVC plus phthalates**   "Toxins, Endocrine Disruptors and Carcinogens That Migrate from the Molecules of Different Plastic Containers to Their Contents," Ecology Center, www.ecologycenter.org/ptf/toxins.html.

70 **bonded to the PVC plastic polymer**   Washington State Nurses Association, www.wsna.org.

70 **below the surface of the Atlantic**   Schettler et al., *Generations at Risk*.

70 **into contact with plastic wrap**   Donald T. Wigle, *Child Health and the Environment* (New York: Oxford University Press, 2003).

71 **Saran Wrap without any phthalates**   See www.besafenet.com/ppc/ archives2006/01/sc_johnson_elim.html.

71 **tucks into her backpack**   Those Nalgene bottles that are made of Lexan,

distinguished by their snazzy colors, are #7 plastic; Nalgene also comes in a relatively benign, less colorful #2 plastic.

71 **as a synthetic estrogen**   John Peterson Myers, "A New View on Toxic Chemicals and How They Impact Our Health," presentation to the Healthy Environment Forum, Seattle, January 24, 2004.

71 **caused by hormone replacement estrogen**   The #1 PET and #7 polycarbonate bottles are made with bisphenol-A; #2 HDPE, #4 LDPE, and #5 PP are not.

71 **most polycarbonate baby bottles**   "Toxic Baby Bottles," February 27, 2007, www.environmentcalifornia.org.

71 **are made of polycarbonate**   *Shopper's Guide to Toxic-Free Kids,* www .environmentcalifornia.org.

71 **plastic resins that line food cans**   "Chemical Linked to Birth Defects Found at Unsafe Levels in Canned Food," March 5, 2007, www.ewg.org/reports/ bisphenola.

71 **the United States, Europe, and Japan**   Travis Madsen, *Growing Up Toxic* (Sacramento: Environment California Research and Policy Center, 2004).

71 **bisphenol-A leaches out**   Myers, "Good Genes Gone Bad."

71 **fourteen billion pounds a year**   See www.vinylinfo.org.

71 **vinyl lunch boxes, for instance**   See www.pvcfree.org; Martha Mendoza, "How the Government Decided Lunch Box Lead Levels," Associated Press, February 18, 2007, www.ap.org. The AP article also disclosed that the regulating agency, the Consumer Product Safety Commission, interpreted lab test results of samples in such a way as to obviate finding lead at hazardous levels.

72 **dumped from barges into the seas**   Plastic items stamped with a #3 or the letter *V* should not be recycled. *PVC, The Poison Plastic* (Arlington, VA: Center for Health, Environment & Justice, 2004), www.besafenet.com/pvc.

72 **families in surrounding neighborhoods**   Citizens' Environmental Coalition, *Building Green Without Going in the Red* (Albany, NY: Citizens' Environmental Coalition, undated).

72 **"highly susceptible populations"**   See www.atsdr.cdc.gov/facts; EPA, *TEACH Chemical Summary,* www.epa.gov/TEACH.

72 **not labeled, you can't tell**   "Flame Retardant Chemicals Pose Serious Risks to Children's Health," fact sheet (Freeland, WA: Institute for Children's Environmental Health, undated), www.iceh.org.

72 **"flabbergasted" by their quick buildup**   Marla Cone, "Cause for Alarm over Chemicals—Fire Retardants," *Los Angeles Times,* April 20, 2003, A1, A30.

72 **"some ingredients combine to release formaldehyde"**   Telephone interview with company consumer affairs spokesperson, January 12, 2005.

73 **other nations, including China**   "Household Antibacterial Products Generate Chloroform," *Environmental Science & Technology Online News,* February 28, 2007, http://pubs.acs.org.

73 **wheeze and eventually develop asthma**   Alison McCook, "Cleaning Products May Affect Babies' Breathing," Reuters, January 11, 2005, www.nlm.hih .gov/medlineplus/news.

73 **formaldehyde and other carcinogenic compounds**   William Nazaroff and C. J. Weschler, "Cleaning Products and Air Fresheners: Exposure to Primary and Secondary Air Pollutants," *Atmospheric Environment* 38, no. 18 (June 2004): 2841–2865.

73 **fifty-two in the air** Ruthann A. Rudel et al., "Phthalates, Alkylphenols, Pesticides, Polybrominated Biphenyl Ethers, and Other Endocrine-Disrupting Compounds in Indoor Air and Dust," *Environmental Science & Technology* 37, no. 20 (September 2003): 4543–4553, http://pubs3.acs.org/acs/journals.

73 **topping the list** Jane Kay, "Study Says Household Dust Holds Dangerous Chemicals," *San Francisco Chronicle,* March 23, 2005, www.sfgate.com.

74 **disintegrated with age** Schettler et al., *In Harm's Way,* 73.

74 **lead levels above health standards** Kirk Johnson, "Lead Peril Lurks Overhead and Underfoot," *New York Times,* November 2, 2003, 1.

74 **pre–industrial age children** Herbert Needleman, "Salem Comes to the National Institutes of Health," *Pediatrics* 90, no. 6 (December 1992): 977; Donald T. Wigle and Bruce P. Lanphear, "Human Health Risks from Low-Level Environmental Exposures," *PloS Medicine* 2, no. 12 (December 2005): 1, www.plosmedicine.org.

74 **discovered in the North Pole** Cone, "Cause for Alarm over Chemicals."

74 **a Lexus or Volvo** *Toxic at Any Speed* (Ann Arbor, MI: Ecology Center, January 2006).

74 **inhaling polluted air** John Wargo, *Children's Exposure to Diesel Exhaust on School Buses* (North Haven, CT: Environment & Human Health, February 2002).

74 **his or her classmate walking to school** Renee Montagne, "Diesel-Burning School Buses May Pose Health Threat to Kids," *Morning Edition,* NPR News, August 26, 2002, http://nl.newsbank.com; Janet Wilson, "Aging U.S. School Buses Still Fouling Air," *Los Angeles Times,* May 25, 2006, www.latimes.com.

74 **particulates than when moving** Wargo, *Children's Exposure to Diesel Exhaust on School Buses.*

75 **nitrogen oxides, in just one year** Anthony Depalma, "Bus Companies Agree to Cut Engine Idling Near Schools," *New York Times,* January 24, 2004, www.nytimes.com.query.nytimes.com/. . . /Reference/Times%20Topics/People/Y/Yaniv,%20Oren.

75 **need not test their drinking water** Paul Schwartz, DC Clean Water Coalition, personal communication, February 6, 2008; a lawsuit under the Safe Drinking Water Act changed school water testing from mandatory to voluntary.

75 **twenty-five cause learning disabilities** Jay Feldman, presentation at Collaborative for Health and the Environment teleconference, October 13, 2004.

75 **dismissed their findings as "scaremongering"** George Clarke, "JAMA Should Seek Second Opinion on Pesticide Exposure at Schools Report," press release, CropLife America, July 26, 2005.

75 **walls and ceilings of many schools** GAO, *School Facilities: Condition of America's Schools* (Washington, DC: Government Accountability Office, GA0/HEHS-95–61, February 1, 1995).

75 **worse shape than our jails** Bill Fischer, ed., "In Worse Shape than Jails," *NEA Today* 16, no. 4 (November 1997): 4–21.

75 **schools in the worst repair** Children's Environmental Health Network, *Administration Report Card, 2001–2004,* www.cehn.org.

75 **failed to pass** "Harkin-Clinton Bill for School Repair Fails," Healthy Schools Campaign, www.healthyschoolscampaign.org, February 3, 2004.

75 **ground-up recycled rubber tires** *Exposures to Recycled Tire Crumbs Used on*

Synthetic Turf Fields, Playgrounds and as Gardening Mulch (Hartford, CT: Environment and Human Health, Inc.), August 2007.

75 **specialize in environmental studies**   Juan Parras, social worker and teacher at Texas Southern University, personal communication, March 15, 2005.

76 **even on top of one**   Lois Gibbs, Center for Health, Environment and Justice, personal communication, December 3, 2004.

76 **radius of a hazardous waste site**   See www.childproofing.org.

76 **some parts of our nation**   The Clean Air Act, passed in 1970 during the Nixon administration after a ten-year battle, identified six pollutants for control—ozone (smog), particulate matter (soot), sulfur dioxide (SO_2, or SO_x), nitrogen dioxide (often called NO_x), carbon monoxide, and lead. Note, by the way, that mercury is missing from that list. The American Trucking Associations, claiming compliance would be too costly, fought the EPA's efforts in 1997 to curb smog and soot all the way to the Supreme Court and delayed them until 2001, when the Court upheld the constitutionality of the EPA's right to set these limits under the Clean Air Act without having to factor in industry expenditures. ("Supreme Court Upholds EPA Position on Smog, Particulate Rules," EPA press release, February 27, 2001, http://yosemite.epa.gov.) Yet 225 counties and the District of Columbia, with a total of 192 million people, more than half our country's population, still do not meet those limits. Michael Janofsky, "Many Counties Failing Fine-Particle Air Rules," *New York Times,* December 18, 2004, www.nytimes.com.

76 **varieties of air pollutants**   See www.scorecard.org.

76 **for adults, not for children**   Woodruff et al., *America's Children and the Environment,* 30–31. This report covers the years 1990–2000.

76 **likely to be playing outdoors**   Needleman and Landrigan, *Raising Children Toxic Free,* 182–183.

76 **additional pollutants as it travels**   Ibid.

77 **power plants, and incinerators**   Woodruff et al., *America's Children and the Environment,* 18.

77 **deep in the lungs if inhaled**   Ruth A. Etzel, "Air Pollution and Bronchitic Symptoms in Southern California Children with Asthma," *Environmental Health Perspectives* 107, no. 9 (September 1999): 757–760.

77 **ingest as well as inhale them**   Committee on Environmental Health, American Academy of Pediatrics, "Ambient Air Pollution: Health Hazards to Children," *Pediatrics* 114, no. 66 (December 2004): 1699–1707.

77 **nitrogen oxide (NO_x) pollution**   Mark Clayton, "In Bid to Cut Mercury, US Lets Other Toxins Through," *Christian Science Monitor,* March 31, 2005, www.csmonitor.com. The utility industry's Edison Power Research Institute claims that cars and trucks, not coal-burning power plants, are responsible for pollution; that claim is refuted by scientists who point out that the study of a normal day in Atlanta on which that claim was based was sponsored by the institute and designed to reach that conclusion. Jeff Nesmith, "Atlanta Sick of Smog," *Atlanta Journal-Constitution,* December 1, 2004, www.ajc.com.

77 **airborne mercury from industry**   Jennifer 8. Lee, "U.S. Proposed Easing Rules on Emissions of Mercury," *New York Times,* December 2, 2003, 20A. The Clean Air Act required power plants and large factories to minimize their emissions of harmful pollutants and established national air quality standards to be met by 1975. Owners were to install pollution control technology if and when they made significant improvements to the plants. For in-

stance, they were to add scrubbers (devices that can cut 95 percent of the pollution) at the top of the smokestacks. But in a concession to industry, since many companies simply patched up their old dirty plants, Congress changed the law in 1977 to let plants phase in upgrades. Mercury emissions from coal-burning power plants are not due for cleanup until 2018, so that a whole generation of children will grow up inhaling or ingesting mercury at levels high enough to harm their brains and nervous systems.

77 **plant owners more up front**   Kenneth J. Stier, "Dirty Secret: Coal Plants Could Be Much Cleaner," *New York Times,* May 22, 2005, 3.

77 **autism and Alzheimer's disease**   Dan R. Laks, "Assessment of Chronic Mercury Exposure and Neurodegenerative Disease," unpublished paper (Berkeley: University of California at Berkeley, 2007), http://adventuresinautism .blogspot.com/2007/06/biological-mechanisms-for-mercury.html.

77 **developing bodies and brains**   *Polluting Our Future.*

78 **major portion of all dioxin releases**   *Toxic Releases and Health: A Review of Pollution Data and Current Knowledge on the Health Effects of Toxic Chemicals* (Washington, DC: U.S. PIRG Education Fund, January 2003).

78 **home to African American populations**   *Polluting Our Future,* 1.

78 **acute lymphocytic leukemia**   Cindy Horswell, "Children Living Near Houston Ship Channel Have Greater Cancer Risk," *Houston Chronicle,* January 18, 2007, www.houstontx.gov/health/UT.html.

79 **any other landscape, including croplands**   Wargo, *Our Children's Toxic Legacy,* 146.

79 **heavy doses of persistent pesticides**   Philip J. Landrigan et al., "Pesticides and Inner-City Children: Exposures, Risks, and Prevention," *Environmental Health Perspectives* 107, Suppl. 3 (June 1999): 431–437.

79 **seldom speak up, fearing retaliation**   Carol Dansereau, Farm Worker Pesticide Project, personal communications, September 23, September 25, and October 10, 2004.

79 **lifetime burden of pesticides**   *Pesticides in the Diet of Infants and Children* (Washington, DC: National Academy of Sciences, 1993).

79 **82 percent of us**   Gina Solomon and Catherine J. Karr, "Health Effects of Common Home, Lawn, and Garden Pesticides," *Pediatric Clinics of North America* 54, no. 1 (February 2007): 63–80.

79 **more than agriculture does**   EPA, *Pesticides Industry Sales and Usage, 2000 and 2001 Market Estimates* (Washington, DC: U.S. Environmental Protection Agency, Office of Prevention, Pesticides, and Toxic Substances, EPA/733/ R-04/001, May 2004).

79 **the nozzle a masculine head**   See http://2001.roundup.com.

79 **world expenditures on insecticides**   See www.epa.gov/oppbead1/pestsales.

80 **cozy names and cute bottles**   Wargo, *Our Children's Toxic Legacy,* 70.

80 **toxic to both pests and humans**   Marquita Kaya Hill, *Understanding Environmental Pollution* (Cambridge: Cambridge University Press, 2004), 372ff.

80 **for some agricultural uses**   See www.epa.gov/REDs/malathion-red.pdf., www.epa.gov/REDs/diazinon-ired.pdf.

80 **every state legislature**   Jay Feldman, Beyond Pesticides, personal communication, December 15, 2004.

80 **model supplied by the coalition**   *What Is State Preemption?* (Washington, DC: Beyond Pesticides, March 2005).

81 **the average American adult**   Philip J. Landrigan et al., "Pesticides and

Inner-City Children: Exposures, Risks and Prevention, *Environmental Health Perspectives* 107, S3 (June 1999): 431–437.

81 **detected even through current technology** Paul Thacker, "Removing Pharmaceuticals, Hormones, and Cosmetics from Tap Water," *Environmental Science & Technology News,* August 4, 2005, http://pubs.acs.org.

81 **high-income suburban communities** Bob Downing, "Medicines, Household Chemicals Flow into Creeks," *Akron Beacon Journal,* February 4, 2005, Document ID: 10812846A8B73CAC, www.ohio.com/archives.

81 **elude understanding and treatment** "Pharmaceutical Data Eludes Environmental Research," *Environmental Science & Technology News,* March 16, 2005, pubs.acs.org/subscribe/journals/esthag-w/2005/mar/science/pt_pharmdata.html.

81 **breakdown products of cigarettes** Thacker, "Removing Pharmaceuticals, Hormones, and Cosmetics from Tap Water."

81 **"sentinels of potential health effects"** Lori Valigra, "How Safe Is the Water?" *Christian Science Monitor,* December 30, 2004, www.csmonitor.com.

81 **found wherever researchers look** David Shaffer, "Former 3M Chemical Is Widespread," *Minneapolis Star Tribune,* August 15, 2004, www.startribune.com.

82 **"the water-quality benchmarks used"** Robert J. Gilliom et al., *Pesticides in the Nation's Streams and Ground Water* (Reston, VA: US Geological Survey, Circular 1291, 2006, revised February 2007).

82 **"these numbers have been arrived at"** Sandra Steingraber, *Living Downstream* (New York: Vintage Books, 1997), 194.

82 **children's intelligence and behavior** Schettler et al., *In Harm's Way,* 90–91; see also Fluoride Action Network, www.fluoridealert.org.

82 **chlorinated drinking water, called trihalomethanes** Frank Bove et al., "Drinking Water Contaminants and Adverse Pregnancy Outcomes: A Review," *Environmental Health Perspectives* 110, Suppl. 1 (February 2002): 61–73.

82 **neural tube defects among babies** *Consider the Source: Farm Runoff, Chlorination Byproducts, and Human Health* (Washington, DC: Environmental Working Group and U.S. Public Interest Research Group, January 8, 2002).

82 **bladder cancer in adults** Cristina M. Villaneuva et al., "Bladder Cancer and Exposure to Water Disinfection By-Products Through Ingestion, Bathing, Showering and Swimming in Pools," *American Journal of Epidemiology* 165, no. 2 (2007): 148–156.

83 **their weight than the average American** Wargo, *Our Children's Toxic Legacy.*

83 **sixteen times more raisins** Environmental Working Group, *They Are What They Eat: Kids' Food Consumption and Pesticides* (Washington, DC: Environmental Working Group, 1999).

83 **citing numerous studies** Woodruff et al., *America's Children and the Environment,* 29.

83 **lead, mercury, and PCBs** Schettler et al., *In Harm's Way,* 253.

83 **the DDT banned decades ago** USDA, *Pesticide Data Program: Annual Summary Calendar Year 2003* (Washington, DC: U.S. Department of Agriculture), www.ams.usda.gov/science/pdp.

83 **infant and toddler menus** Katherine Shea et al., *Reducing Low-Dose Pesticide Exposures in Infants and Children* (Washington, DC: Physicians for Social Responsibility, March 2006).

83 **food processors and grocery chains** "Where Your Food Comes From," editorial, *New York Times,* January 23, 2004, A24.

84 **has not yet been studied** Cynthia Curl and Richard Fenske, "Organophosphorus Pesticide Exposure of Urban and Suburban Preschool Children with Organic and Conventional Diets," *Environmental Health Perspectives* 111, no. 3 (March 6, 2003): 377–382.

84 **you ingest the dioxin** "Transition to PVC/DEHP Free Products," fact sheet (San Francisco: Catholic Healthcare West, 2005).

84 **at all could be recommended** Arnold Schecter, "Intake of Dioxins and Related Compounds for Foods in the US Population," *Journal of Toxicology and Environmental Health,* part A, no. 63 (2001): 1–18.

84 **dioxin-linked health effects** Environmental Working Group, *Greening Hospitals* (Washington, DC: Environmental Working Group and Health Care Without Harm, June 1998), ewg.org/reports/greening.

85 **though at much lower levels** Arnold Schecter et al., "Polybrominated Diphenyl Ethers Contamination of United States Food," *Journal of Environmental Science and Technology* 38 (2004): 5306–5311.

85 **the parents' lifetime accretion** Wargo, *Our Children's Toxic Legacy.*

85 **a Texas study discovered** Purnendu K. Dasgupta et al., "Perchlorate and Iodide in Dairy and Breast Milk," *Environmental Science and Technology* 39, no. 7 (February 2005): 2011–2017.

85 **if found in cow's milk** Schetter et al., *Generations at Risk,* 229.

85 **in three European countries** Charlotte Brody and Stephen Lester, "Protecting Baby's First Food," *Everyone's Backyard* (Spring 2000): 26–28.

85 **breast than in her blood** Bernard Weiss et al., "Pesticides," *Pediatrics* 113, no. 4 (April 2004): 1030–1036.

85 **also found in breast milk** Gina Solomon et al., "Transgeneration Exposures: Persistent Chemical Pollutants in the Environment and Breast Milk," *Pediatric Clinics of North America* 54, no. 1 (February 2007): 81–101.

85 **the lining of popcorn bags** Steve Curwood, "Mother's Milk: A Modern Dilemma," *Living on Earth,* April 20, 2007.

86 **Angela of Gainesville, Florida** This study, *Study Finds Record High Levels of Toxic Fire Retardants in Breast Milk from American Mothers,* was conducted by the Environmental Working Group, Washington, D.C. Results are reported on their website, www.ewg.org/reports/mothersmilk; P. Eriksson et al., "Brominated Flame Retardant: A Novel Class of Developmental Neurotoxicants in Our Environment?" *Environmental Health Perspectives* 109, no. 9 (September 2002): 903–908.

86 **longer time to accumulate them** Schettler et al., *Generations at Risk,* 205.

86 **not present in human milk** Julia R. Barrett, "The Science of Soy: What Do We Know?" *Environmental Health Perspectives* 114, no. 6 (June 2006), www.ehponline.org/members/2006/114–6/focus.html.

86 **effects in female offspring** Wendy N. Jefferson et al., "Adverse Effects on Female Development and Reproduction in CD-1 Mice Following Neonatal Exposure to the Phytoestrogen Genistein at Environmentally Relevant Doses," *Biology of Reproduction* 73 (2005): 798–806.

86 **occurs naturally in breast milk** Francis Crinella et al., "Effects of Neonatal Dietary Manganese Exposure on Brain Dopamine Levels and Neurocognitive Functions," *NeuroToxicity* 23, nos. 4–5 (October 2002): 645–651.

86 **fetal exposure to persistent chemicals** Sonya Lunder and Renee Sharp, *Study Finds Record High Levels of Toxic Fire Retardants* (Washington, DC: Environmental Working Group, July 10, 2003).

86 **$143 billion over the last decade**   Dan Barber, "Stuck in the Middle," *New York Times,* November 23, 2005, 29.

87 **over thousands of acres**   See www.iatp.org/foodandhealth/issues _factoryfarms.cfm.

87 **requires intensive use of pesticides**   For a stomach-turning description of a beef factory farm, read *The Way We Eat* by Peter Singer and Jim Mason (Emmaus, PA: Rodale Press, 2006).

87 **"a top concern"**   Environmental Defense, "Fewer Antibiotics Will Be Used on Hog Farms," *Solutions: Environmental Defense,* newsletter, 36, no. 4 (July–August 2005).

87 **losing money by using them**   Jay P. Graham, Ellen Silbergeld et al., "Growth Promoting Antibiotics in Food Animal Production: An Economic Analysis," *Public Health Reports* 122, no. 1 (January–February 2007): 79–87.

87 **fertilizer on nearby fields**   See www.factoryfarming.com.

88 **"The data indicate arsenic causes cancer"**   John Vandiver, "Chicken Feed Effects Questioned," *Salisbury Daily Times* (Salisbury, MD), January 4, 2004, www.dailytimesonline.com.

88 **"Americans who consume chicken"**   Ellen K. Silbergeld, "Arsenic in Food," *Environmental Health Perspectives* 112, no. 3 (2004): 338–339.

88 **before American chicken consumption tripled**   Tamar Lasky et al., "Mean Total Arsenic Concentrations in Chicken 1989–2000 and Estimated Exposures for Consumers of Chicken," *Environmental Health Policy* 112, no. 1 (January 2004): 18–21. A four-piece portion of McNuggets weighs 2.3 ounces.

88 **demanding arsenic-free chickens**   Libby Lawson, public relations, Tyson Food, personal communication, June 14, 2007.

88 **fast-food chicken products**   David Wallinga, *Arsenic Widespread in Chicken, Testing Finds* (Minneapolis: Institute for Agriculture and Trade Policy, April 5, 2006); Marian Burros, "Chicken with Arsenic? Is That O.K.?" *New York Times,* April 5, 2006, www.nytimes.com. As McDonald's and other fast-food companies have said they no longer accept fowl raised with arsenic, the arsenic found in the samples the Institute for Agriculture and Trade Policy tested might have come from soil or water or air pollution, according to David Wallinga, Institute for Agriculture and Trade Policy, personal communication, April 22, 2007.

89 **genetically engineered grains**   See www.organicconsumer.org.

89 **extra $40 per head**   Janet Raloff, "Hormones: Here's the Beef," *ScienceNews* 161, no. 1 (January 5, 2002): 10.

89 **Steroid . . . reasons for declining male fertility**   A recent study found that the sons of mothers who while pregnant consumed the relatively high level of seven beef meals a week have sperm concentrations below the World Health Organization threshold for subfertility. Swan, Shanna et al., "Semen Quality of Fertile U.S. Males in Relation to Their Mothers' Beef Consumption During Pregnancy," *Human Reproduction* 6 (June 22, 2007): 1497–1502.

89 **organic milk factory farms**   Melissa Allison, "Organic-Milk Fight Takes Aim at Grazing Time," *Seattle Times,* June 6, 2006, http://seattletimes.com.

89 **production of stress-fighting antioxidants**   Michael Pollan, "Mass Natural," *New York Times Magazine,* May 1, 2006, 15ff.

89 **deficient in these omega acids**   Genevieve Young and Julie Conquer, "Omega-3 Fatty Acids and Neuropsychiatric Disorders," *Journal of Reproduction Nutrition Development* 45 (2005): 1–28.

90 **"a uniquely broad palette"** See www.fmcbiopolymer.com.

90 **"Most of the money"** "Obesity: The Silent Environmental Epidemic," conference call, Bolinas, CA: Collaborative on Health and the Environment, September 23, 2004.

90 **and the risk of diabetes** Denise Mann, "Trans Fats: The Science and the Risks," WebMD, www.webmd.com/contents/Article/71/81217.htm.

90 **and probably to diabetes** David Wallinga, personal communication, September 6, 2005.

90 **vanilla/banana, and blueberry flavors** Caroline E. Mayer, "Putting a Healthy Spin on Processed Foods," *Washington Post,* January 10, 2005, www.washingtonpost.com.

90 **among workers at popcorn plants** Stephen Labaton, "OSHA Leaves Worker Safety Largely in Hands of Industry, *New York Times,* April 25, 2007, 1.

91 **no raspberries in this raspberry flavoring** Allan Magaziner and Anthony Zolezzi, *Chemical Free Kids* (New York: Kensington Publishing Company, 2003).

91 **identified as a factor in ADHD** Jim Stevenson et al., "Food Additives and Hyperactive Behaviour in Three-Year-Old and Eight-to-Nine-Year-Old Children in the Community," www.thelancet.com, September 6, 2007.

91 **damaging to nerve cells** K. Lau et al., "Synergistic Interactions Between Commonly Used Food Additives in a Developmental Neurotoxicity Test," *Toxicological Sciences* 90, no. 1 (2006): 178–187.

91 **have vanished or diminished** Jane Hersey, Feingold Association, personal communication, January 13, 2007.

91 **fire false, excitable signals** *Pure Facts,* newsletter (Williamsburg, VA: Feingold Association, February 1995).

91 **aggressive behavior and even crime** Roger D. Masters, "Environmental Pollution, Brain Chemistry, and Violent Crime," in Gerald A. Corey Jr. and Russell Gardner Jr., eds., *The Evolutionary Neuroethology of Paul MacLean* (Westport, CT: Praeger, 2002), 275–296.

91 **investigatory arm of the U.S. Congress** GAO, *School Meal Programs: Competitive Foods Are Available in Many Schools* (Washington, DC: Government Accountability Office, GAO-04-673, April 2004).

91 **$18 a year per child** Michael Jacobson, Center for Science in the Public Interest, personal communication, February 2, 2007.

92 **and some other junk foods** Center for Science in the Public Interest, "Junk Food in Schools Enjoys Bipartisan Support," news release, Center for Science in the Public Interest, May 20, 2004, www.cspinet.org/new/200405201.html.

92 **a year marketing to children** Melanie Warner, "Guidelines Are Urged in Food Ads for Children," *New York Times,* March 17, 2005, www.nytimes.com.

92 **restrictions on the ads** Jeremy Manier and Delroy Alexander, "Scientists Say Kids' Food Ads Are Junk," *Chicago Tribune,* December 7, 2005, www.chicagotribune.com.

92 **"We have created this monster"** Derrick Z. Jackson, "Diabetes and the Trash Food Industry," *Boston Globe,* January 11, 2006, A15.

92 **DNA damage wrought by toxins** Christopher D. Jensen et al., "Maternal Dietary Risk Factors in Childhood Acute Lymphoblastic Leukemia," *Cancer Causes and Control* 15 (2004): 565–566; Marilyn L. Kwan et al., "Food Consumption of Children," paper presented at the International Scientific Con-

ference on Childhood Leukemia, London, September 6–10, 2004, reprinted in the Northern California Childhood Leukemia study materials; Neela Guba et al., "MTHFR Polymorphisms, Maternal Folate," from the Northern California Childhood Leukemia study materials.

92 **deprived of thyroid hormones**   Environmental Working Group, *Thyroid Threat* (Washington, DC: Environmental Working Group, 2006), www.ewg .org/reports/thyroidthreat.

92 **understanding multiple chemical sensitivity**   Nicholas Ashford and Claudia Miller, *Chemical Exposures: Low Levels and High Stakes,* 2nd ed. (New York: John Wiley & Sons, 1998).

93 **in a first such search**   Toxic Use Reduction Institute, Five Chemicals Assessment Study (June 2006); *An Investigation of Alternatives to Mercury Containing Products* (Lowell: University of Massachusetts, Toxic Use Reduction Institute, January 2003). The six chemicals are mercury, lead, perc, hexavalent chromium, formaldehyde, and the DEHP phthalate.

93 **"formaldehyde as a nail hardener"**   "Leading Nail Polish Manufacturer Removes Toxic Ingredients," Campaign for Safe Cosmetics, March 29, 2007, www.safecosmetics.org/newsroom/press.cfm?pressReleaseID=22.

93 **Kaiser Permanente in six states**   Dorothy Kalins, "Fresh Ideas About Food," *Newsweek,* October 16, 2006, www.noharm.org.

93 **products in the United States**   "Kaiser Permanente Turns Green," April 23, 2003, www.greenbiz.com.

93 **and green cleaning products**   Jenn Abelson, "Organic Apparel Gets Hip," *Boston Globe,* April 9, 2007, E1; Ruth La Ferla, "The Cream Is Cleaning Green," *New York Times,* April 22, 2007, "Style" section, 1.

93 **found in any U.S. community**   *Athens Banner-Herald* (Athens, GA), April 17, 2007.

94 **and it even saves money**   William McDonough and Michael Braungart, "Transforming the Textile Industry," green@work, May–June 2002, www .mcdonough.com/writings/transforming_textile.htm.

Six FORENSICS

95 **"the dose makes the poison"**   What he actually said is that "all substances are poisons; there is none which is not a poison. The right dose differentiates a poison from a remedy." Quoted in "The Dose Makes the Poison," issued by ChemSafe, a program of the Chemical Industries Association, http://learn .caim.yale.edu/chemsafe/references/dose.html.

95 **"if the dose is low enough"**   Chem Safe, "The Dose Makes the Poison."

95 **"water will kill you"**   Ann Knef, "Atrazine Poses No Harm, EPA Finds," *Madison-St. Clair Record,* August 8, 2006, www.madisonrecord.com. Alex Avery appears to be continuing the work of his father, Dennis Avery, a longtime advocate for agribusiness and chemical interests whose latest campaign is against organic agriculture, according to SourceWatch, August 15, 2006, www.sourcewatch.org.

96 **add up to devastating consequences**   Steven G. Gilbert, *A Small Dose of Toxicology* (Boca Raton, FL: CRC Press, 2004), vii.

96 **sometimes differently vulnerable**   National Research Council, *Pesticides in the Diet of Infants and Children* (Washington, DC: National Academy Press, 1993).

96 **quadrillion can affect their development** "Pesticides and Aggression," *Rachel's Environment & Health News* 648 (New Brunswick, NJ: Environmental Research Foundation, April 28, 1999).

96 **a critical day of development** Schettler, et al., *In Harm's Way,* 80.

96 **relieved of that restriction** See www.epa.gov.oppsrrd1/REDs/chlorpyrifos_ired.pdf.

97 **IQ later in life** Kropp et al., *Body Burden.*

97 **interfere with different processes** Pauline Mendola et al., "Environmental Factors Associated with a Spectrum of Neurodevelopmental Deficits," *Mental Retardation and Developmental Disabilities Research Reviews* 8 (2002): 188–197.

97 **118 bathtubs of water** Environmental Working Group, "Fiction #1: You'd Have to Drink 500 Bathtubs to Get a Dose That Caused Any Harm in Animal Studies," *Chemical Industry Archives* (Washington, DC: Environmental Working Group), www.chemicalindustryarchives.org/factfiction.

97 **learning disabilities, and ADHD** David Douglas, "ADHD Linked to Mom's Iodine Levels," Reuters, January 17, 2005, www.nlm.nih.gov/medlineplus.html.

97 **day twelve of gestation!** Sadler, *Langman's Medical Embryology.*

97 **researchers subsequently realized** Patricia M. Rodier, "An Embryological Approach to Autism: The Thalidomide Connection," *Narrative* (Princeton, NJ: National Alliance for Autism Research, Summer 1997), 1–18.

97 **of the environmental insult** Ellen K. Silbergeld and Thelma E. Patrick, "Environmental Exposures, Toxicologic Mechanisms, and Adverse Pregnancy Outcomes," *American Journal of Obstetrics and Gynecology* 192 (May 2005): S11–S21.

97 **cardiac defects or cleft lips** Beate Ritz, "Ambient Air Pollution and Risk of Birth Effects in Southern California," *American Journal of Epidemiology* 155, no. 1 (January 2002): 17–25.

98 **more girls than boys** See www.medicinenet.com.

98 **to cancer and asthma** *A Father's Day Report—Men, Boys and Environmental Health Threats,* June 15, 2007, www.healthyenvironmentforkids.ca.

98 **protects the body against oxidative stress** Another possibility for the higher male-to-female ratio in autism lies in the fact that females have two X chromosomes (whereas males have an X and a Y chromosome). If a girl has one X chromosome with a gene that predisposes to autism, she will also have one "good" chromosome, and therefore her chances of expressing the disease are halved. If a boy's copy of X has the marker genes, his chances are high because the Y chromosome has completely different genes and cannot serve as a backup. For a thorough explanation of why boys suffer from many chronic illnesses in greater number than girls, read *A Father's Day Report: Men, Boys and Environmental Health Threats,* Canadian Partnership for Children's Health & Environment, June 15, 2007, www.healthyenvironmentforkids.ca.

98 **PCBs and flame retardants** Environmental Working Group, *Overloaded?: New Science, New Insights* (Washington, DC: Environmental Working Group, December 13, 2004), www.ewg.org/reports.

98 **as do pesticides and air pollution** Ning Li et al., "Ultrafine Particulate Pollutants Induce Oxidative Stress and Mitochondrial Damage," *Environmental Health Perspectives* 111, no. 4 (April 2003), www.ehponline.org.

98 **tadpoles raised in clean water** David Biello, "Mixing It Up," *Scientific American,* May 1, 2006, www.sciam.com.

98 **it was discovered recently**   Russ Hauser et al., "Evidence of Interaction Between Polychlorinated Biphenyls and Phthalates in Relation to Human
Sperm Motibility," *Environmental Health Perspectives* 113, no. 4 (April 2005):
425–430.

99 **low levels of helpful enzymes**   Clement E. Furlong et al., "PON1 Status of
Farmworker Mothers and Children as a Predictor of Organophosphate Sensitivity," *Pharmacogenetics and Genomics* 16, no. 3 (2006): 183–190.

99 **more of the metals than normal**   J. B. Adams et al., "Mercury, Lead, and
Zinc in Baby Teeth of Children with Autism Versus Controls," *Journal of
Toxicology and Environmental Health* 70, no. 12 (June 2007): 1046–1051. Another study, of the urine of children with autism, similarly showed they retain
mercury; Jeff Bradstreet, "A Case-Control Study of Mercury Burden in Children with Autistic Spectrum Disorders," *Journal of American Physicians and
Surgeon* 8, no. 3 (Summer 2003): 76–79. Neuropharmacologist Richard Deth
of Northeastern University has discovered that the mercury in thimerosal inhibits the enzyme methionine synthase that would otherwise clean out the
neurotransmitter glutamate.

99 **the non-ADHD test group**   Schettler et al., *In Harm's Way,* 69; Trinh Tran,
Francis M. Crinella, et al., "Effects of Neonatal Dietary Manganese Exposure
on Brain Dopamine Levels and Neurocognitive Functions," *Neurotoxicology*
23 (2002): 645–651.

99 **have the opposite effect**   Daniel M. Sheehan, "Activity of Environmentally
Relevant Low Doses of Endocrine Disruptors and the Bisphenol A Contro-
versy: Initial Results Confirmed," *Proceedings of the Society for Experimental
Biology and Medicine* 224 (June 2000): 57–60.

99 **"a paradigm shift"**   Ibid.

100 **a study demonstrated**   Rolv T. Lie et al., "A Population-Based Study of the
Risk of Recurrence of Birth Defects." *New England Journal of Medicine* 331
(1994): 1–4.

101 **precipitate cancer at low levels**   Natural estrogens are known human carcinogens; prenatal exposure to nature and synthetic estrogens is associated
with increased breast and vaginal tumors in humans. Linda S. Birnbaum and
Suzanne E. Fenton, "Cancer and Developmental Exposure to Endocrine Disruptors," *Environmental Health Perspectives* 111 (2003): 389–394.

101 **according to lab studies**   Myers, "Good Genes Gone Bad."

101 **carcinogens the EPA has ever studied**   Birnbaum and Fenton, "Cancer and
Developmental Exposures."

101 **only some dioxins are carcinogenic**   See www.vinylinfo.org.

101 **"Pesticides could become"**   "Pesticides May Be the Ultimate Contraceptive:
Why?" Mt. Sinai Center for Children's Environmental Health, ad in *New
York Times,* July 2002; also online, www.childenvironment.org.

101 **infancy and the toddler years**   John Peterson Myers, "Emerging Science on
the Impacts of Endocrine Disruptors on Intelligence and Behavior," undated,
www.ourstolenfuture.org.

101 **sixty-three hormone-disrupting pesticides**   Theo Colborn, "A Case for Revisiting the Safety of Pesticides: A Closer Look at Neurodevelopment," *Environmental Health Perspectives* 114, no. 11 (January 2006): 10–17.

101 **herbicides disrupt hormones**   Ibid.

101 **widely used in agriculture**   Bernard Weiss et al., "Pesticides," *Pediatrics* 113,
no. 4 (April 2004): 1030–1036.

101 **"To date the EPA has never"**   Colborn, "A Case for Revisiting."

102 **from conception until old age**   Gouveia-Vigeant and Tickner, *Toxic Chemicals and Childhood Cancer*.

102 **children subsequently diagnosed with leukemia**   "Progress Towards Understanding the Causes of Childhood Leukemia," paper from a University of California at Berkeley symposium, September 2, 2004, UC Berkeley Northern California Childhood Leukemia Study Program, www.ncc.org.

103 **develop through a multistep process**   Ashford and Miller, *Chemical Exposures*.

103 **autism does as well**   Dr. Robert Hendren, director, M.I.N.D. Institute, personal communication, June 23, 2005.

103 **high blood pressure in adulthood**   Steven Reinberg, American Heart Association's High Blood Pressure Conference, *HealthDay News,* www.healthday .com.

103 **Parkinson's disease among the elderly**   B. K. Barlow et al., "The Gestational Environment and Parkinson's Disease: Evidence for Neurodevelopmental Origins of a Neurodegenerative Disorder," *Reproductive Toxicology* 23, no. 3 (April–May 2007): 457–470.

103 **Dr. Theodore Slotkin predicts**   Theodore A. Slotkin, "Comparative Developmental Neurotoxicity of Organophosphates: Effects on Brain Development Are Separable from Systemic Toxicity," *Environmental Health Perspectives* 114, no. 5 (May 2006): 746–751.

103 **or grandmother and grandfather**   Gouveia-Vigeant and Tickner, *Toxic Chemicals and Childhood Cancer*.

104 **maintain the alteration and are inherited**   Michael Skinner, M. D. Amway, et al., "Epigenetic Transgenerational Actions of Endocrine Disruptors and Male Fertility," *Science* 308, no. 1 (2005): 1466–1469. Dr. Skinner subsequently discovered that female rats preferred males who were not exposed, down through three generations. See Michael Skinner, "Transgenerational Epigenetic Imprints on Mate Preference," *Proceedings of the National Academy of Science* 104, no. 14: 5942–5946.

104 **grandparent's exposure to a pollutant**   Jon Luoma, notes from the Vallombrosa conference on fertility, Menlo Park, CA, October 2005, www .healthandenvironment.org/infertility/Vallombrosa_documents.

104 **responsible for these various problems**   Agency for Toxic Substances and Disease Registry, *ToxFAQs for Trichloroethylene (TCE),* July 2003, www.atsdr .cdc.gov/tfacts19.html. ATSDR, a federal agency, tracks the effect on public health of hazardous substances, monitoring Superfund and other waste sites, responding to community inquiries, and maintaining health registries.

104 **low as 1 ppb**   "Evidence Growing on Health Risks of TCE," www .nationalacademies.org/morenews/20060727.html.

105 **each causing a different disease**   David Carpenter et al., "Understanding the Human Health Effects of Chemical Mixtures," *Environmental Health Perspectives* 110, Suppl. 1 (February 2002): 25–42.

105 **degenerative brain disorders**   Myers, "Good Genes Gone Bad"; Marla Cone, "Chemical in Plastics Is Tied to Prostate Cancer," *Los Angeles Times,* June 1, 2006, www.latimes.org.

105 **harm to the pregnancy itself**   R. Thomas Zoeller, "Environmental Neuroendocrine and Thyroid Disruption," unpublished paper, Biology Department, University of Massachusetts at Amherst, 2006.

105 **disrupt individual organs**   Nicholas A. Ashford and Claudia S. Miller, "Social and Policy Implications of Low-Level Exposures to Chemicals," *Risk, Health, and Environment,* background papers for the Third Ministerial Conference on Environment and Health, Amsterdam, June 16–18, 1999.

105 **only one or two**   Niels Erik Skakkebaek et al., "Testicular Dysgenesis Syndrome: An Increasingly Common Development Disorder with Environmental Aspects," *Human Reproduction* 16 (2001): 972–978.

105 **testicular cancer than her son's**   L. Hardell et al., "Increased Concentrations of Polychlorinated Biphenyls, Hexachlorobenzene, and Chlordane in Mothers of Men with Testicular Cancer," *Environmental Health Perspectives* 111, no. 7 (June 2003): 930–934.

105 **suffer from undescended testicles**   Katharina Maria Main, "Flame Retardants in Placenta and Breast Milk and Cryptorchidism in Newborn Boys," *Environmental Health Perspectives,* May 2007, www.ehponline.org.

106 **abnormalities in autistic brains**   Chloe Silverman and Martha Herbert, "Autism and Genetics: Genes Are Not the Cause of an Emerging Epidemic," *GeneWatch* 16, no. 1 (January 2003), www.gene-watch.org.

106 **mental-behavioral illnesses**   Generally, an autistic child cannot connect socially, engages in repetitive behavior like rocking back and forth, and has difficulty with language.

106 **"The widespread changes we're seeing"**   Martha R. Herbert, "Autism: A Brain Disorder, or a Disorder That Affects the Brain?" *Clinical Neuropsychiatry* 2, no. 6 (2005): 354–379.

106 **may improve brain conditions**   Ibid., 357.

106 **impact of environmental chemicals**   Martha Herbert, "Autism, Biology and the Environment," *Journal of the San Francisco Medical Society* 78, no. 8 (November–December 2005): 13–16.

106 **"New evidence is emerging"**   Herbert and Silverman, "Autism and Genetics."

107 **"The revolution under way"**   John Peterson Myers, "A New View on Toxic Chemicals and How They Impact Our Health," presentation, Seattle Art Museum conference, January 27, 2004, www.iceh.org/presentations.html.

107 **triggers precursors of carcinomas**   Ana Soto, presentation during Collaborative on Health & the Environment teleconference, December 7, 2006, www.healthandenvironment.org.

107 **such as autism and ADHD**   Carol Reinisch, "Clam Embryo Study Shows Pollutant Mixture Adversely Affects Nerve Cell Development," *Environmental Toxicology and Pharmacology* 19 (January 2005): 9–18.

108 **pollutants that have triggered an illness**   Richard Jackson, "Body Burden: The Pollution in Us—A Conversation on Biomonitoring," presentation during Collaborative on Health & the Environment teleconference, September 15, 2005, www.healthandenvironment.org.

108 **powerful tool for prevention**   Frederica Perera, "Molecular Epidemiology: On the Path to Prevention?" *Journal of the National Cancer Institute* 92, no. 8 (2000): 602–612.

108 **one month of exposure**   Bruce Lanphear, "Origins and Evolution of Children's Environmental Health."

108 **in a single sample of blood**   Philip J. Landrigan, "Disease of Environmental Origin in American Children: Prospects for Research and Prevention," testi-

mony before the Committee on Appropriations, U.S. House of Representatives, Washington, DC, May 2, 2000.

109 **multiple routes of exposure**   Evelyn O. Talbott and Gunther F. Craun, eds., *Introduction to Environmental Epidemiology* (Boca Raton, FL: CRC Press Lewis Publishers, 1995), 63.

109 **discipline of children's environmental health**   Lanphear, "Origins and Evolution of Children's Environmental Health."

109 **expanding exponentially over the years**   The first report, in 2001, surveyed 27 different compounds; the second, released in 2003, upped the catalog to 116; the third, released in July 2005, tracked 148. The participants were offered a $2,000 physical exam for free, in return for submitting to extensive blood and urine tests. The next study, not yet completed when this book went to press, will have the ability to look at about 473 different chemical compounds in the blood of people across our nation. See www.cdc.gov/exposurereport.

109 **what the gene normally does**   "Knockout Mice," fact sheets, National Human Genome Research Institute, October 2006, www.genome.gov.

110 **on the developing nervous system**   Patricia M. Rodier, "Environmental Causes of CNS Maldevelopment," *Pediatrics* 113, no. 4 (April 2004): 1076–1083.

110 **exposure to environmental toxins**   Landrigan, "Disease of Environmental Origin" testimony.

111 **a wide range of concentrations**   National Research Council, "Toxicity Testing in the Twenty-first Century: A Vision and a Strategy," National Academy of Science, June 12, 2007, www.nap.edu. The National Academies of Science bring together the National Academy of Science, National Academy of Engineering, Institute of Medicine, and National Research Council.

111 **the M.I.N.D. Institute**   The initials stand for Medical Investigation of Neurodevelopmental Disorders.

113 **daily exposure to pollutants**   *Research into Action* (Seattle: University of Washington Center for Child Environmental Health Risks Research, May 2003).

113 **PCBs, and pesticides**   Philip Landrigan et al., "Health and Environmental Consequences of the World Trade Center Disaster," *Environmental Health Perspectives* 112, no. 6 (May 2004).

113 **other chlorinated materials in the buildings**   Center for Health, Environment and Justice, *PVC, the Poison Plastic,* report (New York: PVC Campaign, Center for Health, Environment, and Justice, undated), www.besafenet .com/pvc.

113 **presented no health risk**   The EPA's own inspector general faulted the agency for its misleading pronouncements and the White House for removing cautionary language from the EPA's press releases. It has further come to light that a White House executive order gave Christie Whitman authority to classify embarrassing documents as "secret." Corky Siemaszko, "Secret 9/11 Lies?" *Daily News,* August 1, 2006, www.nydailynew.com. The Government Accountability Office in 2004 criticized the federal government for failing to study the health effects. The judge presiding over the class action lawsuit brought against the EPA and Ms. Whitman as an individual by students, workers, and residents concluded that "Whitman's deliberate and misleading statements to the press shocks the conscience."

Thomas Zambito, *New York Daily News,* February 3, 2006; Julia Preston,
"Public Misled on Air Quality After 9/11 Attack, Judge Says," *New York
Times,* February 3, 2006; Robin Shulman, "Health Fears for Victims of
Ground Zero's Deadly Dust," *The Guardian,* February 10, 2006,
www.guardian.co.uk. In May 2007, New York City's medical examiner, for
the first time, directly linked a death to exposure to dust from the destruction of the World Trade Center. Anthony DePalma, "For the First Time,
New York Links a Death to 9/11 Dust," *New York Times,* May 24, 2007,
www.nytimes.com.

114 **putting them at high risk**   Rescue and cleanup workers are reported to
have developed cancer, emphysema, heart disease, lung disease, and other
maladies. A collaborative of six research centers is investigating the health effects among the at-risk adults. For information, see Mt. Sinai Hospital Department of Community and Preventive Medicine, www.mssm.edu/cpm/div
_of_epidem.shtml.

114 **The mothers seem unharmed**   Sally Ann Lederman et al., "The Effects of
the World Trade Center Event on Birth Outcomes Among Term Deliveries
at Three Lower Manhattan Hospitals," *Environmental Health Perspectives* 112,
no. 17 (December 2004): 1772–1778; Frederica P. Perera, "Relationships
Among Polycyclic Aromatic Hydrocarbon-DNA Adducts, Proximity to the
World Trade Center, and Effects on Fetal Growth," *Environmental Health
Perspectives* 113, no. 8 (August 2005): 1062–1067.

114 **autism, asthma, and diabetes**   Shelley A. Hearne, *Healthy from the Start:
Why America Needs a Better System to Track and Understand Birth Defects and
the Environment* (Washington, DC: Pew Environmental Health Commission,
undated); Wargo and Evenson Wargo, *State of Children's Health and Environment 2002,* 17, www.healthychild.org; American Diabetes Association, "Low
Weight Babies, High Risk Later in Life," undated, www.diabetes.org.

114 **health problems in childhood**   Lederman et al., "The Effects of the World
Trade Center Event on Birth Outcomes."

114 **especially vulnerable unborn babies**   Among women who had been in their
first and second trimester when the 9/11 tragedy occurred and gave birth between December 2001 and June 2002, the center team enrolled 160 exposed
women and 169 as controls who had not been exposed as they came into three
lower Manhattan hospitals to deliver their babies, explained study leader Dr.
Sally Ann Lederman. That size group can produce statistically significant
analyses. The center took samples of blood from the mother or the umbilical
cord and samples of the baby's meconium (first poop) and ran the samples
through analyses, while the hospitals took the babies' measurements. A similar study through the Mt. Sinai Center for Children's Health and the Environment showed a twofold increase in smaller than normal newborns.

115 **also smaller head circumference**   Perera et al., "Effects of Transplacental
Exposure to Environment Pollutants."

115 **a child's risk of cancer**   Perera et al., "Relationships Among Polycyclic Aromatic Hydrocarbon-DNA Adducts"; and Frederica P. Perera et al., "Biomarkers in Maternal and Newborn Blood Indicate Heightened Fetal
Susceptibility to Procarcinogenic DNA Damage," *Environmental Health Perspectives* 112: 1133–1136; Lederman et al., "The Effects of the World Trade
Center Event on Birth Outcomes."

115 **biomonitoring and epidemiology**   Frederica P. Perera, "Molecular Epi-

demiology: On the Path to Prevention?" *Journal of the National Cancer Institute* 2, no. 8 (April 19, 2000): 602–612.

116 **link that damage to cancer**   Jeff Wheelwright, "Is There Cancer in This Photo?" *Discover,* March 2006, 60–66.

116 **babies in the group were significantly shorter**   Robin M. Whyatt et al., "Prenatal Insecticide Exposures, Birth Weight and Length Among an Urban Minority Cohort," *Environmental Health Perspectives* 112, no. 10 (July 2004): 1125–1132.

116 **birth length also increased to normal**   Ibid.

116 **showered down its brew of pollutants**   Anthony M. Szema et al., "Clinical Deterioration in Pediatric Asthmatic Patients after September 11, 2001," *Journal of Allergy and Clinical Immunology* 113, no. 3 (March 2004), www.911ea.org.

117 **behavioral problems, including ADHD**   Virginia A. Rauh et al., "Impact of Prenatal Chlorpyrifos Exposure on Neurodevelopment in the First 3 Years of Life Among Inner-City Children," *Pediatrics* 118, no. 6 (December 2006): 1845–1859.

117 **"flow through science like a river"**   Douglas L. Weed of the National Cancer Institute in Bethesda, MD, as quoted by Janet Raloff, "Benched Science," *Science* 168, no. 15 (October 8, 2005): 232.

118 **untangle gene-environment interactions**   Susan M. Koger, Ted Schettler, and Bernard Weiss, "Environmental Toxicants and Development Disabilities," *American Psychologist* 60, no. 3 (April 2005): 243–255.

118 **alarm that the chemical does harm**   Schettler et al., *Generations at Risk,* 25.

118 **strategy for protecting human health**   Koger et al., "Environmental Toxicants."

118 **research to which they refer**   Richard Clapp and Polly J. Hoppin, "Using Science to Meet Public Health Goals," *American Journal of Public Health* 95 (July 2005): S8–S13.

119 **paint manufacturers in Rhode Island**   Lanphear, "Origins and Evolution of Children's Environmental Health."

119 **"There is no uniformity of opinion"**   Robert Brent and Michael Weitzman, "The Current State of Knowledge About the Effects, Risks, and Science of Children's Environmental Exposures," *Pediatrics* Supplement 113, no. 4 (April 2004): 1158–1166.

119 **enough to say something does harm**   Richard Clapp, personal communication, October 11, 2006.

120 **which is still the case**   Schettler et al., *Generations at Risk,* 310.

120 **"the occurrence of a greater"**   "Cancer Facts," fact sheet (Washington, DC: National Cancer Institute, December 23, 2003), http://cis.nci.nih.gov/asp/factsheet.

121 **help combat unsafe chemicals**   Martin Griffith, "Gene Variation ID'd in Leukemia Cluster," *Boston Globe,* December 1, 2006, A27.

121 **smoking or sunbathing**   David Robinson, *Cancer Clusters: Findings vs. Feelings* (New York: American Council on Science and Health, undated).

122 **cases of children with leukemia**   S. W. Lagakos and M. Zelen, "An Analysis of Contaminated Well Water and Health Effects in Woburn, Massachusetts," *Journal of the American Statistical Association* 81, no. 395 (September 1986): 583–596.

122 **contamination was the lowest**   E. Scott Bair, "What the Judge, Jury and John Travolta Didn't Know: Beyond a Civil Action," presentation, December 1997, modified October 2003, www.geology.ohio-state.edu/courtroom.

123 **most recent data gathering**   Kevin Costas et al., "A Case-Control Study of Childhood Leukemia in Woburn, Massachusetts: The Relationship Between Leukemia Incidence and Exposure to Public Drinking Water," *Science of the Total Environment* 300 (December 2, 2002): 23–25.

123 **chronic diseases from TCE**   See www.tceblog.com.

123 **"I've had several cases"**   Joan Lowry, "Evidence of Chemical Effects on Children Mounts," Scripps Howard News Service, December 16, 2003, www.sej.org.

123 **University of Wisconsin at Madison**   "Pesticides and Aggression," *Rachel's Environment & Health News* 648 (New Brunswick, NJ: Environmental Research Foundation, April 28, 1999); W. P. Porter et al., "Endocrine, Immune and Behavioral Effects of Aldicarb (Carbamate), Atrazine (Triazine) and Nitrate (Fertilizer) Mixtures at Groundwater Concentrations," *Toxicology and Industrial Health* 15, nos. 1–2 (1999): 133–150.

123 **turns rats hyperactive and aggressive**   Frederick vom Saal et al., "Estrogenic Pesticides: Binding Relative to Estradiol in MCF-7 Cells and Effects on Exposure During Fetal Life on Subsequent Territorial Behavior in Male Mice," *Toxicology Letters* 77 (1995): 343–350.

123 **more territorial than normal**   P. Palanza et al., "Prenatal Exposure to Endocrine Disrupting Chemicals: Effects on Behavioral Development," *Neuroscience and Biobehavioral Reviews* 23, no. 7 (November 1999): 1011–1027.

124 **exposure occurs during fetal life**   Colborn et al., *Our Stolen Future,* chap. 13.

124 **"may affect behavior during puberty"**   David Goodman, "Manganese Madness," www.westonaprice.org/soy/manganese.html.

124 **another study announces**   N. Ribas-Fito et al., "Exposure to Hexachlorobenzene During Pregnancy and Children's Social Behavior at 4 Years of Age," *Environmental Health Perspectives* 115, no. 3 (March 2007): 447–450.

124 **white or African American**   Herbert L. Needleman et al., "Bone Lead Levels in Adjudicated Delinquents," *Neurotoxicology and Teratology* 24, no. 6 (November–December 2002): 711–717.

124 **pesticides such as Dursban do**   Orrin Devinsky et al., "Aggressive Behavior Following Exposure to Cholinesterase Inhibitors," *Journal of Neuropsychiatry and Clinical Neuroscience* 4 (1992): 189–194.

124 **levels of heavy metals**   Roger D. Masters et al., "Environmental Pollution, Neurotoxicity, and Criminal Violence," in J. Rose, ed., *Environmental Toxicology: Current Developments* (London: Gordon & Breach, 1998), 13–48. Factors such as poverty, nutrition, and stress could also be involved.

124 **Columbine and Virginia Tech**   The early life in South Korea of the man responsible for the deaths of the Virginia Tech students is unknown, but it is known that when he lived in the United States before setting off for school, he helped his grandparents in their dry-cleaning store.

125 **"It follows . . . that a community"**   Roger D. Masters, "Environmental Pollution, Brain Chemistry, and Violent Crime."

125 **"blame the victims"**   Roger D. Masters, "Lead, Brain Chemistry, and Educational Failure," unpublished paper, Dartmouth College, Hanover, NH, 2004.

Seven PERPETRATORS

126 **"to make expenditures on reducing"**   Milton Friedman, "A Friedman Doctrine—The Social Responsibility of Business Is to Increase Its Profits," *New York Times Sunday Magazine*, September 13, 1970, 33.

127 **"In an ideal free market"**   Ibid.

129 **thyroid, stomach, and skin**   EPA, Persistent Bioaccumulative and Toxic Chemical Program, "Polychlorinated Biphenyls (PCBs)," undated, www.epa .gov/pbt/pubs/pcbs.htm.

129 **in and around the city**   EPA, Toxics Release Inventory, *Envirofacts Report,* September 9, 2005, www.epa.gov/cgi-bin/epaprintonly.cgi.

130 **GE had inflicted on the city**   "General Electric Agrees to $250 Million Settlement to Clean Up PCBs in Housatonic River," press release, U.S. Department of Justice, October 7, 1999, www.udoi.gov/opa/pr/1999/October/ 471env.htm. This document is but one part of the voluminous documentation for the environmental problems GE caused and left behind in Pittsfield.

131 **"no clear and convincing evidence"**   Renata D. Kimbrough, "Polychlorinated Biphenyls (PCBs) and Human Health: An Update" (Washington, DC: Institute for Evaluating Health Risks, 2005), accessed through PubMed, a service of the National Library of Medicine and the National Institutes of Health, www.ncbi.nlm.nih.gov/entrez/query.fcgi?cmd=Retrieve &db+PubMed&list-uids+761.

131 **"show up in unexpected places"**   Letter from R. Kelly Niederjohn to R. J. Dasgrosfillers, manager of the Pittsfield GE transformer plant, May 15, 1981.

131 **"List of Potential Environmental Liabilities"**   Unsigned GE memo dated February 13, 1992.

132 **state average in all subjects**   Pittsfield Public Schools 2005 MCAS results, www.boston.com/education/mcas/scores2005/districts/0236.htm.

133 **goes on for pages**   "GE: Decades of Misdeeds and Wrongdoing," *Multinational Monitor* 212, nos. 7–8 (July–August 2001), www.multinational monitor.org/mm2001/o1july-august01corp4.html.

133 **"For anyone who understands"**   Chris Anderson, "Little Green Men," *Flow,* July 8, 2005, http://idg.communication.utexas.edu/flow/jot?jot.

134 **populated largely by African Americans**   "History of PCB Manufacturing in Anniston," fact sheet (St. Louis, MO: Solutia, Inc., undated), www.solutia .com/pages/anniston/pcbhistory.asp.

134 **landfills in and around Anniston**   Michael Grunwald, "Monsanto Hid Decades of Pollution," *Washington Post,* January 1, 2002, A1.

135 **"acute PCB intoxication"**   Ellen Berry, "A Neighborhood of Poisoned Dreams," *Los Angeles Times,* April 13, 2004.

135 **"no data ever has confirmed"**   Jessica Centers, "Plaintiffs Still Feel Lost in the Shuffle Associated with PCB Cases," *Anniston Star,* September 27, 2004, 1.

135 **"thousands of pages"**   Grunwald, "Decades of Pollution."

136 **to increase milk production**   Russell Mokhiber, "Monsanto: Life Itself," *Multinational Monitor* 11, no. 12 (December 1990), www.multinationalmonitor .org/hyper/issues/1990/12/mokhiber.html.

136 **"On the basis of these results"**   Morando Soffritti et al., "First Experimental Demonstration of the Multipotential Carcinogenic Effects of Aspartame Administered in the Feed to Sprague-Hawley Rats," *Environmental Health Perspectives* 114, no. 3 (March 2006).

136 **affect brain receptors**   Philip J. Landrigan, Herbert L. Needleman, and Mary Landrigan, *Raising Healthy Children in a Toxic World* (Emmaus, PA: Rodale Organic Living Books, 2001), 54.

136 **did not include children**   Associated Press, "Federal Study Rejects Aspartame Risks," *New York Times,* April 5, 2006, www.nytimes.com/aponline/ usAP-Diet-Aspartame.html.

136 **settled out of court**   "Raging Hormones," *Sierra,* November–December 2003, 15.

137 **higher risk of cancer**   Organic Consumers Association, "Despite Industry Propaganda Monsanto's Bovine Growth Hormone Still Threatens Public Health," February 3, 2005, www.organicconsumers.org/rBGH/milkismilk/20405.cfm.

137 **"higher exposures to IGF-1"**   Consumers Union, "Potential Public Health Impacts of the Use of Recombinant Bovine Somatotropin in Dairy Production," September 1997; www.consumersunion.org/pup/core_food_safety/002274.html; letter from Michael Hansen, Consumers Union, to Steven Rowe, attorney general, state of Maine, re petition to suspend use of the state of Maine quality trademark for milk and milk protein, February 11, 2003.

137 **a lifetime of exposure**   EPA, "Technical Factsheet on: ALACHLOR," fact sheet (Washington, DC: Environmental Protection Agency), www.epa.gov/OGWDW/dwh/t-soc/alachlor.html.

137 **spine of newborn babies**   Susan M. Booker, "Dioxin in Vietnam: Fighting a Legacy of War," *Environmental Health Perspectives* 109, no. 3 (March 2001).

138 **"Dow Chemical, on this level"**   Jack Doyle, *Trespass Against Us: Dow Chemical & the Toxic Century* (Monroe, ME: Common Courage Press, 2004), xviii. It should be noted that one of the authors of this book, Philip Shabecoff, served for several years on Dow's Corporate Environmental Advisory Council. He did so because he felt that if he could help nudge a big ship like Dow Chemical even a few inches toward becoming a more benign presence in the environment, he would be doing the Lord's work. In fact, for much of the time during his tenure on the council, the company did seem to be moving in a new and better direction. Management at the highest levels did seem to be honestly committed to the path toward more responsible environmental citizenship. A number of the council's recommendations, such as annual environmental accounting by the company and full cost accounting for its products and operations, were accepted by management. Dow's reputation for environmental responsibility began to improve. Then, however, the company suffered a less-than-stellar year on its balance sheet, and its competitors in the marketplace gained ground. Top management of the company changed, and the new leaders indicated they were more concerned about the bottom line than about Dow's environmental footprint. The company seemed to be retreating toward some of its past environmental practices. The new leadership was not indifferent to environmental concerns, so perhaps the company's green performance improved as its economic performance improved. Along with several other members, however, Philip left the council.

138 **vinyl chloride, and chlorpyrifos**   Ibid., 435.

138 **threat to the health of children**   "Cheerios Brought to a Halt," *New York Times,* June 17, 1994, http://query.nytimes.com/gst/fullpage.html.?res=940DE6D7173DF934A25755COA962958.

139 **illnesses from other toxic exposures**   Jane Kay, "Pesticide Threat to Babies Linked to Enzyme Levels, Researchers Find Them Much More at Risk than Adults," *San Francisco Chronicle,* March 3, 2006, www.sfgate.com/cgi-bin/article.cgi?file=/c/a/2006/03/03/BAGVGHHV1h1.dtl&type.

139 **"In a letter to the EPA"**   Ashford and Miller, *Chemical Exposures,* 59ff.

141 **resulting from exposure to chlorpyrifos**   "EPA Fines DowElanco for Failure to Report Pesticide Health Effects," press release, Environmental Protection Agency, www.yosemite.epa.gov/opa/admpres.nsf/2efl48c99c9c26a, May 2,

1995; statement of Charles Lewis, chairman and executive director of the Center for Public Integrity (a nonpartisan, nonprofit research organization), announcing the release of the report *Unreasonable Risk: The Politics of Pesticides,* June 30, 1998.

141 **other compounds, many hazardous** New Jersey Department of Health and Senior Services, "Citizen's Guide to the Reich Farm Public Health Assessment (Trenton, NJ: State of New Jersey Department of Health and Senior Services, March 2001), www.state.nj.us/health/eoh/assess/reich_cg.pdf.

142 **levels of volatile organic chemicals** New Jersey Department of Health and Senior Services, Under a Cooperative Agreement with the Agency for Toxic Substances and Disease Registry, Health Consultation, Ciba-Geigy Corporation, Toms River, Ocean County, NJ, March 27, 2003, www.atsdr.cdc.gov/HAC/PHA/cibageigy/cib_toc.html.

144 **workers exposed to it** Gerald Colby, *Du Pont Dynasty* (Secaucus, NJ: Lyle Stuart, 1984). Knowledge of lead's hazards dates back to the days of the Roman Empire; Ben Franklin cautioned about its effects in his writing; it was well known that hatmakers using lead in their work often became mad hatters.

144 **"The addition of lead"** Steven G. Gilbert, *A Small Dose of Toxicology,* 89.

144 **"why make all this fuss"** Colby, *Du Pont Dynasty,* 246–250.

144 **emitters of cancer-causing substances** Ibid., 813.

145 **levels of PFOA in their blood** Tom Pelton, "Teflon Chemical Found in Infants," *Baltimore Sun,* February 6, 2006, http://baltimoresun.com/news/local/bal-te.md.teflon06feb06,1,849991,print.story?col.

145 **"Our story is not"** News Media Update, Reporters Committee for Freedom of the Press, 2004, www.rcfp.org/news/2004/05261eachv.html.

146 **no liability for its conduct** "EPA Settles PFOA Case Against Du Pont for Largest Environmental Administrative Penalty in Agency History," EPA press release, Environmental Protection Agency, December 14, 2005.

146 **as of this writing** Environmental Science & Technology Online News, "Scientists Hail PFOA Reduction Plan," March 15, 2006, http://pubs.acs.org/subscrib/journals/esthag-w/2006/mar/policy/rr_PFOAreduction.html.

146 **a likely cause of cancer** Jeff Montgomery, "EPA Admits C8 May Be Unsafe for Humans," *Wilmington News Journal,* March 9, 2006.

146 **severe tooth enamel damage** Marla Cone, "Panel Fears Too Much Fluoride Getting into Water," *Los Angeles Times,* March 22, 2006, www.latimes.com/news/nationworld/nation/la-032206fluoride_lat,0,7247354,print,st.

147 **indicated they would comply** Marla Cone, "US EPA Calls for End to Releases of Chemical in Teflon Process," *Los Angeles Times,* January 26, 2006.

147 **presented to the health of children** Greenwire, "Child Health Risks to Be Assessed," July 5, 2001, www.eenews.net/Greenwire/searcharchive/test_search-display.cgi?q=DuPont&file=%.

148 **we were unable to do so** After several e-mail messages from us asking for a meeting, the communications director of the council, Tiffany Harrington, agreed to arrange for interviews with council experts at its headquarters in Roslyn, Virginia, directly across the Potomac River from Washington, D.C. We had told her that we would be in the Washington area for two days but would arrange our schedule to meet when convenient for council officials. At her request, we had e-mailed our proposed questions in advance. We arrived at the headquarters, the entire floor of a large office building, with sun-filled,

tastefully furnished offices overlooking the Potomac and the Jefferson Memorial, at the specified time. No one was there to meet with us. We waited for half an hour and then were told by the receptionist that Ms. Harrington had taken a "personal day off." Later, when we checked our voice mail message at our home in Brookline, Massachusetts, we found she had called at 4:30 the previous day to say the meeting was off. We were, of course, already in Washington and had checked our messages an hour earlier that day. We never got an explanation for why the meeting was dropped. In a later e-mail, Harrington raised the possibility of a meeting at some unspecified time in the future.

148 **"for understanding the possible effects"**   Press release, National Institute of Environmental Health Sciences, July 26, 2001, www.niehs.gov/oc/news/accmou.htm.

148 **lack of public oversight**   GAO, *NIH and EPA Need to Improve Conflict of Interest Reviews for Research Arrangements with Private Sector Entities* (Washington, DC: Government Accountability Office, GAO-05–191, February 2005).

148 **"to undermine a European plan"**   Elizabeth Becker, "White House Undermined Chemical Test, Report Says," *New York Times,* April 2, 2004, www.nytimes.com/2004/o4/02/business/02chemical.html.

148 **North American Commission on Environmental Cooperation**   "Taking Stock: A Special Report on Toxic Chemicals and Children's Health in North America," comments of the American Chemistry Council, draft April 13, 2004.

150 **industry emissions since 1988**   See www.americanchemistry.com/s_acc/sec_statistics.asp?CID=176&DID=304.

150 **"Responsible Care was, and remains"**   Chemical Industry Archives, a project of the Environmental Working Group, March 30, 2001, www.chemicalindustryarchives.org/dirtysecrets/responsiblecare/6.asp.

152 **Healthy Child Healthy World**   The organization's previous name was Children's Health Environmental Coalition.

Eight CO-CONSPIRATORS

153 **"Corporations do what they do"**   Quoted in Philip Shabecoff, *Earth Rising* (Washington DC: Island Press, 2000), 89.

154 **delivering cleaner water**   Philip Landrigan, personal communication, January 29, 2004.

155 **"the greatest hoax ever perpetrated"**   Todd Neff, "Inhofe, R-Okla., Is a Global-Warming Critic," *Daily Camera,* March 18, 2006, www.dailycamera.com/bdc/cda/article_print/0.1983,BD.C._2432_4552186_ARTICL.

155 **"makes clear that, in general"**   Children's Environmental Health Network, "Children's Environmental Health Bush Administration Report Card, 2001–2004," hwww.cehn.org/cehn/reportcard2004.html.

156 **while in their mother's uterus**   "EPA Finalizes Power Plant Mercury Rule," *PSR Reports* 27, no. 2 (Spring–Summer 2005): 4.

157 **"This is going to be"**   Alan C. Miller and Tom Hamburger, "Critics Swift to Jump on Rule to Reduce Mercury Emissions," *Los Angeles Times,* March 15, 2005, www.latimes.com/news/printedition/asection/la-na-mercury15mar15,1,5315773.

157 **higher than previous estimates**   Shankar Vedantam, "New EPA Mercury

Rule Omits Conflicting Data," *Washington Post*, March 22, 2005, www .washpost.com/wp-dyn/articles/a55268.2005mar21.html.

157 **"a degree of politicization"**   Union of Concerned Scientists, "Scientific Integrity, Information on Power Plant Mercury Emissions Censored," April 2005, www.ucsusa.org/scientific_integrity/interference/mercury-emissions .html.

157 **preparing the mercury rule**   Felicity Barringer, "E.P.A. Accused of a Predetermined Finding on Mercury," *New York Times,* February 4, 2005, www .nytimes.com/2005/03/04/national/04mercury.html.

158 **the output of those committees**   Union of Concerned Scientists, "Scientific Integrity, Restoring Scientific Integrity in Policymaking," February 18, 2004, www.ucsusa.org/scientific_integrity/interference/scientists-signon-statement .html.

159 **in the environment that harm children**   *Politics and Science in the Bush Administration,* a report by the minority staff of the United States House of Representatives Committee on Government Reform—Special Investigations Division, November 3, 2003, 15.

159 **ties to the lead industry**   Ibid., 22. Among the three industry-associated appointees were Dr. Joyce Tsuji, who worked for two companies with contracts from the lead industry, and Dr. William Banner, who had served as an expert witness for the Sherwin-Williams paint company when it was being sued by the state of Rhode Island for damages to children caused by its lead paint.

159 **protect the industry from lawsuits**   Milan Kecmant, "Is the White House Helping Researchers Reach the 'Right' Conclusions?" Cleveland *Plain Dealer,* May 28, 2005, www.cleveland.com/ssf?/base/news/1117272667306010 .xml&coll=2.

159 **"This stuff would be prime material"**   Donald Kennedy, "An Epidemic of Politics," *Science* 299, no. 5607 (January 31, 2003): 625.

159 **kids as test animals**   David D. Kirkpatrick, "E.P.A. Halts Florida Test on Pesticides," *New York Times,* April 9, 2005, www.nytimes.com/2005/04/09/ politics/09pesticides.html?th&emc-th.

160 **public employees' group**   "Chemical Industry Is Now EPA's Main Research Partner," press release, Public Employees for Environmental Responsibility, October 5, 2005, http://by103fd.bay.103.hotmail.msn.com/cgi-bin/getmsg ?msg=1B7FD7B5–925C-45F3–907.

160 **during the Clinton administration**   Alan C. Miller and Tom Hamburger, "EPA Relied on Industry for Plywood Plant Pollution Rule," *Los Angeles Times,* May 21, 2004, 1.

160 **the company was fired**   Marla Cone, "U.S. Fires Firm over Apparent Conflict," *Los Angeles Times,* April 17, 2007, www.latimes.com.

160 **exception to that conclusion**   Eddy Ball, "Advisory Panel Weighs in on Bisphenol-A," *NIEHS Environmental Factor,* newsletter, September 20, 2007, www.niehs.nih.gov/news/newsletter; and Anila Jacob, letter to CERHR re Interim Draft Report on Bisphenol-A, June 20, 2007, Environmental Working Group.

161 **"The revolving door"**   *A Matter of Trust,* report by the Revolving Door Working Group, www.revolvingdoor.info.

161 **industry and the public**   Tom Hamburger, "EPA Puts Mandated Lead-Paint Rules on Hold," *Los Angeles Times,* May 10, 2005, www.latimes.com/ news/nationworld/nation/la-na-paint10may10,1,99282.

161 **"The numbers point out"**   Juliet Eilperin, "EPA Faulted on Clean-Water Violations," *Washington Post,* March 31, 2004, A08.

162 **"just one more example"**   Jim Jeffords and Julie Fox Gorte, "A Dark Cloud over Disclosure," *New York Times,* March 10, 2006, A19.

163 **expenditures to reduce those levels**   David Corn, "The Other Lies of George Bush," *The Nation,* October 13, 2003, www.thenation.com/doc.mhtml ?i=20021013&s=corn.

163 **"a powerful counterattack"**   Stephen Labaton, " 'Silent Tort Reform' Is Overriding States' Powers," *New York Times,* March 10, 2006, C5.

164 **doubled from the previous year**   Chris Baltimore, "U.S. House Panels Push Bills to Help Energy Industry," Reuters, September 29, 2005, http://enn.com/ today.html?id=8919.

164 **2002 midterm elections**   Jonathan Weisman, "A Homeland Security Whodunit, in Massive Bill, Someone Buried a Clause to Benefit Drug Maker Eli Lilly," *Washington Post*, November 28, 2002, A45.

164 **"there was a great deal"**   Stephan S. Hall, "The Drug Lords," *New York Times Sunday Book Review,* November 14, 2004, 9.

165 **"hundreds of times higher"**   Dina Capiello, "Making It Easy," *Houston Chronicle,* January 16, 2005, www.chron.com/cs/CDA/printstory.mpl/ topstory/2989506.

165 **donations to local elections**   Janet Elliott, "Move to Levy Fines for Air Pollution Is Snuffed Out," *Houston Chronicle,* May 4, 2005, A1. The petrochemical industry alone contributed more than $600,000 to the political accounts of state legislators and officials in 2004, according to the article.

165 **"the method the state of Texas"**   Capiello, "Making It Easy."

167 **radioactive, and other hazardous substances**   Wargo and Evenson Wargo, *State of Children's Health and Environment 2002,* 12.

167 **at low concentrations**   EPA, "Contaminant Focus>Perchlorate," April 12, 2006, http://cluin.org/contaminantfocus/default.focus/sec/perchlorate/cat/ Overview.

168 **mental function in adults**   R. Thomas Zoeller et al., "Thyroid Hormone, Brain Development, and the Environment," *Environmental Health Perspectives* 110, Suppl. 3 (June 2002): 355–361; *Thyroid Threat* and *44 Million Women at Risk of Thyroid Deficiency from Rocket Fuel Chemical* (Washington, DC: Environmental Working Group, October 2006), www.ewg.org/reports/ thyroidthreat.

168 **milk, and other foods**   California Department of Health Services, "Perchlorate in California Drinking Water: Overview and Links," May 3, 2006, www .dhs.ca.gov/ps/ddwem/chemicals/perchl/perchlindex.htm.

169 **given to the National Academy**   The controversy over perchlorate has been well documented in the media. For our summary of the issue, we relied on several sources. They included Peter Waldman, "Inside Pentagon's Fight to Limit Regulation of Military Pollutant," *Wall Street Journal,* December 20, 2005, 1; Jennifer 8. Lee, "Second Thoughts on a Chemical: In Water, How Much Is Too Much?" *New York Times,* March 2, 2004; Daniel Danelski, "Agency Raise Perchlorate Concerns," *Press-Enterprise* (Riverside, CA), July 24, 2005; and a report by Danielski on withholding data from the National Academy reported on the Environmental Working Group website, www .ewg.org/issues/perchlorate/20050603/index.phb?print_ve.

169 **"From Cape Cod in Massachusetts"**   Jon R. Luoma, "Toxic Immunity,"

Mother Jones, November–December 2003, www.motherjones.com/cgi-bin/print_article.pl?url+http://www.motherjones.com/new.

169 **contaminated with TCE**   Dr. Samuel L. Brock, "New Trichloroethylene Cleanup Standards," HQAFCEE/ERS: Air Force, April 17, 2003.

170 **disorders at the camp**   See www.watersurvivors.org.

170 **nearly two decades earlier**   Jennifer Sass, Natural Resources Defense Council, March 23, 2005, comments on the EPA's 2001 *Assessing the Human Health Risk of Trichloroethylene.*

170 **"What happened with TCE"**   Ralph Vartgebedian, "How Environmentalists Lost the Battle over TCE," *Los Angeles Times,* March 29, 2006, www.latimes.com/news/science/environment/la-na-toxic29mar29,0,5610036.story.

170 **"the willingness to set aside science"**   Jennifer Sass, "U.S. Department of Defense and White House Working Together to Avoid Cleanup and Liability for Perchlorate Pollution," *International Journal of Occupational and Environmental Health* 10, no 3. (July–September 2004) 333.

170 **until 2008, if ever**   Lenny Siegel, "The Implications of the NRC Report on TCE," Center for Public Environmental Oversight, August 3, 2006, www.cpeo.org.

Nine WITNESSES FOR THE DEFENSE

172 **disagreements over science**   Wendy Wagner, "The Perils of Relying on Interested Parties to Evaluate Scientific Quality," *American Journal of Public Health* 95, Suppl. 1 (2005): S99–S106, www.defendingscience.org/AmJPublicHealth-Supplement.cfm.

172 **"Doubt is our product"**   David Michaels, "Doubt Is Their Product," *Scientific American,* June 2005, 96–101.

172 **could not be subpoenaed**   Lisa C. Friedman et al., "How Tobacco-Friendly Science Escapes Scrutiny in the Courtroom," *American Journal of Public Health* 95, Suppl. 1 (2005): S16–S20; David Michaels, personal communication. The July 2005 issue of the *American Journal of Public Health* in which this and other cited articles appear is available online without charge at www.defendingscience.org/AmJPublicHealth-Supplement.cfm.

173 **purposely for the tobacco industry**   For the full history of the tobacco industry's strategies, see Stanton A. Glantz et al., *The Cigarette Papers* (Berkeley: University of California Press, 1996); Allan M. Brandt, *The Cigarette Century* (New York: Basic Books, 2007). See also the newly developed website tobaccowiki.org.

173 **assignments from the tobacco industry**   It's all available at http://legacy.library.ucsf.edu, where the paper trail uncovered from years of tobacco litigation is stored and keyworded.

173 **"Our overwhelming objective"**   Elisa K. Ong and Stanton A. Glantz, "Constructing 'Sound Science' and 'Good Epidemiology': Tobacco, Lawyers, and Public Relations Firms," *American Journal of Public Health* 91, no. 11 (November 2001): 1749–1757, www.defendingscience.org/AmJPublicHealth-Supplement.cfm.

173 **created a front group**   The front group was called the Advancement for Sound Science Coalition. See Friedman et al., "How Tobacco-Friendly Science Escapes."

173 **signed on as a sponsor**   Ibid.

173 **federally funded research studies** Annamaria Baba et al., "Legislating 'Sound Science': The Role of the Tobacco Industry," *American Journal of Public Health* 95, no. S1 (July 2005): S20–S22.

173 **staff had time to read** See www.thecre.com/post/page3.html.

174 **hamstring environmental regulation** Jim Tozzi was the author of the Data Quality and Data Access acts. He theorizes that the concept of "sound science" gained traction because the OMB was powerful and in the George W. Bush administration shared the antiregulatory ideology needed to put the ideas into action.

174 **paralysis by analysis** Chris Mooney, "Paralysis by Analysis," *Washington Monthly*, May 2004, www.washingtonmonthly.com.

174 **an internal company memo explains** D. M. Cook et al., "The Power of Paperwork: How Philip Morris Neutralized the Medical Code for Secondhand Smoke," *Health Affairs* 24, no. 4 (July–August 2005): 994–1004; Baba et al., "Legislating 'Sound Science.'"

174 **delay their being put in place** Michael Janofsky, "Environmental Groups Are Praising EPA for Updating Cancer Risk Guidelines," *New York Times*, April 4, 2005, A18.

174 **warnings about smokeless tobacco** See www.thecre.com/post/page3.html. Note that this is the website maintained by Jim Tozzi's organization.

174 **part of our national health costs** Cook et al., "The Power of Paperwork."

174 **into planning this strategy** Baba et al., "Legislating 'Sound Science'"; Jim Tozzi, personal communication, October 10, 2006.

175 ***Washington Post* of August 2004** Rick Weiss, "'Data Quality' Law Is Nemesis of Regulation," *Washington Post*, August 16, 2004, 1.

175 **in at least twenty-three states** Steingraber, *Living Downstream*, 161.

175 **water samples in farming communities** Robert Gilliom et al., "Design of the National Water-Quality Assessment Program: Occurrence and Distribution of Water-Quality Conditions," U.S. Geological Survey Circular 1112, http://pubsusgs.gov/circ/circ1112.

175 **never allowed in Switzerland** Alison Pierce, "Biological Warfare," *SF Weekly*, June 2, 2004, www.sfweekly.com.

175 **vigilant in screening for cancer** Weiss, "'Data Quality' Law Is Nemesis."

175 **contradict Syngenta and the EPA** WuQiang Fan et al., "Atrazine-Induced Aromatase Expression Is SF-1-Dependent," *Environmental Health Perspectives* 115, no. 5 (May 2007): 720–727, www.ehponline.org.docs/2007/9758.

175 **Syngenta and EPA representatives** Jennifer Beth Sass and Aaron Colangelo, "European Union Bans Atrazine, While the United States Negotiates Continued Use," *International Journal of Occupational and Environmental Health* 12, no. 3 (July–September 2006): 260–267.

176 **defend Syngenta products** Goldie Blumenstyk, "The Story of Syngenta & Tyrone Hayes: The Price of Research," *Chronicle of Higher Education* 50 (October 31, 2003).

176 **"We showed that these animals"** Tyrone Hayes et al., "Characterization of Atrazine-Induced Gonadal Malformations in African Clawed Frogs," *Environmental Health Perspectives* 114, no. S-1 (April 2006), www.ehponline.org/docs.2006/8067/abstract.htm; Weiss, "'Data Quality' Law Is Nemesis."

176 **publication of his data** Twenty percent of the tadpoles exposed before metamorphosis to the herbicide atrazine grew up hermaphroditic, even though exposed at a level thirty thousand times lower than the previous low-

est effect of atrazine detected on frogs; the difference is that these tests exposed tadpoles, whereas previous tests used adult frogs. John Peterson Myers, "A New View on Toxic Chemicals and How They Impact Our Health," presentation at the Seattle Art Museum conference, January 27, 2004, www.iceh.org/pdfs/SBLF/SBLFMyersTranscript.pdf.

176 **regulating the product**   T. Hayes letter to A. Hosmer, Ecorisk Panel, November 7, 2000. Hayes terminates his professional relationship with the Ecorisk Panel and with Novartis, citing professional and personal reasons, including concern that his findings will not be published in a timely manner.

176 **Kansas Corn Growers Association**   Blumenstyk, "The Story of Syngenta & Tyrone Hayes."

176 **"commitment to sound science"**   Sarah Hull, vice president, Public Affairs, Syngenta America, letter to the editor, *Washington Post*, September 9, 2004, A26.

177 **"Traditional covert influence"**   James Huff, "Industry Influence on Occupational and Environmental Public Health," *International Journal of Occupational and Environmental Public Health* 13, no. 1 (January–March 2007): 107–117.

178 **"reshape the debate"**   Paul D. Thacker, "The Weinberg Proposal," *Environmental Society and Technology,* February 22, 2006, http//:pubs.acs.org; and www.sourcewatch.org.

178 **"You're not biased"**   Lila Guterman, "Researchers Who Study Workplace and Environmental Safety Debate Just How Close Academe and Industry Should Be," *Chronicle of Higher Education* 51, no. 42 (June 24, 2005): 15–21.

178 **evaluates chemicals suspected of causing cancer**   Valerio Gennaro and Lorenzo Tomatis, "Business Bias," *International Journal of Occupational and Environmental Health* 11 (2005): 356–359.

178 **the chemicals were benign**   Daniel Fagin and M. Lavelle, *Toxic Deception* (Secaucus, NJ: Birch Lane Press, 1997).

178 **absolved the chemical of any harm**   Frederick vom Saal and Claude Hughes, "An Extensive New Literature Concerning Low-Dose Effects of Bisphenol A Shows the Need for a New Risk Assessment," *Environmental Health Perspectives* 113, no. 8 (August 2005): 926–933; www.ourstolenfuture.org/newscience/oncompounds/bisphenola/2006/2006–0101vomsaalandwelshons.html.

179 **refused to sign the report**   Peter Waldman, "Common Industrial Chemicals in Tiny Doses Raise Health Issue," *Wall Street Journal*, July 25, 2005, www.mindfully.org/Pesticide/2005.

179 **"There is sometimes fraud"**   John C. Bailar, "How to Distort the Scientific Record Without Actually Lying," *European Journal of Oncology* 11, no. 4 (2006): 217–224.

179 **a clean bill of health**   Renata Kimbrough et al., "Mortality in Male and Female Capacitor Workers Exposed to Polychlorinated Biphenyls," *Journal of Occupational and Environmental Medicine* 41, no. 9 (September 1999): 161–171; Renata Kimbrough et al., "A Mortality Update of Male and Female Capacitor Workers Exposed to Polychlorinated Biphenyls," *Journal of Occupational and Environmental Medicine* 45, no. 3 (March 2003): 271–282.

179 **untreated tanks of water**   Kerry Tremain, "Hopping Mad: A Frog Biologist Battles an Agrichemical Giant," *Sierra,* July–August 2004, www.sierraclub.org/sierra/200407/profile.asp; www.ourstolenfuture.org/NEWSCIENCE/wildlife/frogs/2004.

179 **failed to reach any conclusions** "White Paper on Potential Developmental Effects of Atrazine on Amphibians: In Support of an Interim Reregistration Eligibility Decision on Atrazine" (Washington, DC: Office of Prevention, Pesticides, and Toxic Substances, Office of Pesticide Programs, Environmental Fate and Effects Division, May 29, 2003); Peter Montague, "Ending Government Regulation by Manufacturing Doubt," *Rachel's Environment & Health News* 824 (August 18, 2005), www.rachel.org.

180 **impacts had been discarded** Ted Schettler, science director, Science & Environmental Health Network, personal communication, May 10, 2007; *NTP-CERHR Expert Panel Report on Di-n-butyl Phthalates* (NTP-CERHR-DBP), October 2000, www.cerhr.niehs.nih.gov/chemicals/phthalates/dbp/dbp -final-inprog.pdf.

180 **"almost all of these scares"** Gilbert Ross, "Chemophobia Looms Again in California," essay, American Council on Science and Health, www.acsh.org/ factsfears/newsID.542/news_detail.asp.

180 **years he chose for the study** J. E. Loughlin and Kenneth Rothman, "Lymphatic and Haematopoietic Cancer Mortality in a Population Attending School Adjacent to Styrene-Butadiene Facilities, 1963–1993," *Journal of Epidemiology and Community Health* 53 (1999): 283–287. Dr. Rothman also worked for Texaco to evaluate whether Texaco's oil exploration practices in Ecuador's Amazon region were responsible for environmental damage and harmful health effects; he concluded they were not. The story of Texaco in Ecuador is told in the best seller by John Perkins, *Confessions of an Economic Hit Man* (New York: Plume Group/Penguin, 2006).

180 **"Industry and government"** Presentation at University of Pittsburgh Health & the Environment conference, April 7–9, 2005.

181 **"good epidemiological practice"** Ong and Glantz, "Constructing 'Sound Science' and 'Good Epidemiology.' "

181 **"However, the risks for community exposure"** Robinson, *Cancer Clusters*.

181 **essential to its illegal enterprise** See www.prwatch.org/node/5296.

181 **and hold them secret** Guterman, "Researchers Who Study Workplace and Environmental Safety."

181 **serve Dow in-house worldwide** Dow scientist, Carol Burns, personal communication, September 4, 2006.

182 **over $1 million a year** See http://swz.salary.com.

182 **auto safety failures** It was founded in 1967 under the corporate name Failure Analysis Associates.

182 **and the Department of Energy** Exponent has pulled the listing of its corporate clients off its website, www.exponent.com/about/clients.html. However, the scientific literature is full of articles by company employees naming clients as they discuss cases and in articles by others discussing cases involving Exponent.

183 **loosening of the California standards** Letter prepared by Joseph K. Lyou, California League of Conservation Voters Education Fund, to the University of California's Blue Ribbon Panel on Hexavalent Chromium and the Apparent Conflicting Interests of Panel Member Dr. Dennis Paustenbach; deposition of Gary G. Praglin, *Aguayo v. PGE*.

183 **except for lung cancer** Brad Rodu and Philip Cole, "The Fifty-Year Decline of Cancer in America," *Journal of Clinical Oncology* 19, no. 1 (January 2001): 239–241.

183 **indication of PG&E involvement**   Peter Waldman, "Common Industrial Chemicals in Tiny Doses Raise Health Issue," *Wall Street Journal,* December 23, 2005; David Michaels, "A Chrome-Plated Controversy," December 7, 2006, http://thepumphandle.wordpress.com.

183 **for his conflict of interest**   Deposition of Tony Ye, December 12, 2002, *Aguayo v. PGE; Chrome-Plated Fraud: How PG&E's Scientists-for-Hire Reversed Findings of Cancer Study* (Washington, DC: Environmental Working Group, undated), http://ewg.org/reports/chromium. Paustenbach oversaw most of the ghostwriting of the study originally written by the Chinese scientist JianDong Zhang, while he (Paustenbach) was with a company he had founded, called ChemRisk, and testified about it while he was with Exponent. The article was published in the *Journal of Occupational and Environmental Medicine;* that journal retracted the paper in July 2006 when it discovered the underlying facts. The coauthor alleges that ChemRisk scientists merely "provided minor editorial consulting and input." See Melissa Lee Phillips, "Chromium Paper Retracted Unfairly," *The Scientist,* December 22, 2006, www.the-scientist.com.

183 **"Of course not"**   Deposition of Dennis J. Paustenbach, August 29, 2002, *Aguayo v. PGE.*

183 **for $295 million**   Exponent and the company Paustenbach founded, ChemRisk, were paid more than $23 million over five years by Ford, General Motors, and DaimlerChrysler to defend the companies in lawsuits brought by autoworkers exposed to asbestos on the job. Andrew Schneider, "Pressure at OSHA to Alter Warning," *Sun Reporter,* November 20, 2006.

184 **whose products generate dioxins**   Eric Pianin, "Dioxin Report by EPA on Hold, Industries Oppose Finding of Cancer Link, Urge Delay," *Washington Post,* April 12, 2001, www.besafenet.com. The dioxin subcommittee met November 1–2, 2000.

184 **a by-product of chlorine**   C. R. Kirman et al., "Is Dioxin a Threshold Carcinogen?: A Quantitative Analysis of the Epidemiological Data Using Internal Dose and Monte Carlo Methods," *Organohalogen Compounds* 48 (2000): 219–222.

184 **academic conflicts of interest**   Sheldon Krimsky, *Science in the Private Interest* (Lanham, MD: Rowman & Littlefield, 2003), x.

184 **sun exposure in the United States.**   See www.hcra.harvard.edu/risk.html.

184 **exposed as children and adolescents**   See www.chernobyl.info; and "Health Effects of the Chernobyl Accident," Fact Sheet No. 303, World Health Organization, April 2006. The World Health Organization, citing a United Nations study, states that five thousand have so far been diagnosed with thyroid cancer, and there may be up to nine thousand excess cancer deaths.

184 **Europe from the fallout**   For instance, a radiation-saturated rain drenched the towns around Italy's Lake Como, where the cancer rates have subsequently soared.

185 **soliciting money from Philip Morris**   "Fundraising Letter from Graham to Philip Morris, October 1991," www.citizen.org.

185 **"an arrangement be made"**   "Fundraising Letter from Graham to Kraft, June 1, 1992," http://legacy.library.ucsf.edu/cgi.

185 **makers of those chemicals**   Krimsky, *Science in the Private Interest,* 39–41; www.cspinet.org/integrity/corp_funding_html.

185 **more research was needed**   Robert Kuttner, "University for Rent," *Ameri-*

can Prospect 12 (May 7, 2001); Steve Toloken, "Industry-Backed Harvard Review Clears Styrene," *Plastic News,* June 24, 2002, www.plasticnews.com.

185 **added deaths from malnutrition a year**   "Harvard Study Shows New Food Regulations Bad for Public Health," *Pesticide Broadcast* 11, no. 1 (February 2000), North Carolina Cooperative Extension Service, http://ipm.ncsu.edu/current_ipm/Broadcasts/broadcast02–00.html#title1; www.prospect.org/print/V12/8/Kuttner-r.2.html.

185 **a history of corporate service**   Professor Delzell's clients have included Phillips 66, Ford Motor Company, United Auto Workers, Shell Oil, Union Oil, Novartis/Syngenta, and the Internal Institute of Synthetic Rubber Producers, among others.

186 **"the small study size necessitates"**   Elizabeth Delzell et al., "A Follow-up Study of Workers at a Dye and Resin Manufacturing Plant," *Journal of Occupational Medicine* 31, no. 3 (March 1989): 273–278; Fabio Barbone, Elizabeth Delzell, et al., "A Case-Control Study of Lung Cancer at a Dye and Resin Manufacturing Plant," *American Journal of Industrial Medicine* 22 (1992): 835–849; Fabio Barbone, Elizabeth Delzell, et al., "Exposure to Epichlorohydrin and Central Nervous System Neoplasms at a Resin and Dye Manufacturing Plant," *Archives of Environmental Health* 49, no. 5 (September–October 1994): 355–358; Elizabeth Delzell and Nalini Sathiakumar, "Summary of Retrospective Follow-up Epidemiology Study of Employees at the Toms River Plant," unpublished article on study for Ciba, July 1998; Nalini Sathiakumar and Elizabeth Delzell, "An Updated Mortality Study of Workers at a Dye and Resin Manufacturing Plant," *Journal of Environmental Medicine* 42, no. 7 (July 2000): 762–771.

186 **the norm when they contracted cancer**   Bob Herbert, "IBM Computer Workers: Sick and Suspicious," *New York Times*, September 4, 2003.

186 **overall death rate was 35 percent lower**   Elizabeth Delzell et al., "Mortality Among Semiconductor and Storage Device-Manufacturing Workers," *Journal of Occupational & Environmental Medicine* 47, no. 10 (October 2005): 996–1014.

186 **to bias a study's outcome**   Trevor Stokes, "IBM Finds No Higher Cancer Rate," *The Scientist,* November 9, 2004, www.the-scientist.com.

187 **he won the right to publish**   Richard Clapp, "Mortality Among US Employees of a Large Computer Manufacturing Company: 1969–2001," *Environmental Health: A Global Access Science Source* 5 (October 19, 2006), www.ehjournal.net.

187 **born with birth defects**   In addition, 1,100 cases of similar Bendectin-related birth defects were filed in a Cincinnati-based trial.

187 **according to Mekdeci**   The plaintiffs' evidence included cell culture tests and live animal studies, analyses of pharmacological similarities between Bendectin and other substances known to cause birth defects, and unpublished reanalyses of negative epidemiological studies.

187 **dismiss a case have risen**   Molly Jacobs et al., *Daubert: The Most Influential Supreme Court Ruling You've Never Heard Of* (Boston: Tellus Institute, June 2003).

187 **"President Eisenhower warned the nation"**   Statement of Honorable Mark I. Bernstein, "Science and Justice," December 13, 1996, www.birthdefects.org/information/env_bendectin.htm.

188 **[Dr. Robert Brent]**   The judge referred to Dr. Robert Brent by name in his statement.

188 **plaintiff's evidence more readily**   Ronald L. Melnick, "A Daubert Motion: A Legal Strategy to Exclude Essential Scientific Evidence in Toxic Tort Litigation," *American Journal of Public Health* 95, no. S1 (July 2005): S30–S34.

188 **"These requirements favor the powerful"**   David Michaels, "Scientific Evidence and Public Policy," *American Journal of Public Health* 95, no. S1 (July 2005): S5–S7; Melnick, "A Daubert Motion"; Friedman et al., "How Tobacco-Friendly Science Escapes Scrutiny."

188 **curb class action lawsuits**   Jeffrey H. Birnbaum, "Chamber of Commerce Helps Bush Agenda," *Washington Post,* February 6, 2005, www.concordmonitor .com; *USA Today*, February 18, 2005, 1B.

188 **Chapter 11 bankruptcy protection**   John Heenan, "Graceful Maneuvering: Corporate Avoidance of Liability Through Bankruptcy and Corporate Law," *Vermont Journal of Environmental Law* (2003), www.vjel.org/roscoe.

189 **nobody would ever see the documents**   Paul D. Thacker, "The Weinberg Proposal," Environmental Society and Technology, American Chemical Society, February 22, 2006, http:acs.org.

189 **an outcome of lawsuits**   David S. Egilman and Susanna Rankin Bohme, "Over a Barrel: Corporation Corruption of Science and Its Effects on Workers and the Environment," *International Journal of Occupational and Environmental Health* 11, no. 4 (October–December 2005): 331–337.

189 **"expedite the peer review process"**   Patricia A. McDaniel and Gina Solomon, "The Tobacco Industry and Pesticide Regulations: Case Studies from Tobacco Industry Archives," *Environmental Health Perspectives* 113, no. 12 (December 2005): 1659–1665; "The Risk of Smoky Finances," *ES&T* online news, September 2006, http://pubs.acs.org.

190 **work for environmental organizations**   Fagin and Lavelle, *Toxic Deception,* 116.

190 **renamed the American Chemistry Council**   Environmental Working Group, *Chemical Industry Archives: The Inside Story* (Washington, DC: Environmental Working Group, undated). The amount is in inflation-adjusted dollars. See www.chemicalindustryarchives.org.

190 **Was there ever any real**   Fact sheet, American Council on Science and Health, undated, www.mindfully.org/pesticide/acsh-koop.htm. See also chapter 3 in Elizabeth Whelan's book *Toxic Terror* (Amherst, NY: Prometheus Books, 1993).

190 **contaminating the Hudson River**   Elizabeth Whelan, "Who Says PCBs Cause Cancer?" *Wall Street Journal,* December 12, 2000, www mindfully.org.

190 **corporate-financed foundations**   See "Tobacco Files" for documentation of its start-up funding, www.syx21e00_p18.png, and the editorial written by American Council on Science and Health president Elizabeth Whelan in the twenty-fifth anniversary report, April 29, 2004, http://acsh.org/news/newsID .852news_detail.asp.

191 **"highly untrustworthy individual"**   "In the Cases of Gilbert Ross, M.D., and Deborah Williams, M.D., vs. The Inspector General, U.S. Department of Health and Human Services," Docket Nos. C-94–368 and C-94–369, Decision No. CR478, Steven T. Kessel, administrative law judge. Ross was convicted of a criminal conspiracy and racketeering offense by the U.S. government in 1994, sentenced to forty-six months in prison, fined $612,855 (later reduced because he could not pay the original amount), excluded for ten years from participating in federally funded health care programs, and

stripped of his license to practice as a physician in New York. The judge accused him of "indefensible practices" and of giving "dishonest testimony at his trial" and termed him "a highly untrustworthy individual." He and his partner "ran sham clinics to manufacture fraudulent Medicaid reimbursement claims," accepted "patients" who were "homeless men and drug addicts," and "turned away real patients," while his clinics "lacked decent medical equipment" and were "very dirty and unsanitary." The scheme was to offer patients prescriptions for expensive drugs that they could resell on the street for cash, in return for allowing Ross and his associates to give medically unnecessary examinations, procedures, and tests, for which they received Medicare reimbursements. After his release from prison, Ross answered an ad in *The New York Times* for a "staff assistant" job at the American Council on Science and Health. Ross told council president Elizabeth Whelan his history, saying he found himself inadvertently participating in a scam. After hiring him, she promoted him to "medical/executive director." In 2004, when filling out a form for the official New York State website for physicians, Ross left blank the field titled "criminal convictions." Bill Hogan, "Paging Dr. Ross," *Mother Jones,* November–December 2005, www.motherjones.com.

191 **interfere with human reproduction**   *House of Mirrors: Burson-Marsteller Brussels Lobbying for the Bromine Industry* (Corporate Europe Observatory, January 2005), www.corporateeurope.org/lobbycracy/houseofmirrors.html.

191 **EPA standard for perchlorate in water**   See www.councilonwaterquality .org.

191 **restaurant, and beverage companies**   See www.sourcewatch.org.

191 **sugar in their soft drinks**   See www.consumerfreedom.com.

191 **"blur the distinction"**   Verlyn Klinkenborg, "The Story Behind a New York Billboard and the Interests It Serves," *New York Times,* July 22, 2005, http://select.nytimes.com.

192 **"triggering endless mischief"**   Elizabeth M. Whelan, "Chemical-Contamination Con," *New York Post* editorial, June 15, 2005, www.acsh.org/ healthissues. Michael Fumento of the Hudson Institute claims: "Ultimately the NRDC [National Resources Defense Council] does not want safe use of pesticides; it wants no use. [Americans will] be forced to eat expensive, ugly, shriveled-looking organic produce and foreign competitors will have our farmers foreclosing at rates not seen since the dust bowl days." Fumento has received payments from Monsanto. See www.sourcewatch.org.

192 **unique in her experience**   Lila Guterman, "Peer Reviewers Are Subject in Cancer Lawsuit Against Chemical Companies," *Chronicle of Higher Education* 51, no. 13 (November 19, 2004): A20, http://chronicle.com.

192 **polluting the water in Woburn, Massachusetts**   Deborah Josefson, "US Journal Embroiled in Another Conflict of Interest Scandal," *British Medical Journal* 316 (January 24, 1998): 247–252, www.bmj.com. The reviewer was Jerry H. Berke, who, as part of his blast against Steingraber, cited Philip Cole's research.

192 **"it cannot be used"**   Sandra Steingraber, "How Mercury-Tainted Tuna Damages Fetal Brains, *In These Times,* December 27, 2004, www.inthesetimes .com.

192 **treatment for head lice**   Curt Guyette, "Of Lice and Libel," *Metro Times* (Detroit, MO), August 23, 2006, www.metrotimes.com.

192 **banned in fifty-two other countries**   Melissa Lee Phillips, "Registering

Skepticism: Does the EPA's Pesticide Review Protect Children," *Environmental Health Perspectives* 114, no. 10 (October 2006): A592–A595, www .ehponline.org; Alaska Community Action, "Submission of Information on Lindane Pursuant to Article 8 of the Stockholm Convention as Specified in Annex E of the Convention," Alaska Community Action on Toxics and Pesticide Action Network of North America, undated, www.oztoxics.org/poprc/ Library/Lindane%20IPEN%20Comments%20for%20POPRC.pdf; FDA, "Public Health Advisory: Safety of Topical Lindane Products for Treatment of Scabies and Lice" (Washington, DC: Center for Drug Evaluation and Research, U.S. Food and Drug Administration, 2003), www.fda.gov/cder/drug/ infopage.lindane.

193 **data sets from several researchers** Jocelyn Kaiser, "Lead Paint Experts Face a Barrage of Subpoenas," *Science* 309 (July 2005): 362–363.

193 **"various analytical and statistical errors"** Baba et al., "Legislating 'Sound Science.' "

193 **"If the results were true"** Natasha Singer, "Does It . . . or Doesn't It?" *New York Times,* August 10, 2006, E1.

193 **"the enemy is not"** See www.acsh.org/healthissues, quoted by Nicols-Dezenhall.

194 **his testimony had never been done** Holcomb B. Noble, "Koop Criticized for Role in Warning on Hospital Gloves," *New York Times,* October 29, 1999, www.nytimes.com.

194 **"It was misinformation"** "The Precautionary Principle: Throwing Out Science with the Bath Water," *Issues Perspective* (McLean, VA: Wirthlin Worldwide, February 2000).

194 **according to its own trade association** Philip Shabecoff, "Apple Scare of '89 Didn't Kill Market," *New York Times,* November 13, 1990, http://select .nytimes.com.

195 **flame retardants must be harmless** Stuart Blackman, " 'Industrial' Pollutants Reveal a Surprising Origin," *The Scientist* 19 (June 6, 2005): 24–25.

195 **FMC at $1.5 million** Samuel S. Epstein, *Unreasonable Risk,* 2nd ed. (Chicago: Environmental Toxicology, 2001), 34.

195 **staffers and federal officials** Bill Hogan, ed., *Unreasonable Risk: The Politics of Pesticides* (Washington, DC: Center for Public Integrity, 1998), 13.

195 **safe cosmetics bills in California** Jordan Rau, "Legislature Targets Toxic Risks in Products," *Los Angeles Times,* May 19, 2005, www.latimes.com.

196 **to influence the outcome** Charles Lewis, press release, Center for Public Integrity, June 30, 1998.

196 **remains in use today** Shelia Hoar Zahm and Aaron Blair, "Pesticides and Non-Hodgkin's Lymphoma," *Cancer Research* Suppl. 52 (October 1, 1992): 5485s–5488s; "Overview of the Effects of 2,4-D," Sierra Club of Canada, January 2005, www.sierraclub/ca. Other researchers conclude, "An increased rate of non-Hodgkin's lymphoma has been repeatedly observed among farmers, but identification of specific exposures that explain this observation has proven difficult." See A. J. De Roos et al., "Integrative Assessment of Multiple Pesticides as Risk Factors for Non-Hodgkin's Lymphoma Among Men," *Journal of Occupational and Environmental Medicine* 60, no 9. (2003), www .oem.bmj.com/cgi/content.

196 **those voting against** See www.opensecrets.org, which contains data on contributions to individual congressmen.

Ten *POSSE COMITATUS*

201 **exposure in fetuses and children**   Kenneth P. Stoller, "The Obfuscation of the Iatrogenic Autism Epidemic," *Pediatrics* 111, no. 5 (May 5, 2006), http://pediatrics.aapublications.org/cgi/letters/117/4/1028.

206 **"gives community members power"**   Louisiana Bucket Brigade, "Mission Statement and Purpose and History of the Louisiana Bucket Brigade," www.labucketbrigade.org/about/index.shtml.

208 **dioxin, and other pollutants**   From Birth Defects Research for Children, Inc., www.birthdefects.org.

209 **"Poisoned Schools Report"**   Center for Health, Environment & Justice, "Child Proofing Our Communities," www.childproofing.org.

210 **Healthy Child Healthy World**   Formerly called the Children's Health Environmental Coalition, the organization is now located in California.

210 **community economic development**   Philip Shabecoff, *Earth Rising* (Washington, DC: Island Press, 2001), 54.

215 **"one of the most successful campaigns"**   Krimsky, *Science in the Private Interest,* 187.

215 **put lead back into gasoline**   Herbert L. Needleman, "Salem Comes to the National Institutes of Health: Notes from Inside the Crucible of Scientific Integrity," *Pediatrics* 90, no. 6 (1992): 977ff.

216 **"If that happens, our craft"**   Ibid.

216 **pursued under subpoena**   Jocelyn Kaiser, "Lead Paint Experts Face a Barrage of Subpoenas," *Science* 309 (July 2005): 362–363.

216 **he was barred from testifying**   Bruce Lanphear, personal communication, June 25, 2007.

217 **members live up to its name**   It should be noted that Greater Boston Physicians for Social Responsibility acted as fiscal sponsor for funding for this book.

Eleven VALUES

221 **protect open space and restore wetlands**   Letter from Representative Jerry Weller, Eleventh District, Illinois, to Mrs. Cynthia Sauer, February 23, 2004.

221 **first session of the 109th Congress**   League of Conservation Voters scorecard, www.capwiz/lev/dbq/vote_info/?sort=results&last=Weller&su.

221 **"are sufficiently protective"**   Letter from J. E. Dyer, director, Office of Nuclear Reactor Regulation, United States Nuclear Regulatory Commission, to Mrs. Cynthia Sauer, April 16, 2004.

222 **"as with all ionizing radiation"**   EPA, *Radiation Information—Tritium* (Washington, DC: U.S. Environmental Protection Agency, March 8, 2006), www.epa.gov/radiation/radionuclides/tritium.htm.

222 **"about one-eighth of the dose"**   Eli Port, Tritium fact sheet, undated, www.exeloncorp.com/nr/rdonlyres.

222 **those without such facilities**   Illinois Department of Public Health, "Pediatric Cancer Incidence and Proximity to Nuclear Facilities in Illinois," *Health and Hazardous Substances Registry Newsletter* (Fall 2000).

222 **"the smallest dose of radiation"**   Needleman and Landrigan, *Raising Children Toxic Free,* 140.

223 **"the method of operations"**   "Madigan, Glasgow File Suit for Radioactive

Leaks at Braidwood Nuclear Plant," press release from the Illinois Attorney General's Office, March 16, 2006.

223 **"no specific action"**   United States Nuclear Regulatory Commission, NRC Notice 2006–13: "Ground-Water Contamination Due to Undetected Leakage of Radioactive Water," July 10, 2006.

223 **"It is outrageous"**   "Nuclear Regulatory Commission Ignores Public Demands to Stop Water Contamination, Backroom Deal with Industry Group Alleged," news release from the Union of Concerned Scientists, July 20, 2006.

223 **any member of Congress**   Thomas B. Edsall and Justin Blum, "Rep. Barton Faces Energy Challenge," *Washington Post,* April 14, 2005, A25.

226 **"It is not just a few"**   David S. Egilman and Susanna Rankin Bohme, "Over a Barrel: Corporate Corruption of Science and Its Effects on Workers and the Environment," *Journal of International Environmental Health* 11, no. 4 (October–December 2005): 331.

226 **"inappropriate or unethical behavior"**   Catherine D. DeAngelis, "The Influence of Money on Medical Science," *Journal of the American Medical Association* 296 (2006): 996–998.

227 **"bears a huge burden of guilt"**   Lynn White Jr., "The Historical Roots of Our Ecological Crisis," *Science* 155 (March 10, 1967): 1203–1207.

228 **"religion is now a leading voice"**   Roger S. Gottlieb, *A Greener Faith* (New York: Oxford University Press, 2006), 9.

228 **"The ecological crisis has assumed"**   Jennifer Parmelee, "Pope Says Environmental Misuse Threatens World Stability," *Washington Post,* December 6, 1989.

228 **"The Church's commitment to life"**   Pope John Paul II, "Address of October 29, 1983, to the 35th General Assembly of the World Medical Association," quoted in Paul Gorman, Maida Galvez, and Philip J. Landrigan, "Environmental Threats to the Unborn Child," a collaborative project of the National Religious Partnership for the Environment and the Mt. Sinai Center for Children's Health and the Environment, October 2003, 4.

228 **"has undertaken a broad program"**   Ibid., 5.

230 **"We and our children face"**   "On the Care of Creation: An Evangelical Declaration on the Care of Creation," Evangelical Environmental Network and *Creation Care* magazine, een@creationcare.org.

230 **"This is God's world"**   "Evangelical Leaders Call Global Warming a Moral and Spiritual Crisis," press release, Evangelical Climate Initiative, February 8, 2006, www.christiansandclimate.org/statement.

231 **the chief greenhouse gas**   Michael Janofsky, "When Cleaner Air Is a Biblical Obligation," *New York Times, November* 7, 2005, www.nytimes.com/2005/11/07/politics/07air.html.

231 **"Bible-believing evangelicals"**   Alan Cooperman, "Evangelicals Will Not Take Stand on Global Warming," *Washington Post,* February 2, 2006, www.washingtonpost.com/wp/dyn/content/article/2006/02/01/AR2006020102 132_p.

231 **"biblical values ... thereby reversing"**   Website of Concerned Women for America, www.cwfa.org/about.asp.

232 **"that homosexual activists"**   *Dr. Dobson's Newsletter,* June 2006, www.family.org/docstudy/newsletters/a0040893.cfm.

232 **"If the Green Revolution"**   Bill Moyers, "A Question for Journalists: How

Do We Cover Penguins and Politics of Denial?" keynote speech to the Society of Environmental Journalists convention, Austin, TX, October 1, 2005.

232 **"with biblical worldviews"**   Kevin Phillips, *American Theocracy* (New York: Viking, 2006), 64.

232 **"may not believe in end times"**   Ibid., 67.

233 **"left-leaning environmentalism"**   Ibid., 238.

Twelve JUSTICE

237 **"stumble toward a future"**   Michael Lerner, "The Age of Extinction and the Emerging Environmental Health Movement," unpaginated essay, available from commonweal@commonweal.org.

237 **"Human activity is putting"**   Statement of the Board, "United Nations Millennium Ecosystem Assessment," 2005, 7, www.maweb.org.

241 *Premum non nocere*   "The origin of the phrase is not precisely known; contrary to popular belief, the phrase is not in the Hippocratic Oath. However, it is often described as a Latin paraphrase by Galen of a Hippocratic aphorism . . . yet no specific mention in Galen's writings has been reported. The closest approximation to the phrase that can be found in the Hippocratic Corpus is 'to help, or at least to do no harm,' taken from *Epidemics,* Bk. I, Sect. V." Wikipedia.

241 **"When an activity raises threats"**   "Wingspread Statement on the Precautionary Principle," reproduced in Carolyn Raffensperger and Joel Tickner, eds., *Protecting Public Health and the Environment* (Washington, DC: Island Press, 1999), appendix A, 353.

241 **"a requirement of absolute 'proof' "**   Ted Schettler, Katherine Barrett, and Carolyn Raffensperger, "The Precautionary Principle: Protecting Public Health and the Environment," Science and Environmental Health Network, www.sehn.org.

241 **Joel Tickner, a scientist**   Joel Tickner, "A Map Toward Precautionary Decision Making," in *Protecting Public Health and the Environment,* 163.

242 **"Humans and the environment"**   Ibid.

243 **book about Pittsfield, Massachusetts**   Janna G. Kippe and Jane Keye, "PCBs and the Precautionary Principle," in Poul Harremoes et al., eds., *The Precautionary Principle: Late Lessons from Early Warnings* (London: Earthscan Publications Ltd., 2002), 64ff.

244 **until many years later**   Philip J. Landrigan et al., "Children's Health and the Environment: A New Agenda for Prevention Research," *Environmental Health Perspectives Supplements* 106, no. S3 (June 1998).

244 **"The precautionary principle"**   Bailey Ronald, "Invasion of the Fear-mongers (How Executives Can Manage Negative Publicity Perpetuated by Media and a Misinformed Public)," *Chief Executive (U.S.),* December 1, 2000.

244 **"When studies in the early 90's"**   *The Precautionary Principle: Throwing Science Out with the Bath Water,* Wirthlin World Wide Issues Perspective, a special report prepared in cooperation with Nichols-Dezenhall, February 2000, http://209.204.197.52/publcns/report/ppfinal.pdf.

244 **1.3 million cases of illness**   Quoted in Peter Montague and Tim Montague, "Urban Legend: Precaution and Cholera in Peru," *Rachel's Democracy and Health News,* August 4, 2005.

245 **decision by the Peruvian government**   Ibid.

245 **"the Peruvian cholera epidemic"**   Joel Tickner and Tami Gouveia-Vigeant, "The 1991 Cholera Epidemic in Peru: Not a Case of Precaution Gone Awry," *Risk Analysis* 25, no. 3 (June 2005): 495.

245 **"with appropriate research"**   Bernard D. Goldstein, "The Precautionary Principle and Scientific Research Are Not Antithetical," *Environmental Health Perspectives* 107, no. 12 (December 1999).

245 **"There always will be scientists"**   Schettler et al., *Generations at Risk,* 38.

246 **"where we've created a mess"**   Elizabeth Weise, "Green Chemistry Takes Root," *USA Today, November* 21, 2004, www.usatoday.com/news/science/2001–11–21-green_x.htm.

248 **a wide variety of uses**   Jon Brodkin, "Study Finds Toxic Chemicals More Expensive," *MetroWest Daily News,* July 11, 2006.

248 **PVC and many other products**   Franc Ackerman and Rachel Massey, "The Economics of Phasing Out PVC" (Waltham, MA: Tufts University, 2003).

249 **"a tragedy of our own making"**   Remarks at the Women's Health and Environment Conference, Boston, MA, October 24, 2005.

249 **"a world of abundance"**   See www.mcdonough.com.

249 **"could have very adverse"**   Polly J. Hoppin and Richard Clapp, "Science and Regulation: Current Impasse and Future Solutions," *American Journal of Public Health* 95, no. S1 (2005): 510.

250 **toxic chemicals in commerce**   Michael Wilson, "EPA Chemical Management System," testimony before the U.S. Senate Committee on Environment and Public Works, *Congressional Quarterly,* August 2, 2006.

251 **"Frankenstein's monster"**   Quoted by Jane Anne Morris in "Fixing Corporations—Part 1: Legacy of the Founding Parents," *Rachel's Environment and Health Weekly,* April 4, 1996, www.ejnet.org/rachel/rehw488.htm.

252 **"corporations, which should be"**   Ibid.

255 **"be the greatest unifying force"**   Lerner, "The Age of Extinction and the Emerging Environmental Health Movement."

256 **"a world of scarce health resources"**   Barry R. Bloom, *Harvard Public Health Review* (Fall 2004).

257 **reduction in ADHD symptoms**   Frances E. Kuo and Andrea Faber Taylor, "A Potential Natural Treatment for Attention-Deficit/Hyperactivity Disorder: Evidence from a National Study," *American Journal of Public Health* 94, no. 9 (September 2004): 1580–1586.

257 **"The way we build our communities"**   Frumkin comment during teleconference of the Collaborative on Health and the Environment, September 15, 2005.

259 **"smooth superhighway on which we progress"**   Rachel Carson, *Silent Spring* (Boston: Houghton Mifflin Company, 1962), 277–278.

Ashford, Nicholas, and Claudia Miller. *Chemical Exposures, Low Levels and High Stakes,* 2nd ed. New York: John Wiley & Sons, Inc., 1998.

Brent, Robert, et al., eds. *The Vulnerability, Sensitivity, and Resiliency of the Developing Embryo, Infant, Child, and Adolescent to the Effects of Environmental Chemicals, Drugs, and Physical Agent as Compared to the Adult.* American Academy of Pediatrics Supplement to *Pediatrics* 113, no. 4 (April 2003).

Carson, Rachel. *Silent Spring,* 25th anniversary ed. Boston: Houghton Mifflin Company, 1987.

Colby, Gerard. *Du Pont Dynasty.* Secaucus, NJ: Lyle Stuart, Inc., 1984.

Cooper, Kathleen. *Child Health and the Environment—A Primer.* Toronto: Canadian Partnership for Children's Health & Environment, 2005.

Davis, Devra. *The Secret History of the War on Cancer.* New York: Basic Books, 2007.

————. *When Smoke Ran Like Water.* New York: Basic Books, 2002.

Doyle, Jack. *Trespass Against Us.* Monroe, ME: Common Courage Press, 2004.

Fagin, Dan, Marianne Lavelle, and the Center for Public Integrity. *Toxic Deception.* Secaucus, NJ: Birch Lane Press, 1996.

Farquhar, Doug, and Glen Anderson. *Healthy Environment, Healthy Kids.* Denver: National Conference of State Legislatures, 2004.

Etzel, Ruth A., ed. *Pediatric Environmental Health,* 2nd ed. Elk Grove, IL: American Academy of Pediatrics, 2003.

Geiser, Kenneth. *Materials Matter.* Cambridge, MA: MIT Press, 2001.

Gilbert, Steven G. *A Small Dose of Toxicology.* Boca Raton, FL: CRC Press, 2004.

Gillick, Linda. *For the Love of Mike.* Waco, TX: WRS Publishing, 1994.

Gottlieb, Roger S. *A Greener Faith.* New York: Oxford University Press, 2006.

Harding, Ronnie, ed. *Environmental Decision Making.* Sydney: Federation Press, 1998.

————. *Perspectives on the Precautionary Principle.* Sydney: Federation Press, 1999.

Harremoes, Poul, et al., eds. *The Precautionary Principle in the 20th Century.* London: Earthscan Publications Ltd., 2002.

Kirby, David. *Evidence of Harm.* New York: St. Martin's Press, 2005.

Krimsky, Sheldon. *Science in the Private Interest.* Lanham, MD: Rowman & Littlefield Publishers, Inc., 2003.

Lakoff, George. *Don't Think of an Elephant!* White River Junction, VT: Chelsea Green Publishing, 2004.

Landrigan, Philp J., Herbert L. Needleman, and Mary Landrigan. *Raising Healthy Children in a Toxic World.* Emmaus, PA: Rodale, 2001.

Markowitz, Gerald, and David Rosner. *Deceit and Denial.* Berkeley: University of California Press, 2002.

Michaels, David. *Doubt Is Their Product: How Industry Scientists Manufacture Uncertainty and Threaten Your Health.* New York: Oxford University Press, 2007.

Moore, Colleen F. *Silent Scourge: Children, Pollution, and Why Scientists Disagree.* New York: Oxford University Press, 2003.

National Research Council. *Environmental Epidemiology,* vol. 1: *Public Health and Hazardous Wastes.* Washington, DC: National Academy Press, 1991.

———. *Pesticides in the Diets of Infants and Children.* Washington, DC: National Academy Press, 1993.

National Research Council and Institute of Medicine. *Children's Health, the Nation's Wealth.* Washington, DC: National Academy Press, 2004.

Needleman, Herbert L., and Philip J. Landrigan. *Raising Children Toxic Free.* New York: Farrar, Straus & Giroux, 1994.

Nestle, Marion. *Food Politics.* Berkeley: University of California Press, 2002.

Phillips, Kevin. *American Theocracy.* New York: Viking, 2006.

Raffensperger, Carolyn, and Joel Tickner, eds. *Protecting Public Health and the Environment.* Washington, DC: Island Press, 1999.

Rampton, Sheldon, and John Stauber. *Trust Us, We're Experts.* New York: Jeremy P. Tarcher/Putnam, 2001.

Sadler, T. W. *Langman's Medical Embryology,* 9th ed. Philadelphia: Lippincott, Williams & Wilkins, 2004.

Satterthwaite, David, et al. *The Environment for Children.* London: Earthscan Publications Ltd., 1996.

Schettler, Ted, et al. *Generations at Risk.* Cambridge, MA: MIT Press, 1999.

Schettler, Ted, et al., eds. *In Harm's Way: Toxic Threats to Child Development.* Cambridge, MA: Greater Boston Physicians for Social Responsibility, 2001.

Schlosser, Eric. *Fast Food Nation.* Boston: Houghton Mifflin Company, 2001.

Schneider, Dona, and Natalie Freeman. *Children's Environmental Health.* Washington, DC: American Public Health Association, 2000.

Setterberg, Fred, and Lonny Shavelson. *Toxic Nation.* New York: John Wiley & Sons, Inc., 1993.

Shabecoff, Philip. *Earth Rising.* Washington, DC: Island Press, 2000.

———. *A Fierce Green Fire,* rev. ed. Washington, DC: Island Press, 2003.

Steingraber, Sandra. *Living Downstream.* New York: Vintage Books, 1997.

Talbott, Evelyn O., and Gunther F. Craun, eds. *Introduction to Environmental Epidemiology.* Boca Raton, FL: CRC Press, Inc., 1995.

Tickner, Joel A., ed. *Precaution, Environmental Science and Preventive Public Policy.* Washington, DC: Island Press, 2003.

United Nations Environmental Programme, United Nations Children's Fund, and World Health Organization. *Children in the New Millennium: Environmental Impact on Health,* 2002.

Wargo, John. *Our Children's Toxic Legacy,* 2nd ed. New Haven, CT: Yale University Press, 1998.

Wigle, Donald T. *Child Health and the Environment.* New York: Oxford University Press, 2003.